GLYCOIMMUNOLOGY 2

ADVANCES IN EXPERIMENTAL MEDICINE AND BIOLOGY

Recent Volumes in this Series

GLYCOIMMUNOLOGY 2

Edited by

John S. Axford

St. George's Hospital Medical School
University of London
London, United Kingdom

SPRINGER SCIENCE+BUSINESS MEDIA, LLC

Library of Congress Cataloging-in-Publication Data

On file

Proceedings of the Fourth Jenner International Glycoimmunology Meeting, held November 12 – 15, 1996, in Loutraki, Greece

ISBN 978-1-4613-7457-2   ISBN 978-1-4615-5383-0 (eBook)
DOI 10.1007/978-1-4615-5383-0

© 1998 Springer Science+Business Media New York
Originally published by Plenum Press, New York in 1998

http://www.plenum.com

10 9 8 7 6 5 4 3 2 1

PREFACE

The Jenner International Glycoimmunology Meetings have charted the rapid development of glycobiology within the field of inflammation. In less than a decade, the science has grown from basically being involved in carbohydrate analysis to the understanding of how sugars are associated with inflammation and how they have potential as anti-inflammatory therapeutics. The 4th Jenner International Glycoimmunology Meeting was recently held in Loutraki, Greece, and set the scene for what promises to be an exciting future for the speciality. Discussion reflected the rapid advances glycobiology is making and ranged from the basic biochemistry of carbohydrate physiology to therapeutic trials utilizing synthetic sugars designed to block inflammatory responses.

The meeting is summarized in considerable detail in this book which will provide the interested scientist and clinician with the essential up-to-date facts within the field of glycoimmunology.

Acknowledgments

Many people have been involved in ensuring the success of the Jenner Glycoimmunology Meetings but none more so than my secretary Susan Henderson who has borne the brunt of all four meetings and is currently preparing for the 5th.

Dr. John S. Axford
Academic Rheumatology Unit
St. George's Hospital Medical School
Cranmer Terrace, London SW17 ORE, UK
Email: j.axford@sghms.ac.uk

CONTENTS

GLYCOBIOLOGY: THE BASICS

OLIGOSACCHARIDES AND PROTEIN RECOGNITION

OLIGOSACCHARIDES AND BIOLOGICAL FUNCTION

GLYCOSYLATION AND INFLAMMATION

GLYCOSYLATION AND DISEASE

GLYCOTHERAPEUTICS

GLYCOBIOLOGY: THE BASICS

NOVEL PATHWAYS IN COMPLEX-TYPE OLIGOSACCHARIDE SYNTHESIS
New Vistas Opened by Studies in Invertebrates

Dirk H. Van den Eijnden[1], Alex P. Neeleman[2], Hans Bakker[1]
and Irma Van Die[1]

[1]Department of Medical Chemistry
[2]Department of Medical Microbiology
Vrije Universiteit
Van der Boechorststraat 7
1081 BT Amsterdam, The Netherlands

Protein- and lipid-linked complex-type oligosaccharide chains typically contain N-acetyllactosamine (Galβ1→4GlcNAc, LacNAc) units as building blocks of their so called backbone regions. These backbones are the carriers of the terminal structures such as Lewis^X and sialyl-Lewis^X, that confer specific properties on the glycoconjugates carrying them.

In an increasing number of cases, however, a N,N'-diacetyllactosediamine (GalNAcβ1→4GlcNAc, LacdiNAc) unit is found in stead of the LacNAc unit in these backbones (Van den Eijnden *et al.*, 1995 and references cited therein). These backbone structures may carry terminal sialic acid or fucose residues in structures that are analogous to those occurring on LacNAc chains. Interestingly, termination by α2→6-linked NeuAc or sulfation at carbon 4 (of GalNAc in LacdiNAc) is confined to the glycans of mammalian glycoconjugates. By contrast, terminal α1→3-linked Fuc (to form the LacdiNAc analog of the Lewis^X epitope) occurs in vertebrate as well as in invertebrate glycans (Table 1). On the other hand, several other terminal motifs are only found in invertebrates.

As LacdiNAc based chains occur relatively abundant in invertebrate glycoconjugates we attempted to study the biosynthesis of this unit in lower animals. We have characterized novel UDP-GalNAc:GlcNAcβ-R β1→4-N-acetylgalactos-aminyltransferases (β4-GalNAcTs) in cercariae of the schistosome *Trichobilharzia ocellata* (Neeleman *et al.*, 1994), the albumen gland (Mulder *et al.*, 1995) and the prostate gland of the snail *Lymnaea stagnalis*, and in several insect cell lines (Van Die *et al.*, 1996) that appear to be involved in the biosynthesis of such LacdiNAc-type oligosaccharide chains. By others this enzyme has also been found in the adult worms of the schistosome *Schistosoma mansoni* (Srivatsan *et al.*, 1994). These β4-GalNAcTs control the LacdiNAc pathway of complex-type oligosaccharides in these organisms.

Table 1.

Occurrence in different species of various terminal substitution motifs on
N,N′-diacetyllactosediamine (lacdiNAc) units of protein- and lipid-linked glycans

Terminal structure on lacdiNAc-type glycan	Occurrence in
NeuAcα2→3 GalNAcβ1→4GlcNAc..	reptiles
NeuAcα2→6 GalNAcβ1→4GlcNAc..	mammals
SO₄⁻-4 GalNAcβ1→4GlcNAc..	mammals
GalNAcβ1→4[**Fucα1→3**]GlcNAc..	mammals, reptiles, amphibians, lower fishes, insects, schistosomes
Tyv1→3 GalNAcβ1→4[**Fucα1→3**]GlcNAc..	nematodes
***Galβ1→3**GalNAcβ1→4GlcNAc..	molluscs
***GalNAcα1→4**GalNAcβ1→4GlcNAc..	insects
Galα1→3**GalNAcβ1→4[Fucα1→3**]GlcNAc..	schistosomes
Fucα1→3**GalNAcβ1→4[Fucα1→3**]GlcNAc..	schistosomes

The sugars marked by an * may be further substituted

They act analogously to mammalian β4-galactosyltransferase (β4-GalT) (Figure 1) and share with this enzyme its acceptor properties and, sometimes, the responsiveness to α-lactalbumin (Neeleman and Van den Eijnden, 1996). Unlike the β4-GalNAcT that has been described in mammalian pituitary gland (Smith and Baenziger, 1992) the invertebrate enzymes are glycoprotein hormone unspecific and act on any acceptor structure that carries a β-linked GlcNAc at its non-reducing end regardless of the underlying structure.

In view of these results we anticipated that mammalian β4-GalT and invertebrate β4-GalNAcT would show a molecular similarity that could be exploited to isolate DNA sequences coding for β4-GalNAcT by cross hybridization using β4-GalT cDNA as a probe. Indeed a cDNA (clone P11.4) could be isolated from a snail prostate gland

invertebrate β4-*N*-acetylgalactosaminyltransferase

$$Mn^{2+}$$

UDP-GalNAc + GlcNAc-R ————————> GalNAcβ1→4GlcNAc-R + UDP

mammalian β4-galactosyltransferase

$$Mn^{2+}$$

UDP-Gal + GlcNAc-R ————————> Galβ1→4GlcNAc-R + UDP

Fig. 1. β4-*N*-acetylgalactosaminyltransferase of cercariae of the schistosome *Trichobilharzia ocellata* and the albumen gland of the snail *Lymnaea stagnalis*, and mammalian β4-galactosyltransferase catalyze analogous reactions.

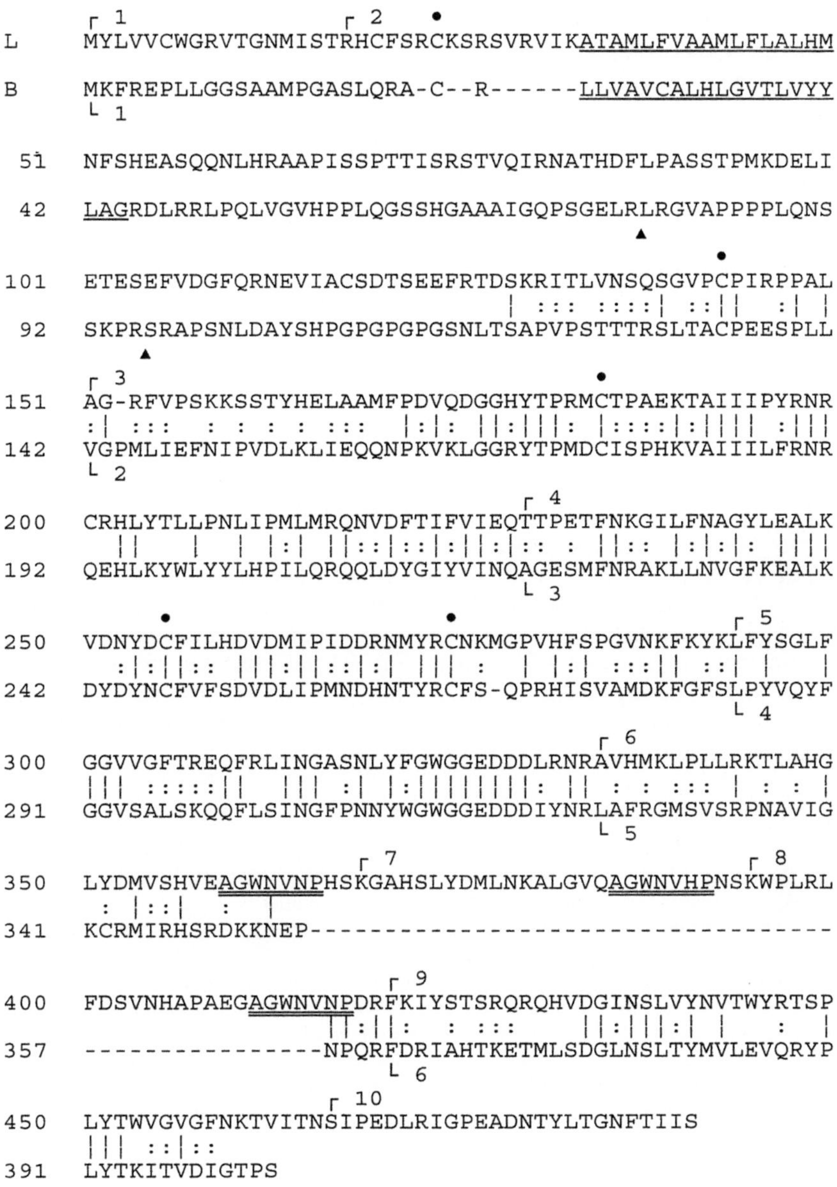

Fig. 2. Comparison of the predicted amino acid sequences of *L.stagnalis* P11.4 (L) and bovine (B) β4-galactosyltransferase. ⌐ and ∟, start of exon; |, Conserved amino acid; :, conservative substitution; _ , membrane spanning domain; ═ , repeat; -, empty place to optimize alignment; ▲, proteolytic cleavage site to yield soluble, fully active enzyme. • indicate conserved Cys residues.

library coding for a protein with a domain structure typical for a glycosyltransferase (GT) and showing considerable similarity with β4-GalT (Figure 2). Unexpectedly, expression of this cDNA resulted in the production of an enzyme that catalyzes the transfer of GlcNAc, rather than Gal or GalNAc, in β1→4-linkage to β-*N*-acetyl-glucosaminides (Bakker *et al.*, 1994). This novel β4-*N*-acetylglucosaminyltransferase (β4-GlcNAcT) has not been described before and its existence could not be predicted from known oligosaccharide structural data. Activity of a naturally ocurring form of this enzyme was found in the prostate gland of the snail.

The intron-exon partition of the β4-GlcNAcT gene shows remarkable similarities to that of β4-GalT (cf. Figure 2). Interestingly, two exons in the β4-GlcNAcT show a high degree of similarity to part of the preceeding exon suggesting that they have originated by exon duplication. The corresponding 'single-copy' exon in β4-GalT has been implicated in nucleotide-sugar binding. Deletion of the two repeated exons in the snail gene resulted in a β4-GlcNAcT with slightly altered sugar-donor specificity (Bakker *et al.*, 1997a) The full length β4-GlcNAcT shows essentially no chitin synthase activity, but rather acts on β6-linked GlcNAc residues on O-linked core 2 and 4 glycans and branching points of blood-group I-active substances (Bakker *et al.*, 1997b). Therefore, it may function in another variant of complex-type oligosaccharide synthesis, the 'chitobio pathway', which might be common in invertebrates. More specifically, because the β4-GalNAcT acts on the product formed by the β4-GlcNAcT, the formation of a GalNAcβ1→4GlcNAcβ1→4GlcNAc structural element might be the result of the combined action of these enzymes. The extensive sequence similarity found for β4-GalT and β4-GlcNAcT cDNAs and the properties shared by β4-GalT and β4-GalNAcT suggest that these three enzymes might be evolutionary related and represent members of a novel GT family: 'the β4-galactosyltransferase family' (Van den Eijnden and Joziasse, 1993; Bakker *et al.*, 1997c). Interestingly, several species have appeared to contain two or more different transcripts that may encode candidate members of this family.

REFERENCES

Bakker, H., Agterberg, M., Van Tetering, A. Koeleman, C.A.M., Van den Eijnden, D.H. and Van Die, I. (1994) A *Lymnaea stagnalis* Gene, with sequence similarity to that of mammalian β1→4-galactosyltransferases, encodes a novel UDP-GlcNAc: GlcNAcβ-R β1→4-*N*-acetylglucosaminyl-transferase. J. Biol. Chem. <u>269</u>, 30326-30333.

Bakker, H., Van Tetering, A., Agterberg, M., Smit, A.B., Van den Eijnden D.H. and Van Die, I. (1997a) Deletion of two exons from the *Lymnaea stagnalis* β1→4-N-acetylglucosaminyl-transferase gene affects the UDP-sugar donor specificity of the encoded enzyme. submitted.

Bakker, H., Schoenmakers, P.S., Koeleman, C.A.M., Joziasse, D.H., Van Die, I. and Van den Eijnden, D.H. (1997b) The substrate specificity of the snail *Lymnaea stagnalis* UDP-GlcNAc:GlcNAcβ-R β4-*N*-acetylglucosaminyltransfer-ase reveals a novel variant pathway of complex-type oligosaccharide synthesis. Glycobiology, in press.

Bakker, H., Van den Eijnden, D.H. and Van Die, I. (1997c) Mammalian β4-galactosyltransferase and *Lymnaea stagnalis* β4-*N*-acetylglucosaminyl-transferase: members of a large and widespread gene family? submitted.

Mulder, H., Spronk, B.A., Schachter, H., Neeleman, A.P., Van den Eijnden, D.H., De Jong-Brink, M., Kamerling, J.P. and Vliegenthart, J.F.G. (1995) Identification of a novel UDP-GalNAc:GlcNAcβ-R β1-4-*N*-acetylgalactosaminyltransferase from connective tissue and the albumen gland of the snail *Lymnaea stagnalis*. Eur. J. Biochem. <u>227</u>, 175-185.

Neeleman, A.P., Van der Knaap, W.P.W. and Van den Eijnden, D.H. (1994) Identification and characterization of an UDP-GalNAc: GlcNAcβ-R β1→4-*N*-acetylgalactosaminyltransferase from cercariae of the schistosome *Trichobilharzia ocellata*. Catalysis of a key step in the synthesis of *N,N*'-diacetyllactosediamino (lacdiNAc)-type glycans. Glycobiology <u>4</u>, 641-651.

Neeleman, A.P. and Van den Eijnden, D.H. (1996) α-Lactalbumin affects the acceptor specificity of *Lymnaea stagnalis* albumen gland UDP-GalNAc:GlcNAcβ-R β1→4-*N*-acetylgalactosaminyl-transferase. Synthesis of GalNAcβ1→4Glc. Proc. Natl. Acad. Sci. USA, <u>93</u>, 10111-10116.

Smith, P.L. and Baenziger, J.U. (1992) Molecular basis of recognition by the glycoprotein hormone-specific *N*-acetylgalactosaminetransferase. Proc. Natl. Acad. Sci. USA, <u>89</u>, 329-333.

Srivatsan, J., Smith, D.F. and Cummings, R.D. (1994) Demonstration of a novel UDP-GalNAc:GlcNAc β1-4-N-acetylgalactosaminyltransferase in extracts of *Schistosoma mansoni*. J. Parasitol., **80**, 884-890.

Van den Eijnden, D.H. and Joziasse, D.H. (1993) Enzymes associated with glycosylation. Curr. Opin. Struct. Biol. <u>3</u>, 711-721.

Van den Eijnden, D.H., Neeleman, A.P., Van der Knaap, W.P.W., Bakker, H., Agterberg, M. and Van Die, I. (1995) Novel glycosylation routes for glycoproteins: The lac**di**NAc pathway. Biochem. Soc. Trans. <u>23</u>, 175-179.

Van Die, I., Van Tetering, A., Bakker, H. Van den Eijnden, D.H. and Joziasse, D.H. (1996) Glycosylation in lepidopteran insect cells: idetification of a β1→4-*N*-acetylgalactosaminyl-transferase involved in the synthesis of complex-type oligosaccharide chains. *Glycobiology*, **6**, 157-164.

DEFECTIVE GLYCOSYLTRANSFERASES ARE NOT GOOD FOR YOUR HEALTH

Harry Schachter,[1] Jenny Tan,[1] Mohan Sarkar,[1]
Betty Yip,[1] Shihao Chen,[1], James Dunn,[2] and Jaak Jaeken[3]

[1]Hospital for Sick Children, 555 University Avenue, and
Dept. of Biochemistry, Univ. of Toronto, Toronto, Ont. Canada
[2]Visible Genetics Inc., Toronto, Ont. Canada
[3]Univ. Hospital Gasthuisberg, Univ. of Leuven, Leuven, Belgium

INTRODUCTION

The history of biochemical research on the structure and function of glycoproteins and glycolipids (glycoconjugates) dates back to the early 19th century (Montreuil, 1995). The variety of these macromolecules, their wide-spread distribution among all forms of life and their quantitative contribution to the bio-mass rank with those of proteins and nucleic acids. Nevertheless for a long period of time the lack of decisive information on the functions of glycoconjugates (Varki, 1993) discouraged most biochemists from entering this field. A series of discoveries over the past 25-30 years has slowly changed this attitude, e.g., the demonstration of mammalian lectins and of the association of defects in glycoconjugate metabolism with human diseases such as cancer, inflammatory and infectious diseases and various congenital diseases (Montreuil et al., 1996). Five autosomal recessive diseases have been reported to date (Table 1) in which a defect in the synthesis of asparagine-linked carbohydrate (N-glycans) has been clearly demonstrated. This presentation will focus on three of these diseases, carbohydrate-deficient glycoprotein syndromes Types I and II (CDGS I and II) and hereditary erythroblastic multinuclearity with a positive acidified serum lysis test (HEMPAS, Congenital Dyserythropoietic Anemia Type II). The study of these diseases should provide valuable information on the roles of N-glycans in human development.

CARBOHYDRATE-DEFICIENT GLYCOPROTEIN SYNDROME TYPE I

The carbohydrate-deficient glycoprotein syndromes (CDGS) are a group of congenital multisystemic diseases characterized by defective glycosylation of N-glycans and moderate to severe abnormalities of the nervous system and most other organs (Blennow et al., 1991; Jaeken et al., 1991; Hagberg et al., 1993; Jaeken and Carchon, 1993; Jaeken et al., 1993a; Stibler et al., 1994). CDGS I patients were first described in 1980 (Jaeken et al., 1980) and over 200 cases are now known world-wide. Three other variants, types II (Ramaekers et al., 1991; Jaeken et al., 1993b; Jaeken et al., 1994; Charuk et al., 1995; Tan et al., 1996), III (Stibler et al., 1993) and IV (Stibler et al., 1995), have been reported but these are rare and represented by only two families each.

Table 1. Autosomal recessive disorders with defective N-glycan synthesis

Name of disease	Type of disease	Enzyme defect	Reference (footnote)
Inclusion Cell Disease (I-cell disease) (mucolipidosis II)	Lysosomal glycoprotein storage disease with severe developmental abnormalities	Phospho-N-acetyl glucosaminyl-transferase required for synthesis of Man-6-phosphate targeting signal	1
Leukocyte Adhesion Deficiency Type II (LAD II)	Immunodeficiency and severe develop-mental abnormalities	Defective synthesis of GDP-Fucose	2
Carbohydrate-Deficient Glycoprotein Syndrome Type I (CDGS I)	Severe developmental abnormalities	Phosphomanno-mutase with defective synthesis of GDP-Mannose (85% of cases)	3
Carbohydrate-deficient Glycoprotein Syndrome Type II (CDGS II)	Severe developmental abnormalities	GlcNAc-transferase II deficiency	4
HEMPAS (Congenital Dyserythropoietic Anemia Type II)	Relatively mild disease with mild anemia and haemosiderosis	α-mannosidase II deficiency. Other factors?	5

1. (Kornfeld and Mellman, 1989; Kornfeld, 1990)
2. (Etzioni et al., 1992; Von Andrian et al., 1993; Etzioni et al., 1995; Phillips et al., 1995)
3. (Jaeken et al., 1980; Wada et al., 1992; Jaeken et al., 1993a; Yamashita et al., 1993a; Yamashita et al., 1993b; Knauer et al., 1994; Powell et al., 1994; Yasugi et al., 1994; Panneerselvam and Freeze, 1995; Van Schaftingen and Jaeken, 1995; Panneerselvam and Freeze, 1996a)
4. (Ramaekers et al., 1991; Jaeken et al., 1993b; Jaeken et al., 1994; Charuk et al., 1995; Tan et al., 1996)
5. (Fukuda et al., 1987; Fukuda, 1990; Fukuda et al., 1990)

Clinical Features and Clinical Biochemistry of CDGS I

CDGS I patients show mental retardation, cerebellar hypoplasia, olivopontocerebellar atrophy (OPCA), cerebellar ataxia, hypotonia, polyneuropathy, oculomotor disturbance, retinal pigmentary degeneration, skeletal deformities, developmental delay, failure to thrive, hepatic dysfunction, dysmorphic features and a distinctive lipodystrophy (Ohno et al., 1992; Hagberg et al., 1993; Jaeken and Carchon, 1993; Jaeken et al., 1993a; Yamashita and Ohno, 1996). The onset and course of these features are variable. Both sexes are equally affected and 20% of the patients die within the first year. Computerized tomography shows a hypoplastic cerebellum and a small brain stem. Language and motor development are severely delayed.

Clinical biochemistry analysis in infancy shows elevated serum transaminases and lactate dehydrogenase, hypoalbuminemia, hypocholesterolemia and low serum cholinesterase activity. Significant abnormalities are observed in the levels of plasma proteins which inhibit coagulation and in endocrine status (De Zegher and Jaeken, 1995).

The most valuable diagnostic test is either isoelectric focusing or chromatofocusing of serum glycoproteins such as transferrin or α_1-antitrypsin in order to demonstrate the presence of under-glycosylated and under-sialylated glycoforms. Normal transferrin has two occupied N-glycan sites and 95% of these transferrin molecules have a disialylated, biantennary N-glycan at each site (this transferrin glycoform is termed S4 since it has 4 sialic acid residues per mole) ; the remaining 5% of molecules have at least one trisialylated, triantennary N-glycan. Analysis of the transferrin glycoforms in CDGS I serum by isoelectric focusing or chromatofocusing reveals about 50% reduction in S4 with concomitant appearance of glycoforms S2 and S0 (bearing 2 and 0 sialic acid residues per mole respectively).

The N-glycan structures on transferrin glycoforms S4, S2 and S0 from Japanese CDGS I patients were analyzed by lectin affinity chromatography and gel filtration chromatography before and after glycosidase digestion, by electrospray ionization mass spectrometry (ESI/MS) (Wada et al., 1992; Yamashita et al., 1993a; Yamashita et al., 1993b; Yamashita and Ohno, 1996) and by matrix-assisted laser desorption time-of-flight mass spectrometry (MALDI-TOF MS) (Wada et al., 1994). The only N-glycans detected on S4 and S2 were disialylated, biantennary N-glycans while S0 was completely devoid of carbohydrate. The data indicate an "all-or-none" situation in which each glycosylation site on transferrin is either occupied by a normal N-glycan or is not occupied. Sugar analyses of serum glycoprotein N-glycans from two CDGS I patients showed no qualitative differences between CDGS I and control samples although the total hexose content of the patient samples showed a 30% reduction relative to protein (Powell et al., 1994) . Metabolic labeling of fibroblasts using incorporation of radioactive Man showed no qualitative differences in N-glycan composition between control and CDGS I fibroblasts (Powell et al., 1994). These data support the "all-or-none" model suggested by the Japanese workers.

The serum protein glycoform patterns are normal in the CDGS I fetus and therefore pre-natal diagnosis is not possible by this method. Abnormal glycoform patterns appear in the second to third weeks of post-natal life.

Linkage analysis of DNA from 25 CDGS I pedigrees of Scandinavian, German and Belgian origin has localized the CDGS I gene to chromosome 16p13.3-p13.12 (Martinsson et al., 1994).

The Biochemical Defect in CDGS I

The "all-or-none" pattern of transferrin glycosylation described above suggests that the defect in CDGS I involves the oligosaccharyltransferase-catalyzed transfer of oligosaccharide from dolichol-pyrophosphate-oligosaccharide (Dol-PP-oligosaccharide) to the asparagine residue of the protein core. Since oligosaccharyltransferase activity was shown to be normal in fibroblasts from four CDGS I patients (Knauer et al., 1994), attention was focused on the synthesis of Dol-PP-oligosaccharide.

Extracts of fibroblasts from three Japanese CDGS I patients had normal amounts of dolichyl phosphate and normal levels of UDP-GlcNAc:dolichyl phosphate GlcNAc-1-phosphate transferase, the enzyme which makes Dol-PP-N-acetylglucosamine, the first step in the synthesis of Dol-PP-oligosaccharide (Yasugi et al., 1994).

Radioactive mannose incorporation into fibroblasts showed significant reduction of both protein-linked and dolichol-linked oligosaccharides in CDGS I cells relative to controls (Powell et al., 1994; Krasnewich et al., 1995; Panneerselvam and Freeze, 1996a). The dolichol-linked oligosaccharides in CDGS I fibroblasts were more heterogeneous and contained significant amounts of oligosaccharides smaller than the $Glc_3Man_9GlcNAc_2$ structure found in normal cells. The following enzymes in the metabolic pathways from Man and Glc to GDP-Man and dolichol-phospho-mannose (Dol-P-Man) were found to be normal or slightly reduced in CDGS I fibroblasts: hexokinase, phosphoglucose isomerase and phosphomannose isomerase (Panneerselvam and Freeze, 1995; Van Schaftingen and Jaeken, 1995). Recently, it has been shown that fibroblast extracts from 85% of CDGS I patients show less than 10% of normal phosphomannomutase activity (conversion of Man-6-phosphate to Man-1-phosphate) (Van Schaftingen and Jaeken, 1995); phosphoglucomutase activity was normal. It has also been observed that CDGS I fibroblasts show considerably lower incorporation of exogenous radioactive Man into total and free intracellular Man, GDP-Man/GDP-Fuc, Man-1-phosphate and Man-6-phosphate (Krasnewich et al., 1995; Panneerselvam and Freeze, 1996a). Surprisingly, incorporation into Dol-P-Man was reported to be normal (Panneerselvam and Freeze, 1996a).

Phosphomannomutase deficiency can be reconciled with the "all-or-none" model as follows. The enzyme deficiency should result in the accumulation of Man-6-phosphate and Man and in a reduced synthesis of Man-1-phosphate, GDP-Man and Dol-P-Man. The high level of free Man will reduce the specific activity of administered radioactive Man thereby accounting for the observed reduced incorporation of label into free and phosphorylated Man. The reduced levels of GDP-Man and Dol-P-Man will result in a reduced rate of synthesis of Dol-PP-oligosaccharide as well as the production of truncated dolichol-linked oligosaccharides. The oligosaccharyltransferase that transfers oligosaccharide from Dol-PP-oligosaccharide to protein can use as donor not only the physiological Dol-PP-Glc$_3$Man$_9$GlcNAc$_2$ but also shorter lipid-linked oligosaccharides, including Dol-PP-GlcNAc$_2$ (Struck and Lennarz, 1980; Verbert, 1995). However, the enzyme shows a marked preference for glucosylated Dol-PP-oligosaccharides and therefore most of the oligosaccharides transferred to protein from both full-length and truncated donors will contain the Man(α1-3)Man(β-) arm of the N-glycan core. These protein-bound oligosaccharides will be processed to a substrate for GlcNAc-transferase I either by the usual pathway or by an "alternate pathway" independent of α-mannosidase II if the Man(α1-6) arm is truncated (Kornfeld and Kornfeld, 1985). The low levels of Dol-PP-oligosaccharides will result in a lowered rate of activity of the oligosaccharyltransferase thereby leaving some Asn-X-Ser/Thr sequons unglycosylated. Any site that is glycosylated will be processed normally, e.g., to a disialylated, biantennary N-glycan in the case of transferrin.

Addition of exogenous Man, but not Glc, to CDGS I fibroblasts corrects both the lowered incorporation of radioactive Man into glycoproteins and the synthesis of short lipid-linked oligosaccharides (Panneerselvam and Freeze, 1996a). Since all the enzymes which convert Glc to Man-6-phosphate are normal in CDGS I fibroblasts, the preferential effect of exogenous Man is difficult to explain. It is possible that intracellular Man in these cells is derived primarily from sources other than the presumed conversion from Glc (Glc - Glc-6-P - Fru-6-P - Man-6-P). Although Man can utilize the Glc transporter, a distinct Man-specific Glc-independent transporter has been demonstrated in mammalian cells (Panneerselvam and Freeze, 1996b). Since the high levels of Glc in serum would prevent entry of Man through the Glc transporter, serum Man may enter through the Man-specific transporter and be a major precursor of N-glycan synthesis. The Glc transporter was shown to be normal in CDGS I cells but there was a significant reduction in Man entry into these cells (Panneerselvam and Freeze, 1995; Panneerselvam and Freeze, 1996a; Panneerselvam and Freeze, 1996b). The role of the Man transporter in the pathogenesis of CDGS I remains to be established but it is an exciting possibility that feeding of Man to CDGS I patients may have some therapeutic value.

Since only 85% of CDGS I patients show phosphomannomutase deficiency, it is clearly a heterogeneous disease. Further evidence for this was obtained in a study of the N-glycan composition of serum α1-antitrypsin (α1-AT) from twin Danish CDGS I patients (Krasnewich et al., 1995). This protein has three N-glycosylation sites all of which were occupied in the CDGS I patients. These N-glycans showed reduced sialylation, increased ratios of Man to both Gal and GlcNAc and insensitivity to endo-β-N-acetylglucosaminidase H suggesting the presence of truncated oligomannose and hybrid N-glycans. This pattern of glycosylation is clearly different from the "all-or-none" model described above. The fibroblasts from these patients show defective incorporation of Man into phosphorylated Man and lipid-linked oligosaccharides but the exact biochemical lesion remains to be determined. Yet another variation was recently reported (Yamashita et al., 1996) in which radioactive Man incorporation into Man-6-phosphate, Man-1-phosphate and GDP-Man of synchronized CDGS I fibroblasts was higher than normal but in which there was deficient Dol-PP-oligosaccharide synthesis due probably to a deficiency in dehydrodolichol reductase.

CONTROL OF COMPLEX N-GLYCAN SYNTHESIS

Both CDGS II and HEMPAS (Table 1) are diseases in which there is a defect in the synthesis of complex N-glycans. The first committed step in the conversion of oligomannose to complex and hybrid N-glycans is catalyzed by UDP-GlcNAc:α-3-D-mannoside β-1,2-N-acetylglucosaminyltransferase I (GnT I; EC 2.4.1.101, Fig.1). The

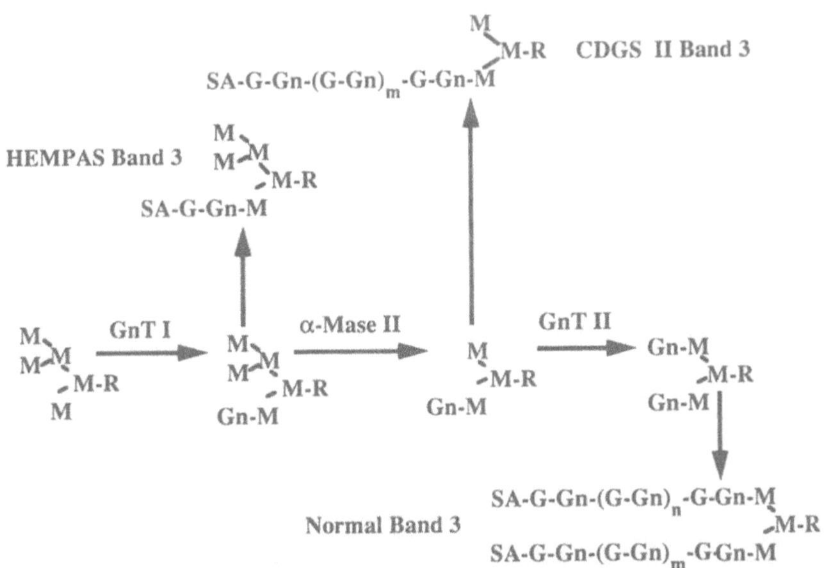

Figure 1. Part of the N-glycan synthesis pathway for erythrocyte Band 3 showing oligosaccharides which accumulate with α-mannosidase II (α-Mase II) deficiency in HEMPAS and GlcNAc-transferase II (GnT II) deficiency in CDGS II. SA, sialic acid; G, Gal; GN, GlcNAc; M, Man; R, N-glycan core attached to Band 3.

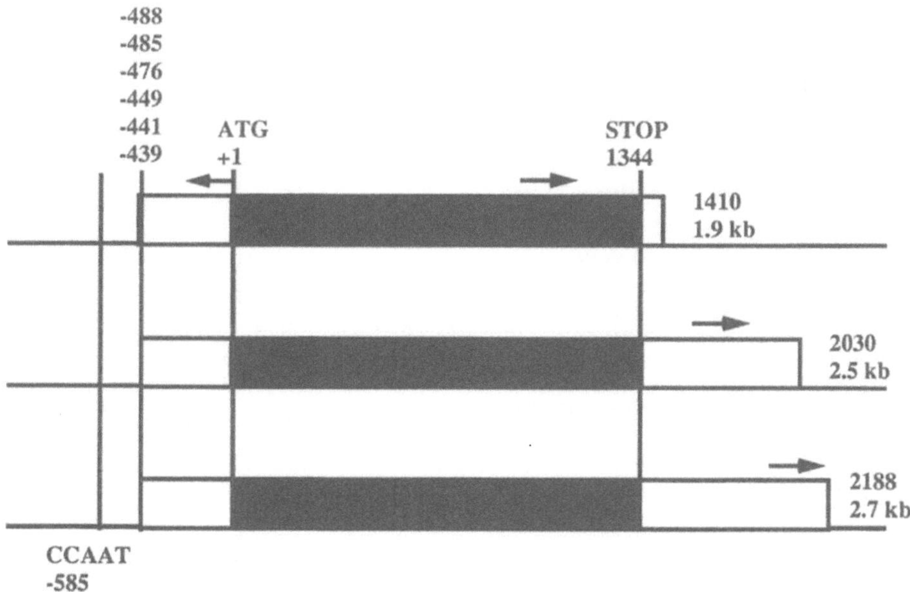

Figure 2. Organization of the single exon human GnT II gene showing the three transcription termination end-points at 1410, 2030, and 2188 nucleorides downstream from the ATG translation start codon, the multiple transcription start sites, and the CCAAT box. The three message sizes (1.9, 2.5, and 2.7 kb) are shown. The boxes show the three mRNAs; the solid boxes represent the open reading frame.

addition of a GlcNAc residue to the Man(α1-3) arm of the N-glycan core is an essential pre-requisite for the action of several enzymes leading to complex N-glycans, α-3/6-mannosidase II, GlcNAc-transferases II, III and IV and core α-1,6-fucosyltransferase (Schachter, 1986; Schachter, 1995).

The GlcNAc-transferase I gene

The human GnT I gene (*MGAT1*) has been mapped to chromosome 5q35 (Kumar et al., 1990; Hull et al., 1991; Tan et al., 1995). The gene encodes a type II transmembrane protein. There are at least two exons in the human GnT I gene, a 2.5 kb exon containing 127 bp of the 5'-untranslated region and the complete coding and 3'-untranslated regions, and a non-coding up-stream exon with the remainder of the 5'-untranslated region. The two exons are between 5.6 and 15 kb apart. There are multiple transcription start sites for the gene compatible with the expression by several human cell lines and tissues of two transcripts, a broad band ranging in size from 2.7 to 3.0 kb and a sharper band at 3.1 kb (Yip et al., 1996). The 5'-flanking region of the upstream exon has a GC content of 81% and has no canonical TATA or CCAAT boxes but contains potential binding sites for transcription factors that recognize GC-rich sequences. CAT expression was observed on transient transfection into HeLa cells of a fusion construct containing the CAT gene and a genomic DNA fragment from the 5'-flanking region of the up-stream exon. It is concluded that *MGAT1* is a typical housekeeping gene.

Similar to previous findings in mouse (Kumar et al., 1992; Pownall et al., 1992) and rat (Fukada et al., 1994), human brain expressed only the larger 3.1 kb transcript (Yip et al., 1996). Data on the mouse and human GnT I genes differ in two respects: there are at least two non-coding up-stream exons in the mouse gene whereas only one such exon has to date been identified in the human gene and the mouse transcripts are at about 2.9 and 3.3 kb respectively. The increased length of the mouse brain 3.3 kb transcript was shown to reside entirely in the 5'-untranslated region and evidence was obtained for the presence of two distinct promoters for the short and long mouse transcripts (Yang et al., 1994). A probe specific for the mouse 3.3 kb transcript was developed (Yang et al., 1994). Although the evidence for a human tissue-specific GnT I transcript (Yip et al., 1996) is not as conclusive as that obtained for the mouse (Yang et al., 1994), the data suggest that there are tissue-specific promoters for the longer transcript in both species. Since the proteins produced by the two transcripts are the same, it is likely that their purpose is to allow differential regulation of enzyme levels.

Mice lacking a functional GnT I gene die at about 10 days after fertilization with multiple developmental abnormalities particularly in the brain (Ioffe and Stanley, 1994; Metzler et al., 1994) indicating that *N*-acetyllactosamine- and hybrid-type N-glycans are important in cell-cell interactions.

The GlcNAc-transferase II gene

UDP-GlcNAc:α-6-D-mannoside β-1,2-*N*-acetylglucosaminyltransferase II (GnT II; EC 2.4.1.143) catalyzes incorporation of a GlcNAc residue in β-1,2 linkage to the Man(α1-6) arm of the N-glycan core and is therefore an essential step in the biosynthetic pathway leading from hybrid to complex N-glycans (Fig.1). The genes encoding rat (D'Agostaro et al., 1995) and human (Tan et al., 1995) GnT II have been cloned and encode typical type II transmembrane proteins. Southern blot analysis indicated only a single copy of the GnT II gene (*MGAT2*) in the human genome. The gene was mapped to human chromosome 14q21. The 1341 bp open reading frame is flanked by a GC-rich 5'-untranslated region and a long AT-rich 3'-untranslated region containing three canonical polyadenylation signals (AATAAA) at 68, 688 and 846 bp downstream from the translation stop codon. Northern analysis has shown a major transcript at 2.0 kb and a minor band at ~2.9 kb in five different human cell lines (Chen et al., 1996) and a major transcript at ~3.0 kb in several human tissues (Tan et al., 1995). RACE analysis of human cell line LS180 (Fig.2) showed that all three AATAAA sequences are utilized for transcription termination, that there are multiple transcription initiation sites within a 50 bp region and that the entire GnT II gene is on a single exon (Chen et al., 1996). The gene has a CCAAT box and lacks a TATA box but an initiator (Inr)-like element (Pugh and Tjian, 1990; Madden et al., 1993) was identified at

-417 bp. The GC-rich 5´-untranslated region contains consensus sequences suggestive of multiple binding sites for Sp1 and other transcription factors.

Chimeric constructs containing different lengths of the 5'-flanking region fused to the CAT reporter gene were tested in transient transfection experiments using HeLa cells (Chen et al., 1996). A region between -636 and -553 bp relative to the ATG start codon (+1) was identified as the main promoter region of the gene. Deletion of the 5'-flanking region from ~-1,000 to -680 bp resulted in a small reduction in CAT activity. There was no significant difference in CAT activity between -680 and -636 bp. A further deletion of 83 bp resulted in a 7-fold reduction in CAT activity.

An unusual feature of this GnT II promoter region is the presence of proto-oncogene transcription factor targeting sites for c-Myb and Est proteins. Ets proteins have been implicated in development and differentiation (Wasylyk et al., 1993). Since GnT II action is required for GnT V action, it is of interest that binding sites for c-Myb and Ets have also been reported in the promoter region of GnT V whose product is related to malignant transformation (Saito et al., 1995).

The absence of a TATA box, the multiple transcription start sites and the GC-rich nature of the promoter region all indicate that GnT II is a typical housekeeping gene. However, some of the properties of both the 5'- and 3'-flanking regions suggest that regulation of GnT II gene expression may play a role in development and differentiation.

The α-mannosidase II gene

Processing of the product of GnT I by α-3/6-mannosidase II (EC 3.2.1.114, MII) to form the specific substrate for GnT II (Fig.1) is the first committed step in the conversion of hybrid to complex N-glycans (Tulsiani et al., 1977; Harpaz and Schachter, 1980; Kornfeld and Kornfeld, 1985; Schachter, 1991). Like GnT I, and probably also GnT II, rat liver MII is a type II integral membrane glycoprotein facing the lumen of the Golgi membrane (Velasco et al., 1993). A 110 kDa catalytically active protein was derived from the 124 kDa membrane protein by proteolysis, purified as a disulfide-linked dimer (Moremen and Touster, 1986; Moremen et al., 1991) and the corresponding cDNA was partially cloned (Moremen, 1989). Northern analysis of rat liver RNA showed a message of about 8 kb, a size approximately 4.7 kb larger than the size of the predicted open reading frame. The function of this disproportionately large untranslated region is not clear. Murine MII cDNA was later cloned (Moremen and Robbins, 1991) encoding the full length open reading frame and most of the 5'- and 3'-untranslated regions. MII is a large type II transmembrane protein containing a short cytoplasmic tail (5 amino acids), a single transmembrane domain (21 amino acids) and a large COOH-terminal catalytic domain (1124 amino acids). This domain organization, which is shared with all the Golgi-localized glycosyltransferases cloned to date, suggests that the common structural motif may have a functional role in Golgi enzyme function or localization. MII cDNA was transiently expressed in COS cells and cross-reactive material was observed in a perinuclear membrane array consistent with Golgi localization. Partial human MII cDNA clones were also isolated and the gene was localized to human chromosome 5q21-22 (Moremen and Robbins, 1991; Misago et al., 1995).

Processing α-mannosidases have been divided into two classes on the basis of catalytic properties and protein sequence homologies (Daniel et al., 1994; Moremen et al., 1994). Class 1 is comprised of calcium -dependent α1,2-mannosidases and class 2 is a more heterogeneous group which includes MII, a lysosomal acidic α-mannosidase and the mammalian cytosolic/endoplasmic reticulum α-mannosidase.

During the isolation of human MII genomic DNA clones (Misago et al., 1995), genes related to that encoding MII were isolated. One such gene was found to encode an isozyme designated α-MIIx. A 5 kb cDNA α-MIIx clone was isolated encoding a truncated polypeptide with 796 amino acid residues. Alternative splicing of the α-MIIx transcript resulted in an additional transcript encoding a 1139-amino acid polypeptide. Northern analysis showed expression of the gene is many tissues suggesting that the α-MIIx gene is a housekeeping gene. COS cells transfected with α-MIIx cDNA containing the full-length open reading frame showed an increase of enzyme activity. The α-MIIx gene was mapped to human chromosome 15q25.

MII is strongly inhibited by swainsonine (Tulsiani et al., 1982) and mannostatin (Elbein, 1987; Tropea et al., 1990; Pan and Elbein, 1995). Inhibition of MII has been shown to reverse the transformed phenotype of NIH 3T3 cells *in vitro* (De Santis et al., 1987) and to inhibit tumour cell metastasis *in vivo* (Dennis, 1986; Humphries and Olden, 1989).

Since at least some patients with HEMPAS have a deficiency of MII (see below), it is important to point out that there are several alternate processing pathways which are independent of MII action. A deficiency in Dol-P-Man synthesis will result in synthesis of a truncated lipid-linked oligosaccharide, Glc(α1-2)Glc(α1-3)Glc(α1-3)Man(α1-2)Man(α1-2)Man(α1-3)[Man(α1-6)]Man(β1-4)GlcNAc(β1-4)GlcNAc(α1-)-pyrophosphate-dolichol which will be transferred to protein and processed to normal complex N-glycans without the need for MII action (Kornfeld et al., 1979; Kornfeld and Kornfeld, 1985). Under conditions of glucose starvation, Chinese hamster ovary cells accumulate similar truncated lipid-linked oligosaccharides (Rearick et al., 1981). Energy deprivation causes the same effect (Kornfeld and Kornfeld, 1985). The level of MII is very low in rat brain but this tissue has a membrane-bound α1.2/1,3/1,6-mannosidase which can cleave protein-bound Man$_{9-4}$GlcNAc$_2$ to Man$_3$GlcNAc$_2$ thereby providing an MII-independent alternate pathway (Tulsiani and Touster, 1985; Moremen et al., 1994); the enzyme is not inhibited by swainsonine. Similar enzymes have been described in rat liver (Bonay and Hughes, 1991; Bonay et al., 1992) and other tissues (Moremen et al., 1994). The physiological roles of these enzymes are not known but they can provide protection against mutations or environmental factors which interfere with the MII-dependent processing pathway.

CARBOHYDRATE-DEFICIENT GLYCOPROTEIN SYNDROME TYPE II

Two CDGS patients were described recently which differed in part from the classic picture of CDGS I described above, an Iranian girl (Ramaekers et al., 1991) and a Belgian boy (Jaeken et al., 1993b). Both of these patients have been shown to have inactivating point mutations in the GnT II gene *MGAT2* .(Tan et al., 1996)

Clinical Features and Clinical Biochemistry of CDGS II

In contrast to CDGS I, both patients have a more severe psychomotor retardation, no peripheral neuropathy and a normal cerebellum on Magnetic Resonance Imaging. Biochemical differences from CDGS I are the absence of proteinuria, a deficiency of clotting factors IX and XII (in addition to the clotting factor deficiencies found in CDGS I), decreased β-glucuronidase, and normal serum glutamic pyruvic transaminase activity, albumin and arylsulfatase A. The CDGS II serum transferrin isoelectric focusing pattern differs markedly from that in CDGS I (see above) in that there are no detectable hexasialo-, pentasialo- nor asialo-transferrrins, low amounts of tetrasialo- and monosialo-transferrins, a moderate amount of trisialotransferrin and a marked increase of disialotransferrin (Jaeken et al., 1994; Jaeken et al., 1996). Carbohydrate analysis, ESI/MS and high resolution proton NMR spectroscopy of CDGS II serum transferrin glycoforms (Jaeken et al., 1994; Jaeken et al., 1996) showed that both glycosylation sites on the disialotransferrin glycoform were occupied by the same truncated bi-antennary N-glycan:
NeuNAc(α2-6)Gal(β1-4)GlcNAc(β1-2)Man(α1-3)[Man(α1-6)]Man-R
where R is -(β1-4)GlcNAc(β1-4)GlcNAc-Asn-X. This pattern is clearly different from the "all-or-none" model described above for CDGS I. The major disialotransferrin glycoform S2 and the minor glycoforms S1, S3 and S4 in CDGS II serum can be explained as follows.

The two glycosylation sites of the normal transferrin glycoform may be occupied by
S-G-Gn-Man(α1-3)[R2-Man(α1-6)]Man-R1
where S-G-Gn- is NeuNAc(α2-6)Gal(β1-4)GlcNAc(β1-2)- and
R2 can be either one or two antennae consisting of either S-G-Gn- or G-Gn-
Lack of GnT II converts this glycan to
S-G-Gn-Man(α1-3)[Man(α1-6)]Man-R

Or these sites may be occupied by
G-Gn-Man(α1-3)[R2-Man(α1-6)]Man-R1 or
R3-Gn(β1-2)[R3-Gn(β1-4)]Man(α1-3)[R2-Man(α1-6)]Man-R1
where R3 can be either S-G- or G-.

Lack of GnT II converts these glycans respectively to
G-Gn-Man(α1-3)[Man(α1-6)]Man-R1 or
R3-Gn(β1-2)[R3-Gn(β1-4)]Man(α1-3)[Man(α1-6)]Man-R1

It is of interest to point out that the presence of the normal transferrin glycoform S4 in CDGS I serum at about 50% of normal levels indicates that this mutation is quite "leaky". In contrast, the ESI/MS data indicate that there is very little normal S4 in CDGS II serum and that the small amount of S4 seen on isoelectric focusing is probably a truncated glycoform in which both sites are occupied by S-G-Gn(β1-2)[S-G-Gn(β1-4)]Man(α1-3)[Man(α1-6)]Man-R (Jaeken et al., 1996).

The Biochemical Defect in CDGS II

GnT II activity was reduced by over 98% in fibroblast extracts from the two unrelated CDGS type II patients (Iranian patient A and Belgian patient B) (Jaeken et al., 1994) and there was no detectable GnT II activity in mononuclear cell extracts from the Belgian patient (Charuk et al., 1995). Direct sequencing in both directions of the entire GnT II coding region from the two patients identified two point mutations in the catalytic domain of GnT II, Ser290Phe (TCC to TTC) in patient A and His262Arg (CAC to CGC) in patient B (Tan et al., 1996). Both mutations occur in the C-terminal catalytic domain (Schachter, 1994; Schachter, 1995) at locations which are conserved between rat and human GnT II (Tan et al., 1995). Both patients are homozygous for their respective mutations and have therefore inherited the same allele from each parent. The father, mother and brother of patient B show one normal allele and one allele with the same mutation as patient B.

The CDGS II mutations were introduced into the normal *MGAT2* gene followed by expression in the baculovirus/Sf9 system (Tan et al., 1996). No enzyme activity was detected in cells transfected by either of the two mutant genes (less than 1% of the control value). Western blot analysis showed that both mutant proteins were expressed at about 8% of the level of normal GnT II expression indicating that the mutations either interfere with transcription-translation or lower the stability of the protein in the baculovirus/Sf9 system. Northern blot analysis showed normal transcription of the GnT II gene in CDGS II fibroblasts. It remains to be established whether there is a defect of either translation or protein stability in human cells. The data suggest that the mutations reduce protein expression and also inactivate GnT II enzyme activity. It is of interest that although the mutations in the two patients are different, they cause the identical phenotype.

Restriction endonuclease analysis of DNA from 23 blood relatives of one of these patients showed that 13 donors were heterozygotes; the other relatives and 21 unrelated donors were normal homozygotes. All heterozygotes showed a significant reduction (33 to 68%) in mononuclear cell GnT II activity. Analysis of patient B's family proves that CDGS II is a recessive autosomal disease located at chromosome 14q21 and that the H262R mutation is not due to polymorphism. The inactivation of GnT II by the S290F mutation in patient A suggests that this mutation is also not due to polymorphism.

"Null" mouse embryos lacking an active GnT I (Ioffe and Stanley, 1994; Metzler et al., 1994) do not survive past 10.5 days of embryonic life and show many developmental defects particularly in the central nervous system. GnT II is essential for the biosynthesis of complex Asn-linked glycans (Schachter, 1986; Schachter, 1995) and CDGS II patients with inactivating mutations of the GnT II gene also develop severe multisystemic developmental abnormalities, especially in the nervous system. Studies with both mice and humans therefore indicate the importance of N-glycans in normal morphogenesis.

HEMPAS

Clinical Features

Ineffective and morphologically abnormal erythropoiesis is called dyserythropoiesis and results in anemia if the output of erythrocytes by the bone marrow is significantly reduced. In 1966 in Toronto, Crookston et al (Crookston et al., 1966) reported a young man (MF) with a life-long history of dyserythropoietic anemia. His bone marrow showed marked hyperplasia and multinuclear erythroblasts. His erythrocytes were lysed by acidified sera from other donors but not by his own serum. The disorder was later found in several other patients, including two sisters (CL, LF) and was determined to be a Congenital Dyserythropoietic Anemia (CDA). The disease was named Hereditary Erythroblastic Multinuclearity with a Positive Acidified-Serum Lysis Test (HEMPAS) (Crookston et al., 1969). Three major types of CDA have been classified (Heimpel and Wendt, 1968). HEMPAS is Type II CDA and is distinguished from Types I and III CDA primarily on the basis of multinuclear erythroblasts and a positive serum lysis test.

In 1987, there were over 120 known cases of HEMPAS (Nathan and Oski, 1987). HEMPAS consists of a group of autosomal recessive disorders, the age of diagnosis varies from infancy to old age and the anemia varies from slight to severe. Jaundice, hepatosplenomegaly, diabetes and gall-stones are common. Intravascular hemolysis with hemoglobinuria has not been reported (Crookston et al., 1972). The body iron stores in HEMPAS are usually increased; hepatic hemosiderosis is common and hepatic cirrhosis has been reported. Electron microscopy shows that the red cell membrane is "duplicated" and covered with pits and plaques (Crookston et al., 1972; Verwilghen et al., 1973). HEMPAS manifests itself primarily as a disease of erythroblasts and erythrocytes although some patients express biochemical defects in leukocytes and other organs.

Structural Studies on HEMPAS Glycans

About 30 per cent of normal individuals have in their serum an antibody which when acidified can lyse HEMPAS red cells. HEMPAS serum does not contain this antibody. HEMPAS red cells also react strongly with anti-i and anti-I sera (Crookston et al., 1969; Crookston et al., 1972; Verwilghen et al., 1973; Nathan and Oski, 1987) suggesting that the abnormal antigen may be carbohydrate in nature. The blood group i and I epitopes are, respectively, linear and branched poly-N-acetyllactosamines (PLs), i.e., [Galβ1-4GlcNAcβ1-3Galβ-]$_n$ and [Galβ1-4GlcNAcβ1-6{Galβ1-4GlcNAcβ1-3}Galβ1-]$_n$. Several laboratories have reported abnormal erythrocyte membrane protein patterns (Anselstetter et al., 1977; Baines et al., 1982; Harlow and Lowenthal, 1982) and decreased membrane protein glycosylation (Scartezzini et al., 1982; Mawby et al., 1983; Zdebska et al., 1987) in HEMPAS red cells.

These findings prompted Fukuda et al (Fukuda et al., 1984b) to carry out chemical studies using galactose oxidase/NaB[^{3}H]$_4$ to label cell surface glycans followed by endo-β-galactosidase to detect PLs. Red cells from three HEMPAS patients (KB, MB, CD) showed the same changes on SDS-PAGE, i.e., no detectable PLs on erythrocyte Bands 3 and 4.5 which are major protein constituents of the red cell membrane. There was concomitant appearance of large amounts of a PL-bearing compound identified as a macroglycolipid (Fukuda et al., 1984b; Fukuda et al., 1986a; Fukuda et al., 1986b; Zdebska et al., 1987). Normal red cells have Bands 3 and 4.5 that are rich in PL (Fig.1) and have a negligible quantity of macroglycolipid. This suggested that the genetic defect in HEMPAS caused a block in the synthesis of protein-bound PL with a shift of these structures to lipid. The accumulation of this lipid may explain the increased reactivity of HEMPAS erythrocytes with anti-i and anti-I sera and/or the susceptibility of the erythrocytes to lysis. Underglycosylated HEMPAS Band 3 aggregates within the red cell membrane and this probably accounts for the abnormal membrane structure of HEMPAS red cells, abnormal erythroblast cell division leading to multinuclearity and dyserythropoiesis with resultant anemia (Fukuda et al., 1986b).

Protein-bound carbohydrate structures on the erythrocytes from four HEMPAS patients (TO, BD, MP, SF) were determined by a combination of chromatography on lectin columns, mass spectrometry, methylation analysis and sequential glycosidase digestions (Fukuda et al., 1987). The major Band 3 oligosaccharide from one of the patients (TO) had the following truncated N-glycan structure (HEMPAS Glycopeptide 1): NeuNAc(α2-6)Gal(β1-4)GlcNAc(β1-2)Man(α1-3)[Man(α1-6)]Man-R where R is -(β1-4)GlcNAc(β1-4)GlcNAc-Asn-X. The other three patients showed a different pattern. The major Band 3 N-glycans were mono- and di-sialylated biantennary complex N-glycans; however, the total erythrocyte membranes also contained some Glycopeptide 1 as well as some of the following hybrid glycopeptide (HEMPAS Glycopeptide 2): NeuNAc(α2-6)Gal(β1-4)GlcNAc(β1-2)Man(α1-3)[Man(α1-6){Man(α1-3)}Man(α1-6)]Man-R (Fig.1). Since normal erythrocytes do not contain truncated or hybrid structures but rather biantennary complex N-glycans with or without PL repeats, it was concluded that the block in TO was at GnT II (Fig. 1). The defect in the other three HEMPAS patients was not evident from the data. These patients had the enzymes required for PL synthesis (since they made the macroglycolipid described above) and were able to make di-sialylated biantennary complex N-glycans but did not make any of the normal PL-containing biantennary N-glycans.

Analysis of erythrocyte membrane glycopeptides from HEMPAS patient GC (Fukuda et al., 1990) showed HEMPAS Glycopeptide 2 as a major component suggesting a block at α-mannosidase II (MII) (Fig. 1) but the presence of some HEMPAS Glycopeptide 1 and of biantennary complex N-glycans indicated that the block was not complete.

The glycans present on transferrin purified from the sera of two HEMPAS patients (TM, BR) were shown to consist of structures ranging from oligomannose to di-sialylated biantennary N-glycans whereas normal transferrin contains only biantennary glycans (Fukuda et al., 1992). Neither normal nor HEMPAS transferrin contains PL structures. The data show that N-glycan assembly by the liver of these HEMPAS patients is defective. HEMPAS transferrin was more rapidly cleared from serum by the liver than normal transferrin due probably to clearance by liver lectins which recognize terminal Man, GlcNAc and Gal residues. Rapid transferrin clearance may play a role in the severe hemosiderosis shown by many HEMPAS patients.

Analysis of HEMPAS lymphocytes and lymphoblasts

Fresh peripheral blood mononucleated cells and Epstein-Barr virus (EBV)-transformed B lymphoblasts have both been used to assay various enzymes involved in N-glycan processing (Fukuda et al., 1987; Fukuda et al., 1990). Lymphocytes from HEMPAS patients TO and BD showed GnT II levels 10% and 30% of normal, respectively. The findings for TO confirm the block at GnT II suggested by the accumulation of Glycopeptide 1 in TO Band 3 (above).

GC lymphoblasts have almost no detectable MII enzyme activity, in agreement with the accumulation of Glycopeptide 2 in GC red cells (above). GC lymphocytes and lymphoblasts show normal levels of GnT I, GnT II and the β1,4-GalT and β1,3-GnT required for PL synthesis. Northern blot analysis of mRNA from cultured EBV-transformed lymphoblasts using a cDNA probe for human MII showed that GC had levels of message less than 10% of normal cells (Fukuda et al., 1990). These results suggest that GC cells contain a mutation in the MII gene that results in inefficient expression of MII mRNA, either through reduced transcription or message instability. The site of the GC mutation has not as yet been elucidated. Recently a second MII gene called α-MIIx (see above) has been described but its role in HEMPAS remains to be elucidated (Misago et al., 1995).

Can HEMPAS be caused by GlcNAc-transferase II deficiency?

Normal adult human erythrocyte Band 3 carries a single biantennary N-glycan with PL chains on both antennae (Fig.1) (Fukuda et al., 1984a) whereas Band 3 from all HEMPAS patients studied to date lacks PL chains (Fukuda et al., 1984b; Fukuda et al., 1987; Fukuda, 1990; Fukuda et al., 1990; Fukuda, 1993). Since there are PL chains on the Man(α1-3) arm of the normal Band 3 N-glycan (Fig.1), a defect in either Gn T II or MII should not result in complete absence of PL chains. Indeed, in contrast to the complete absence of PL on HEMPAS erythrocyte Band 3, PL on CDGS II Band 3 is reduced only by about 50% (Charuk et al., 1995). There is no preferential attachment of PL chains to either of the arms of a biantennary N-glycan although preferential attachment to the Man(α1-6) arm has been reported for tri- and tetra-antennary N-glycans (van den Eijnden et al., 1988). Thus it is expected that GnT II deficit would cause not a complete absence of PL but a 50% reduction as is found in CDGS II. Furthermore, the relatively benign clinical picture of HEMPAS is completely different from the severe abnormalities found in CDGS II (see above) and CDGS II erythrocytes show neither the positive lysis test nor reactivity with anti-i antibodies characteristic of HEMPAS (Charuk et al., 1995). It is concluded that lack of functional GnT II, even if expressed only in erythroblasts, cannot cause HEMPAS; a tissue-specific GnT II defect limited to hematopoietic cells cannot explain the total absence of erythrocyte PL in HEMPAS. The reduced level of lymphocyte GnT II in HEMPAS

patients TO and BD remains an unsolved paradox. Studies on the expression of the
MGAT2 gene in these patients have not as yet been reported.

The abnormality of MII in HEMPAS patient GC has been established at the gene level
suggesting that many of the other HEMPAS patients may also have some sort of defect in
the MII gene. However, studies on the lymphocytes and lymphoblasts of some of these
patients have shown normal levels of MII enzyme activity and normal expression of the
MII message. It is possible that the MII defect in these patients may be expressed only in
the erythroid lineage. Preliminary data from the "null" mutation mouse model lacking a
functional MII gene shows abnormal glycan synthesis only in the erythroid line suggesting
the activation of MII-independent alternate pathways in the other mouse tissues (Jamey
Marth, personal communication). This mouse model cannot explain HEMPAS cases with
normal MII activity in lymphocytes and lymphoblasts; perhaps another MII activity, such
as the α-MIIx discussed above, is active in non-erythroid tissues of these patients. If MII is
in fact defective in HEMPAS erythrocyte biogenesis, why is there no PL incoporation on
the Man(α1-3) arm of the HEMPAS Band 3 glycan? One possibility presently being tested
is that PL addition cannot occur on the Man5 compound which accumulates with MII
deficiency although it clearly occurs on the Man3 compound which accumulates with GnT
II deficiency in CDGS II (Fig.1).

In conclusion, CDGS I and II and HEMPAS, as well as other congential diseases
involving defects in glycan synthesis, provide an excellent approach to the study of glycan
function.

Acknowledgements

These studies were supported by a Medical Research Council of Canada grant to HS and by
a Nationaal Fonds voor Wetenschappelijk Onderzoek of Belgium grant to JJ.

REFERENCES

Anselstetter, V., Horstmann, H.-J. and Heimpel, H., 1977, Congenital dyserythropoietic
anaemia, types I and II; aberrant pattern of erythrocyte membrane proteins in CDA II,
as revealed by two-dimensional polyacrylamide gel electrophoresis, *British Journal of
Haematology*. 35: 209.

Baines, A.J., Banga, J.P.S., Gratzer, W.B., Linch, D.C. and Huehns, E.R., 1982, Red cell
membrane anomalies in congenital dyserythropoietic anaemia, type II (HEMPAS),
British Journal of Haematology. 50: 563.

Blennow, G., Jaeken, J. and Wiklund, L.M., 1991, Neurological findings in the
carbohydrate-deficient glycoprotein syndrome, *Acta Paediatr Scand*. 80: 14.

Bonay, P. and Hughes, R.C., 1991, Purification and characterization of a novel broad-
specificity (alpha1-->2, alpha1-->3 and alpha1-->6) mannosidase from rat liver, *Eur J
Biochem*. 197: 229.

Bonay, P., Roth, J. and Hughes, R.C., 1992, Subcellular distribution in rat liver of a novel
broad-specificity (alpha1-->2, alpha1-->3 and alpha1-->6) mannosidase active on
oligomannose glycans, *Eur J Biochem*. 205: 399.

Charuk, J.H.M., Tan, J., Bernardini, M., Haddad, S., Reithmeier, R.A.F., Jaeken, J. and
Schachter, H., 1995, Carbohydrate-deficient glycoprotein syndrome type II - An
autosomal recessive N-acetylglucosaminyltransferase II deficiency different from
typical hereditary erythroblastic multinuclearity, with a positive acidified-serum lysis
test (HEMPAS), *Eur J Biochem*. 230: 797.

Chen, S., Tan, J. and Schachter, H., 1996, Transcriptional regulation of the human UDP-GlcNAc:α-6-D-mannoside β1-2-*N*-acetylglucosaminyltransferase II gene (*MGAT2*) which controls complex N-glycan synthesis, *Glycoconjugate J*. Submitted.:

Crookston, J.H., Crookston, M.C., Burnie, K.L., Francombe, W.H., Dacie, J.V., Davis, J.A. and Lewis, S.J., 1969, Hereditary erythroblastic multinuclearity associated with a positive acidified-serum test; a typical congenital dyserythropoietic anaemia, *Brit.J.Haematol*. 17: 11.

Crookston, J.H., Crookston, M.C. and Rosse, W.F., 1972, Red-Cell Abnormalities in HEMPAS (Hereditary Erythroblastic Multinuclearity with a Positive Acidified-Serum Test), *Brit.J.Haematol*. 23 (supplement): 83.

Crookston, J.H., Godwin, T.F., Wightman, K.J.R., Dacie, J.V., Davis, J.A., Lewis, S.M. and Patterson, M.J.L. (1966). Congenital Dyserythropoietic Anaemia. International Society of Haematology, XIth Congress, Sydney, Australia,

D'Agostaro, G.A.F., Zingoni, A., Moritz, R.L., Simpson, R.J., Schachter, H. and Bendiak, B., 1995, Molecular cloning and expression of cDNA encoding the rat UDP-N-acetylglucosamine:alpha-6-D-mannoside beta-1,2-N-acetylglucosaminyltransferase II, *J Biol Chem*. 270: 15211.

Daniel, P.F., Winchester, B. and Warren, C.D., 1994, Mammalian alpha-mannosidases-multiple forms but a common purpose?, *Glycobiology*. 4: 551.

De Santis, R., Santer, U.V. and Glick, M.C., 1987, NIH 3T3 cells transfected with human tumor DNA lose the transformed phenotype when treated with swainsonine, *Biochem. Biophys. Res. Communs*. 142: 348.

De Zegher, F. and Jaeken, J., 1995, Endocrinology of the carbohydrate-deficient glycoprotein syndrome type 1 from birth through adolescence, *Pediatr Res*. 37: 395.

Dennis, J.W., 1986, Effects of swainsonine and polyinosinic:polycytidylic acid on murine tumor cell growth and metastasis, *Cancer Res*. 46: 5131.

Elbein, A.D., 1987, Inhibitors of the biosynthesis and processing of N-linked oligosaccharide chains, *Ann. Rev. Biochem*. 56: 497.

Etzioni, A., Frydman, M., Pollack, S., Avidor, I., Phillips, M.L., Paulson, J.C. and Gershoni-Baruch, R., 1992, Brief report: Recurrent severe infections caused by a novel leukocyte adhesion deficiency, *N Engl J Med*. 327: 1789.

Etzioni, A., Phillips, L.M., Paulson, J.C. and Harlan, J.M., 1995, Leukocyte adhesion deficiency (LAD) II, *in*: "Cell Adhesion and Human Disease", J. Marsh and J.A. Goode, ed., John Wiley & Sons Ltd, Baffins Lane, Chichester, England PO19 7UD, 51.

Fukada, T., Iida, K., Kioka, N., Sakai, H. and Komano, T., 1994, Cloning of a cDNA Encoding N-Acetylglucosaminyltransferase I from Rat Liver and Analysis of Its Expression in Rat Tissues, *Biosci Biotechnol Biochem*. 58: 200.

Fukuda, M., Dell, A., Oates, J.E. and Fukuda, M.N., 1984a, Structure of branched glycosaminoglycan, the carbohydrate moiety of band 3 isolated from adult human erythrocytes, *J.Biol.Chem*. 259: 8260.

Fukuda, M.N., 1990, HEMPAS disease: genetic defect of glycosylation, *Glycobiology*. 1: 9.

Fukuda, M.N., 1993, Congenital dyserythropoietic anaemia type II (HEMPAS) and its molecular basis, *Bailliere's Clinical Haematology.* 6: 493.

Fukuda, M.N., Bothner, B., Scartezzini, P. and Dell, A., 1986a, Isolation and characterization of poly-N-acetyllactosaminylceramides accumulated in the erythrocytes of congenital dyserythropoietic anemia type II patients, *Chem.Physics Lipids.* 42: 185.

Fukuda, M.N., Dell, A. and Scartezzini, P., 1987, Primary defect of congenital dyserythropoietic anemia type II. Failure in glycosylation of erythrocyte lactosaminoglycan-proteins caused by lowered N-acetylglucosaminyltransferase II, *J. Biol. Chem.* 262: 7195.

Fukuda, M.N., Gaetani, G.F., Izzo, P., Scartezzini, P. and Dell, A., 1992, Incompletely processed N-glycans of serum glycoproteins in congenital dyserythropoietic anaemia type II (HEMPAS), *Br J Haematol.* 82: 745.

Fukuda, M.N., Klier, G., Yu, J. and Scartezzini, P., 1986b, Anomalous clustering of underglycosylated band 3 in erythrocytes and their precursor cells in congenital dyserythropoietic anemia type II, *Blood.* 68: 521.

Fukuda, M.N., Masri, K.A., Dell, A., Luzzatto, L. and Moremen, K.W., 1990, Incomplete synthesis of N-glycans in congenital dyserythropoietic anemia type II caused by a defect in the gene encoding alpha-mannosidase II, *Proc Natl Acad Sci Usa.* 87: 7443.

Fukuda, M.N., Papayannopoulou, T., Gordon-Smith, E.C., Rochant, H. and Testa, U., 1984b, Defect in glycosylation of erythrocyte membrane proteins in congenital dyserythropoietic anaemia type II (HEMPAS), *British J. Haematology.* 56: 55.

Hagberg, B.A., Blennow, G., Kristiansson, B. and Stibler, H., 1993, Carbohydrate-Deficient Glycoprotein Syndromes - Peculiar Group of New Disorders, *Pediat Neurol.* 9: 255.

Harlow, R.W.H. and Lowenthal, R.M., 1982, Erythrocyte membrane proteins in an unusual case of congenital dyserythropoietic anaemia, type II (CDA II), *British Journal of Haematology.* 50: 35.

Harpaz, N. and Schachter, H., 1980, Control of glycoprotein synthesis. V. Processing of asparagine-linked oligosaccharides by one or more rat liver Golgi α-D-mannosidases dependent on the prior action of UDP-N-acetylglucosamine:α-D-mannoside β-2-N-acetylglucosaminyltransferase I, *J Biol Chem.* 255: 4894.

Heimpel, H. and Wendt, F., 1968, Congenital dyserythropoietic anemia with karyorrhexis and multinuclearity of erythroblasts, *Helvetica Medica Acta.* 34: 103.

Hull, E., Sarkar, M., Spruijt, M.P.N., Höppener, J.W.M., Dunn, R. and Schachter, H., 1991, Organization and localization to chromosome 5 of the human UDP-N-acetylglucosamine:alpha-3-D-mannoside beta-1,2-N-acetylglucosaminyltransferase I gene, *Biochem Biophys Res Commun.* 176: 608.

Humphries, M.J. and Olden, K., 1989, Asparagine-linked oligosaccharides and tumor metastasis, *Pharmacol Ther.* 44: 85.

Ioffe, E. and Stanley, P., 1994, Mice Lacking N-Acetylglucosaminyltransferase I Activity Die at Mid-Gestation, Revealing an Essential Role for Complex or Hybrid N-Linked Carbohydrates, *Proc Natl Acad Sci USA.* 91: 728.

Jaeken, J. and Carchon, H., 1993, The Carbohydrate-Deficient Glycoprotein Syndromes - An Overview, *J Inherited Metab Dis.* 16: 813.

Jaeken, J., Carchon, H. and Stibler, H., 1993a, The Carbohydrate-Deficient Glycoprotein Syndromes - Pre-Golgi and Golgi Disorders?, *Glycobiology*. 3: 423.

Jaeken, J., Decock, P., Stibler, H., Vangeet, C., Kint, J., Ramaekers, V. and Carchon, H., 1993b, Carbohydrate-Deficient Glycoprotein Syndrome Type II, *J Inherited Metab Dis*. 16: 1041.

Jaeken, J., Hagberg, B. and Stromme, P., 1991, Clinical presentation and natural course of the carbohydrate-deficient glycoprotein syndrome, *Acta Paediatr Scand*. 80: 6.

Jaeken, J., Schachter, H., Carchon, H., Decock, P., Coddeville, B. and Spik, G., 1994, Carbohydrate deficient: Glycoprotein syndrome type II: A deficiency in Golgi localised N-acetyl-glucosaminyltransferase II, *Arch Dis Child*. 71: 123.

Jaeken, J., Spik, G. and Schachter, H., 1996, Carbohydrate-deficient glycoprotein syndrome Type II: an autosomal recessive disease due to mutations in the N-acetylglucosaminyltransferase II gene, *in*: "Glycoproteins and Disease", J. Montreuil, J.F.G. Vliegenthart and H. Schachter, ed., Elsevier, Amsterdam, The Netherlands, 457.

Jaeken, J., Vanderschueren-Lodeweyckx, M., Casaer, P., Snoeck, L., Corbeel, L., Eggermont, E. and Eeckels, R., 1980, Familial psychomotor retardation with markedly fluctuating serum prolactin, FSH and GH levels, partial TBG deficiency, increased serum arylsulphatase A and increased CSF protein: a new syndrome?, *Pediatric Res*. 14: 179.

Knauer, R., Lehle, L., Hanefeld, F. and Vonfigura, K., 1994, Normal N-oligosaccharyltransferase activity in fibroblasts from patients with carbohydrate-deficient glycoprotein syndrome, *J Inherited Metab Dis*. 17: 541.

Kornfeld, R. and Kornfeld, S., 1985, Assembly of asparagine-linked oligosaccharides, *Ann. Rev. Biochem*. 54: 631.

Kornfeld, S., 1990, Lysosomal enzyme targeting, *Biochem Soc Trans*. 18: 367.

Kornfeld, S., Gregory, W. and Chapman, A., 1979, Class E Thy-1 negative mouse lymphoma cells utilize an alternate pathway of oligosaccharide processing to synthesize complex-type oligosaccharides, *J.Biol.Chem*. 254: 11649.

Kornfeld, S. and Mellman, I., 1989, The biogenesis of lysosomes, *Annu Rev Cell Biol*. 5: 483.

Krasnewich, D.M., Holt, G.D., Brantly, M., Skovby, F., Redwine, J. and Gahl, W.A., 1995, Abnormal synthesis of dolichol-linked oligosaccharides in carbohydrate-deficient glycoprotein syndrome, *Glycobiology*. 5: 503.

Kumar, R., Yang, J., Eddy, R.L., Byers, M.G., Shows, T.B. and Stanley, P., 1992, Cloning and expression of the murine gene and chromosomal location of the human gene encoding N-acetylglucosaminyltransferase I, *Glycobiology*. 2: 383.

Kumar, R., Yang, J., Larsen, R.D. and Stanley, P., 1990, Cloning and expression of N-acetylglucosaminyltransferase I, the medial Golgi transferase that initiates complex N-linked carbohydrate formation, *Proc Natl Acad Sci Usa*. 87: 9948.

Madden, M.J., Morrow, C.S., Nakagawa, M., Goldsmith, M.E., Fairchild, C.R. and Cowan, K.H., 1993, Identification of 5' and 3' sequences involved in the regulation of transcription of the human mdr1 gene in vivo, *J.Biol.Chem*. 268: 8290.

Martinsson, T., Bjursell, C., Stibler, H., Kristiansson, B., Skovby, F., Jaeken, J., Blennow,

G., Stromme, P., Hanefeld, F. and Wahlstrom, J., 1994, Linkage of a locus for carbohydrate-deficient glycoprotein syndrome type I (CDG1) to chromosome 16p, and linkage disequilibrium to microsatellite marker D16S406, *Hum Mol Genet.* 3: 2037.

Mawby, W.J., Tanner, M.J.A., Anstee, D.J. and Clamp, J.R., 1983, Incomplete glycosylation of erythrocyte membrane proteins in congenital dyserythropoietic anaemia type II (CDA II), *British Journal of Haematology.* 55: 375.

Metzler, M., Gertz, A., Sarkar, M., Schachter, H., Schrader, J.W. and Marth, J.D., 1994, Complex asparagine-linked oligosaccharides are required for morphogenic events during post-implantation development, *EMBO J.* 13: 2056.

Misago, M., Liao, Y.F., Kudo, S., Eto, S., Mattei, M.G., Moremen, K.W. and Fukuda, M.N., 1995, Molecular cloning and expression of cDNAs encoding human alpha-mannosidase II and a previously unrecognized alpha-mannosidase IIx isozyme, *Proc Natl Acad Sci USA.* 92: 11766.

Montreuil, J., 1995, The history of glycoprotein research, a personal view, *in:* "Glycoproteins", J. Montreuil, J.F.G. Vliegenthart and H. Schachter, ed., Elsevier, Amsterdam, The Netherlands, 1.

Montreuil, J., Vliegenthart, J.F.G. and Schachter, H., 1996, "Glycoproteins and Disease", Elsevier, Amsterdam, The Netherlands.

Moremen, K.W., 1989, Isolation of a rat liver Golgi mannosidase II clone by mixed oligonucleotide-primed amplification of cDNA, *Proc.Natl.Acad.Sci.USA.* 86(14): 5276.

Moremen, K.W. and Robbins, P.W., 1991, Isolation, characterization, and expression of cDNAs encoding murine alpha-mannosidase II, a Golgi enzyme that controls conversion of high mannose to complex N-glycans, *J Cell Biol.* 115: 1521.

Moremen, K.W. and Touster, O., 1986, Topology of mannosidase II in rat liver Golgi membranes and release of the catalytic domain by selective proteolysis, *J.Biol.Chem.* 261: 10945.

Moremen, K.W., Touster, O. and Robbins, P.W., 1991, Novel purification of the catalytic domain of Golgi alpha-mannosidase II. Characterization and comparison with the intact enzyme, *J Biol Chem.* 266: 16876.

Moremen, K.W., Trimble, R.B. and Herscovics, A., 1994, Glycosidases of the asparagine-linked oligosaccharide processing pathway, *Glycobiology.* 4: 113.

Nathan, D.G. and Oski, F.A., 1987, "Hematology of Infancy and Childhood", W.B.Saunders Company, Philadelphia, PA.

Ohno, K., Yuasa, I., Akaboshi, S., Itoh, M., Yoshida, K., Ehara, H., Ochiai, Y. and Takeshita, K., 1992, The carbohydrate deficient glycoprotein syndrome in three Japanese children, *Brain Dev.* 14: 30.

Pan, Y.T. and Elbein, A.D., 1995, How can N-linked glycosylation and processing inhibitors be used to study carbohydrate synthesis and function, *in:* "Glycoproteins", J. Montreuil, J.F.G. Vliegenthart and H. Schachter, ed., Elsevier, Amsterdam, The Netherlands, 415.

Panneerselvam, K. and Freeze, H.H., 1995, Enzymes involved in the synthesis of mannose-6-phosphate from glucose are normal in carbohydrate deficient glycoprotein syndrome fibroblasts, *Biochem Biophys Res Commun.* 208: 517.

Panneerselvam, K. and Freeze, H.H., 1996a, Mannose corrects altered N-glycosylation in carbohydrate-deficient glycoprotein syndrome fibroblasts, *J Clin Invest.* 97: 1478.

Panneerselvam, K. and Freeze, H.H., 1996b, Mannose enters mammalian cells using a specific transporter that is insensitive to glucose, *J Biol Chem.* 271: 9417.

Phillips, M.L., Schwartz, B.R., Etzioni, A., Bayer, R., Ochs, H.D., Paulson, J.C. and Harlan, J.M., 1995, Neutrophil adhesion in leukocyte adhesion deficiency syndrome type 2, *J Clin Invest.* 96: 2898.

Powell, L.D., Paneerselvam, K., Vij, R., Diaz, S., Manzi, A., Buist, N., Freeze, H. and Varki, A., 1994, Carbohydrate-deficient glycoprotein syndrome: Not an N-linked oligosaccharide processing defect, but an abnormality in lipid-linked oligosaccharide biosynthesis?, *J Clin Invest.* 94: 1901.

Pownall, S., Kozak, C.A., Schappert, K., Sarkar, M., Hull, E., Schachter, H. and Marth, J.D., 1992, Molecular cloning and characterization of the mouse UDP-N-acetylglucosamine: alpha-3-D-mannoside beta-1,2-N-acetylglucosaminyltransferase I gene, *Genomics.* 12: 699.

Pugh, B.F. and Tjian, R., 1990, Mechanism of transcriptional activation by Sp1: evidence for coactivators, *Cell.* 61: 1187.

Ramaekers, V.T., Stibler, H., Kint, J. and Jaeken, J., 1991, A new variant of the carbohydrate deficient glycoproteins syndrome, *J.Inher.Metab.Dis.* 14: 385.

Rearick, J., Chapman, A. and Kornfeld, S., 1981, Glucose starvation alters lipid-linked oligosaccharide biosynthesis on Chinese hamster ovary cells, *J.Biol.Chem.* 256: 6255.

Saito, H., Gu, J.G., Nishikawa, A., Ihara, Y., Fujii, J., Kohgo, Y. and Taniguchi, N., 1995, Organization of the human N-acetylglucosaminyltransferase V gene, *Eur J Biochem.* 233: 18.

Scartezzini, P., Forni, G.L., Baldi, M., Izzo, C. and Sansone, G., 1982, Decreased glycosylation of band 3 and band 4.5 glycoproteins of erythrocyte membrane in congenital dyserythropoietic anaemia type II, *Brit.J.Haematol.* 51: 569.

Schachter, H., 1986, Biosynthetic controls that determine the branching and microheterogeneity of protein-bound oligosaccharides, *Biochem. Cell Biol.* 64: 163.

Schachter, H., 1991, The "yellow brick road" to branched complex N-glycans, *Glycobiology.* 1: 453.

Schachter, H., 1994, Molecular Cloning of Glycosyltransferase Genes, *in*: "Molecular Glycobiology", M.Fukuda and O.Hindsgaul, ed., Oxford University Press, Oxford, UK, 88.

Schachter, H., 1995, Glycosyltransferases involved in the synthesis of N-glycan antennae, *in*: "Glycoproteins", J.Montreuil, J.F.G.Vliegenthart and H.Schachter, ed., Elsevier, Amsterdam, The Netherlands, 153.

Stibler, H., Blennow, G., Kristiansson, B., Lindehammer, A. and Hagberg, B., 1994, Carbohydrate-deficient glycoprotein syndrome: Clinical expression in adults with a new metabolic disease, *J Neurol Neurosurg Psychiatry.* 57: 552.

Stibler, H., Stephani, U. and Kutsch, U., 1995, Carbohydrate-deficient glycoprotein syndrome - A fourth subtype, *Neuropediatrics.* 26: 235.

Stibler, H., Westerberg, B., Hanefeld, F. and Hagberg, B., 1993, Carbohydrate deficient glycoprotein (CDG) syndrome - a new variant, type III, *Neuropediatrics.* 24: 51.

Struck, D.K. and Lennarz, W.J., 1980, The function of saccharide-lipids in synthesis of glycoproteins, *in*: "The Biochemistry of Glycoproteins and Proteoglycans", W.J. Lennarz, ed., Plenum Press, New York, N.Y., 35.

Tan, J., D'Agostaro, G.A.F., Bendiak, B., Reck, F., Sarkar, M., Squire, J.A., Leong, P. and Schachter, H., 1995, The human UDP-N-acetylglucosamine: alpha-6-D-mannoside-beta-1,2-N-acetylglucosaminyltransferase II gene (MGAT2) - Cloning of genomic DNA, localization to chromosome 14q21, expression in insect cells and purification of the recombinant protein, *Eur J Biochem*. 231: 317.

Tan, J., Dunn, J., Jaeken, J. and Schachter, H., 1996, Mutations in the *MGAT2* gene controlling complex N-glycan synthesis cause Carbohydrate-Deficient Glycoprotein Syndrome Type II, an autosomal recessive disease with defective brain development, *American J.Human Genetics*. 59: 810.

Tropea, J.E., Kaushal, G.P., Pastuszak, I., Mitchell, M., Aoyagi, T., Molyneux, R.J. and Elbein, A.D., 1990, Mannostatin A, a new glycoprotein-processing inhibitor, *Biochemistry*. 29: 10062.

Tulsiani, D.R., Opheim, D.J. and Touster, O., 1977, Purification and characterization of alpha-D-mannosidase from rat liver golgi membranes., *J.Biol.Chem*. 252: 3227.

Tulsiani, D.R. and Touster, O., 1985, Characterization of a novel α-D-mannosidase from rat brain microsomes, *J.Biol.Chem*. 260: 13081.

Tulsiani, D.R.P., Harris, T.M. and Touster, O., 1982, Swainsonine inhibits the biosynthesis of complex glycoproteins by inhibition of Golgi mannosidase II, *J.Biol.Chem*. 257: 7936.

van den Eijnden, D.H., Koenderman, A.H.L. and Schiphorst, W.E.C.M., 1988, Biosynthesis of blood group i-active polylactosaminoglycans. Partial purification and properties of an UDP-GlcNAc:N-acetyllactosaminide β1->3-N-acetylglucosaminyltransferase from Novikoff tumor cell ascites fluid, *J. Biol. Chem*. 263: 12461.

Van Schaftingen, E. and Jaeken, J., 1995, Phosphomannomutase deficiency is a cause of carbohydrate-deficient glycoprotein syndrome type I, *FEBS Lett*. 377: 318.

Varki, A., 1993, Biological roles of oligosaccharides: All of the theories are correct, *Glycobiology*. 3: 97.

Velasco, A., Hendricks, L., Moremen, K.W., Tulsiani, D., Touster, O. and Farquhar, M.G., 1993, Cell type-dependent variations in the subcellular distribution of alpha-mannosidase I and II, *J Cell Biol*. 122: 39.

Verbert, A., 1995, From $Glc_3Man_9GlcNAc_2$-protein to $Man_5GlcNAc_2$-protein: transfer 'en bloc' and processing, *in*: "Glycoproteins", J. Montreuil, J.F.G. Vliegenthart and H. Schachter, ed., Elsevier, Amsterdam, The Netherlands, 145.

Verwilghen, R.L., Lewis, S.M., Dacie, J.V., Crookston, J.H. and Crookston, M.C., 1973, HEMPAS: congenital dyserythropoietic anaemia (type II), *Quarterly Journal of Medicine N.S*. 42: 257.

Von Andrian, U.H., Berger, E.M., Ramezani, L., Chambers, J.D., Ochs, H.D., Harlan, J.M., Paulson, J.C., Etzioni, A. and Arfors, K.E., 1993, In vivo behavior of neutrophils from two patients with distinct inherited leukocyte adhesion deficiency syndromes, *J Clin Invest*. 91: 2893.

Wada, Y., Gu, J.G., Okamoto, N. and Inui, K., 1994, Diagnosis of Carbohydrate-Deficient Glycoprotein Syndrome by Matrix-Assisted Laser Desorption Time-of-Flight Mass Spectrometry, *Biol Mass Spectrom.* 23: 108.

Wada, Y., Nishikawa, A., Okamoto, N., Inui, K., Tsukamoto, H., Okada, S. and Taniguchi, N., 1992, Structure of serum transferrin in carbohydrate-deficient glycoprotein syndrome, *Biochem Biophys Res Commun.* 189: 832.

Wasylyk, B., Hahn, S.L. and Giovane, A., 1993, The Ets family of transcription factors, *Eur.J.Biochem.* 211: 7.

Yamashita, K., Ideo, H., Ohkura, T., Fukushima, K., Yuasa, I., Ohno, K. and Takeshita, K., 1993a, Sugar chains of serum transferrin from patients with carbohydrate deficient glycoprotein syndrome. Evidence of asparagine-N-linked oligosaccharide transfer deficiency, *J Biol Chem.* 268: 5783.

Yamashita, K., Ohkura, T. and Fukushima, K., 1996, A partial deficiency of dehydrodolichol reduction is a cause of carbohydrate-deficient glycoprotein syndrome type I, *Abstracts International Symposium on Molecular and Cell Biology pf Glycoconjugate Expression Switzerland.*

Yamashita, K., Ohkura, T., Ideo, H., Ohno, K. and Kanai, M., 1993b, Electrospray Ionization-Mass Spectrometric Analysis of Serum Transferrin Isoforms in Patients with Carbohydrate-Deficient Glycoprotein Syndrome, *J Biochem Tokyo.* 114: 766.

Yamashita, K. and Ohno, K., 1996, Carbohydrate-Deficient Glycoprotein Syndrome Type I, *in*: "Glycoproteins and Disease", J. Montreuil, J.F.G. Vliegenthart and H. Schachter, ed., Elsevier, Amsterdam, The Netherlands, 445.

Yang, J., Bhaumik, M., Liu, Y. and Stanley, P., 1994, Regulation of N-linked glycosylation. Neuronal cell-specific expression of a 5' extended transcript from the gene encoding N-acetylglucosaminyltransferase I, *Glycobiology.* 4: 703.

Yasugi, E., Nakasuji, M., Dohi, T. and Oshima, M., 1994, Major Defect of Carbohydrate-Deficient-Glycoprotein Syndrome Is Not Found in the Synthesis of Dolichyl Phosphate or N-Acetylglucosaminyl-Pyrophosphoryl-Dolichol, *Biochem Biophys Res Commun.* 200: 816.

Yip, B., Mulder, H., Chen, S., Höppener, J.W.M. and Schachter, H., 1996, Organization of the human β1-2 *N*-acetylglucosaminyltransferase I gene (*MGAT1*) which controls complex and hybrid N-glycan synthesis, *Biochemical J.* In press.:

Zdebska, E., Anselsetter, V., Pacuszka, T., Krauze, R., Chelstowska, A., Heimpel, H. and Koscielak, J., 1987, Glycolipids and glycopepties of red cell membranes in congenital dyserythropoietic anemia type II (CDAII), *British J.Haematol.* 66: 385.

PROBING CARBOHYDRATE-PROTEIN INTERACTIONS BY HIGH- RESOLUTION NMR SPECTROSCOPY

S. W. Homans, R. A. Field, M. J. Milton,
M. Probert and J. M. Richardson

Centre for Biomolecular Sciences,
The Purdie Building,
University of St. Andrews,
St. Andrews, Fife
KY16 9ST UK

INTRODUCTION

An important requirement for a detailed understanding of the molecular basis of the interaction of a carbohydrate with its protein receptor is a high-resolution three dimensional structure of the complex. Historically, such structural information has derived from crystallographic studies which can illustrate in detail the precise nature of certain carbohydrate-protein interactions in the solid state (reviewed by Cambillau (1995)). In contrast, few high-resolution structural studies of glycan-protein interactions in solution using nuclear magnetic resonance have been reported. The solution structure of the complex is of importance since a comparison with the solution structure of the free ligand may be more meaningful, and moreover the dynamics of the system are accessible from relaxation time measurements.

The chosen NMR techniques for probing carbohydrate-protein interactions depend upon the affinity of the carbohydrate ligand for the protein. If the affinity is low $(K_d > 1\mu M)$ and the 'off' rate of the ligand is within the appropriate timescale, the bound state conformation of the ligand can be determined in principle by measurement of nuclear

Overhauser effects in the exchanging system. While these measurements, commonly known as transferred nuclear Overhauser effects (TRNOEs) (Clore and Gronenborn, 1982, 1983; Ni, 1994), have been applied to a variety of glycan-protein interactions (Glaudemans et. al., 1990; Bevilacqua et. al., 1992; Bundle et. al., 1994; Weimar and Peters, 1994; Andrews et. al., 1995; Asensio et. al., 1995; Scheffler et. al., 1995), they do not in general provide any information on the nature of the interaction of the ligand with the protein, since ligand-protein TRNOEs are observable with difficulty (Arepalli et. al., 1995) or not at all. A further complication is that TRNOEs can only be quantified accurately with knowledge of the architecture of the protein binding site, since the presence of spin-diffusion in the molecular complex requires a multi-spin full relaxation matrix analysis (London et. al., 1992). Exclusion of nuclear spins derived from the protein can lead to erroneous conclusions regarding the bound state conformation of the ligand (Glaudemans et. al., 1990; Arepalli et. al., 1995). Determination of the bound-state conformation of the ligand using TRNOEs is however experimentally very straightforward. Since the exchange rate is fast on the NMR timescale, separate resonances for the ligand in the bound and free states are not observed, and a single resonance is observed at the weighted average of the chemical shifts (figure 1).

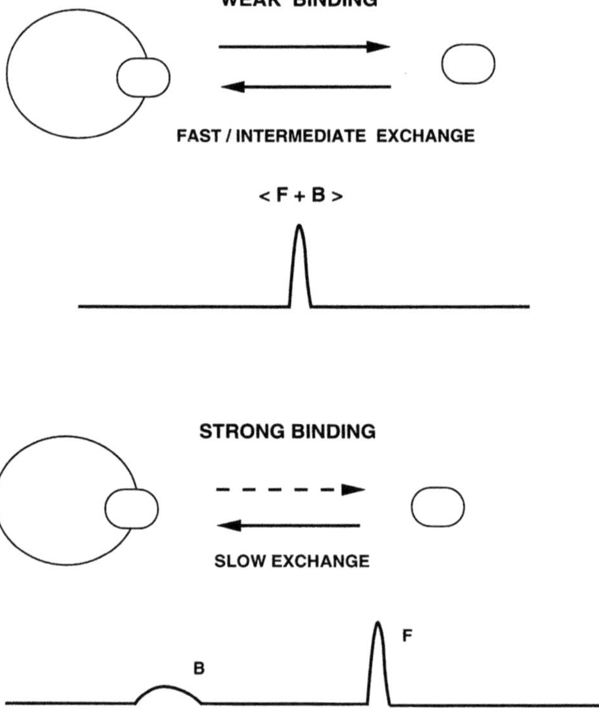

Figure 1. Diagrammatic representation of typical NMR spectra for a ligand in fast and slow exchange with a macromolecule.

These resonances typically have a much narrower linewidth than resonances from the slowly tumbling protein, and are much more intense since it is usual to work at relatively high ligand:protein ratios (10-20:1). There are thus no particular problems in discerning the averaged ligand resonances from background protein.

In situations where the affinity of the carbohydrate for the receptor is high (K_d < 1µM), the TRNOE becomes vanishingly small as the 'off' rate becomes very slow. However, in this exchange regime substantial information on the nature of the ligand-protein interaction is available using conventional NOE measurements, because the carbohydrate ligand is effectively bound for 100% of the time. As a result, separate resonances are observed for the ligand in the bound and free states (figure 1). The bound state ligand resonances now have a linewidth and intensity similar to that of the protein resonances, and usually cannot be directly observed due to the severe resonance overlap which is typical of macromolecules. A suitable method whereby resonances derived from the ligand can be distinguished from those of the protein is by enrichment of either the ligand or protein with a stable isotope, together with application of isotope-editing (Weber et. al., 1991) or isotope-filtration techniques (Petros et. al., 1992), respectively. Here, we describe case-studies which illustrate the application of TRNOE and isotope-edited nuclear Overhauser effect spectroscopy (NOESY) experiments in probing two carbohydrate-protein complexes of topical biological interest. The first concerns a study of the interaction of the carbohydrate-binding B subunit homopentamer (VTB) derived from verotoxin 1 of enterohemorrhagic *E. coli* with the carbohydrate receptor globotriaosylceramide (Gb3). The second study concerns delineation of the bound-state conformation of a small glycoconjugate in association with an antibody Fv fragment.

SOLUTION STRUCTURE OF VTB IN ASSOCIATION WITH GB3

Certain pathogenic *E.coli*, in common with a variety of other pathogenic bacteria, produce cytotoxins (verotoxins, VT) associated with diarrhoeal diseases. Their morphology comprises an enzymatic A subunit in which the toxic activity resides, associated with a B oligomer which binds to specific cell-surface carbohydrate receptors.

Recently, the crystal structure of the B subunit of verotoxin VT-1 from *E.coli* has been reported (Stein et. al., 1992). By comparison of invariant residues in the sequences of members of the Shiga toxin family, a putative conserved carbohydrate binding site was identified in a cleft formed by the β-sheet interaction between adjacent monomers, suggesting five potential binding sites per pentamer. However, the structure of the carbohydrate-protein complex was not determined. We therefore chose to attempt the structural characterisation of the complex using high-resolution NMR methods.

Solution Structure of VTB

Initially, we sought to determine whether the solution structure of VTB parallels that observed in the crystal. The determination of the solution structure of a complex of this size (37 kDa) is not trivial. This is because the molecule tumbles very slowly in solution, giving rise to very broad resonance lines. As a result, conventional homonuclear (^{1}H-^{1}H) NMR methods cannot be used in view of their very low sensitivity. Instead, it is necessary to use efficient multidimensional, multinuclear techniques which have recently been developed for larger proteins (Ikura et. al., 1990). This in turn requires that the protein of interest is uniformly enriched with the stable isotopes ^{13}C and ^{15}N, which can be achieved by growth of the organism which over-expresses the protein in media enriched with ^{13}C and ^{15}N. Using this approach, $\sim 90\%$ of ^{1}H, ^{13}C and ^{15}N resonance assignments were obtained, which were used in the interpretation of three dimensional NOESY-type experiments to generate an initial set of proton-proton through-space distance constraints. These were applied in structure calculations by dynamical simulated annealing. In total some 700 restraints were obtained per monomer, giving some 3500 constraints for the VTB pentamer.

The overall structure of the homopentamer in solution is generally similar to the crystal structure, in that all of the secondary structural features observed in the crystal are present in solution. Thus, adjacent monomers interact via three antiparallel β-sheets, and the central pore is lined with five α-helices, one from each monomer (figure 2).

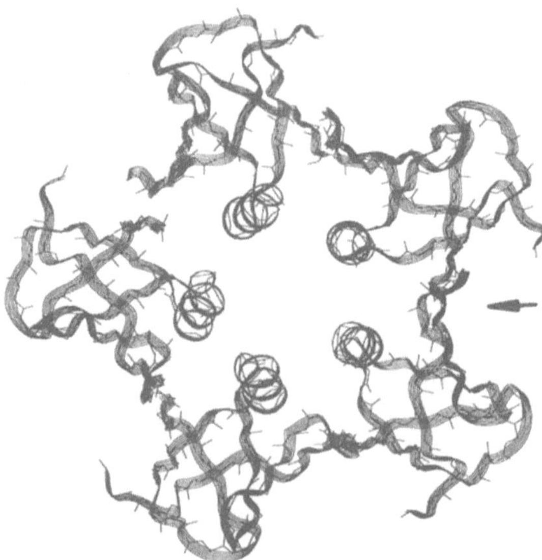

Figure 2. Solution structure of VTB homopentamer. The carbohydrate-binding site (five per pentamer) is arrowed.

Bound-State Conformation of Oligosaccharide and Oligosaccharide Binding Site.

In their original study, Stein *et al* (1992) suggested a putative binding site for the Gb3 ligand in a cleft located between adjacent monomers. These observations have been further developed in the recent molecular modeling study of Nyholm et al., (1996) leading to a model of the VTB-Gb3 complex. Recently we examined the solution structure of the complex between VTB and the oligosaccharide moiety (Galα1-3Galβ1-4Glc) of Gb3 by high-resolution NMR in order to obtain experimental support for this model . In view of the low affinity of the monovalent interaction ($K_d \sim$ 1mM), the exchange of ligand between the free and bound states is fast on the NMR time-scale, and as described above the bound-state conformation of the ligand can be delineated from TRNOE measurements. A potential difficulty with this approach is that carbohydrate-protein interactions described to date involve stacking between an aromatic sidechain and the hydrophobic face of a sugar residue (Cambillau, 1995). The aromatic ring is usually in van-der-Waals contact with the sugar residue, and aromatic ring protons are sufficiently proximal to the ligand spin-system to contribute to the observed TRNOE (Low et al., 1996). Importantly, we observed small but significant ($\sim$ 15 Hz) titratable upfield shifts and broadening of the C-1 proton of the Galβ residue of the oligosaccharide ligand on adding VTB . Extrapolation to 100% bound ligand suggested a bound-state shift for this proton of $\sim$ 0.7 ppm, consistent with a substantial ring-current shift (Perkins, 1982) arising from a stacking interaction. This compares favourably with a ring-current shift of 0.59 ppm predicted from 'site 1' of the theoretical model (Nyholm et. al., 1996) which involves stacking of Galβ on Phe 30, and the observed TRNOEs compare favourably with those predicted from a full relaxation matrix TRNOE simulation of the complex using the theoretical model , after minor adjustments to the φ,ψ angles of the Galα1-4Gal glycosidic linkage of the bound-state conformation of the oligosaccharide (φ,ψ = -48°, -4° in the theoretical model versus -56°, -1° obtained experimentally). Taken together, the available NMR data are consistent with the stacking interaction in "site 1" proposed by Nyholm et al., (1996).

The bound-state conformation of the ligand is in good agreement with one of the two major conformers (φ, ψ = -40°, -5° and -5°, 40°) about the Galα1-4Gal glycosidic linkage determined for Gb3 in dimethylsulphoxide solution and for Gb3 oligosaccharide in aqueous solution (Poppe et. al., 1990). Hence, in common with another toxin of the AB5 class, namely heat-labile toxin, one of the low-energy conformers of a flexible carbohydrate ligand is selected upon binding (Richardson et. al., 1995). In the context of rational inhibitor design, an appropriate strategy might therefore be to utilise the disaccharide Galα1-4Gal or trisaccharide Galα1-4Galβ1-4Glc as a scaffold from which to covalently attach pendant groups with the capacity to make favourable interactions in the vicinity of the binding site. There appear to be several hydroxyl groups on the sugar which do not partake in fruitful hydrogen bonds which might be utilised for this purpose.

SOLUTION STRUCTURE OF A GLYCOCONJUGATE BOUND TO AN ANTIBODY FV FRAGMENT

This example demonstrates the practical value of isotope-edited NMR techniques in probing the bound state conformation of the glycoconjugate estrone-3-glucuronide (E3G, figure 3) in association with an anti-E3G antibody Fv fragment. Furthermore, we assess the extent to which the architecture of the binding-site of the protein can be predicted with these methods in the absence of high-resolution structural data on the protein moiety.

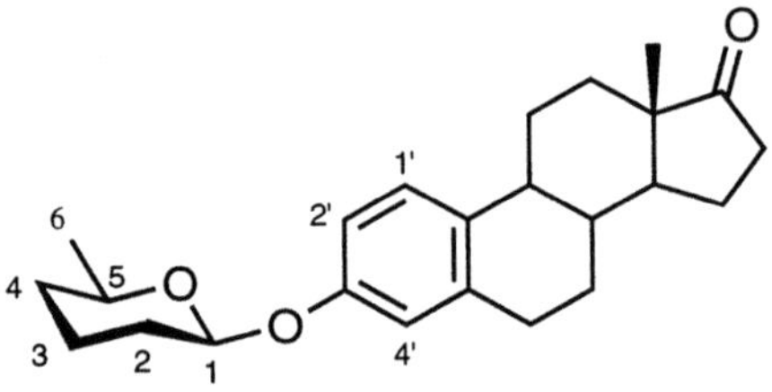

Figure 3. Structure of estrone-3-glucuronide.

Solution Structure of E3G

Initially, we sought to determine the solution structure and dynamics of E3G , using conventional rotating-frame Overhauser effect measurements in combination with restrained dynamical simulated annealing and molecular dynamics simulations (Homans and Forster, 1992). Proton resonance assignments for the glycan, a prerequisite for these studies, could not however be obtained from conventional ^{1}H-^{1}H correlation methods due to extreme overlap of resonances corresponding to the glucuronic acid moiety. The ^{13}C NMR spectrum of the glycan showed that all of the carbon resonances for the glucuronic acid moiety were well resolved. The ^{13}C spectrum of the glucuronic acid moiety was therefore assigned by use of a ^{13}C-^{13}C COSY experiment on estrone-3-U-^{13}C-glucuronide, from which proton resonance assignments were derived by conventional ^{13}C-^{1}H correlation methods. These assignments were then utilised in the interpretation of ^{1}H-^{1}H ROESY measurements on E3G. Two inter-residue ROEs were observable, from the C-1 proton of the glucuronic acid moiety to the C-2' and C-4' protons of the estrone residue, together with several intra-residue ROEs. The inter-residue ROEs were quantified together with the intra-residue ROE between the C-1 and

C-5 protons of the glucuronic acid moiety by measurement of crosspeak volumes. A simple model-building study indicated that no single conformation about the glycosidic linkage of the glycan was consistent with the inter-residue ROEs of the observed intensity, suggesting considerable motional averaging about this linkage. Molecular dynamics simulations on the free glycan were consistent with this observation.

Bound-State Conformation of E3G

Study of the bound-state conformation of E3G was addressed by use of ^{13}C isotope-edited NOESY experiments (Weber et al., 1991). This approach allows the selective observation of NOEs to and from protons directly bonded to a ^{13}C nucleus. By use of estrone-3-U-^{13}C-glucuronide, the technique therefore potentially enables the direct observation of inter-residue NOEs from the C-1 through C-5 protons of the glucuronic acid moiety to the steroid and to amino-acid residues within the binding site. A prerequisite for these studies is the need to obtain resonance assignments for the glycan in the bound state. This is not trivial, but can be achieved by use of a technique known as HCCH-COSY originally designed for the resonance assignment of sidechains in proteins (Bax et. al., 1990), which relies upon much larger one-bond ^{13}C-^{13}C and ^{1}H-^{13}C couplings. The resulting assignments were then utilised in the interpretation of a ^{13}C isotope-edited NOESY experiment on the complex. Crosspeaks in this spectrum corresponding to intra-residues NOEs within the glucuronic acid moiety could readily be assigned, but however the assignment of inter-residue NOE connectivities was not straightforward due to the absence of resonance assignments for the steroid moiety in the bound state and for amino-acid sidechains in the binding site. In order to overcome this difficulty, various NOE connectivities to the steroid were therefore simulated in terms of the possible bound-state configurations of E3G arising from torsional variations about the glycosidic linkage, using a full-relaxation matrix analysis. No single conformation could be found which satisfied the observed NOE connectivities either quantitatively or qualitatively. We therefore surmised that certain of the observed inter-residue connectivities were between the glucuronic acid moiety and protein. We were guided in this assumption by the very large proton resonance shifts of the glucuronic acid moiety observed in the bound state in comparison with the free state. Shifts of this magnitude can only arise from ring currents (Perkins, 1982), and the particularly large upfield shifts of the C-1, C-3 and C-5 protons of the glucuronic acid moiety strongly suggested a stacking interaction between the C-1, C-3 and C-5 face and an aromatic sidechain. In order to determine the disposition of this sidechain, the observed shifts were interpreted in terms of the distance of each proton in glucuronic acid from the centre of a simple phenolic ring, using a gridsearch and optimisation procedure coupled with program RCCAL (Perkins, 1982) for computation of the solution to the Johnson-Bovey equation for the various geometries. The steroid moiety was then added to this optimised geometry, and NOEs were simulated for this complex for a variety of geometries obtained by varying the torsion angles φ and ψ independently in 30° increments. Good qualitative agreement was obtained between simulated and observed NOEs for a geometry $\varphi,\psi = +30°, +30°$, and this geometry was further optimised manually to obtain the best quantitative fit with experimental data, giving a predicted bound-state conformation of the ligand with $\varphi,\psi=+45°, +24°$ (figure 4).

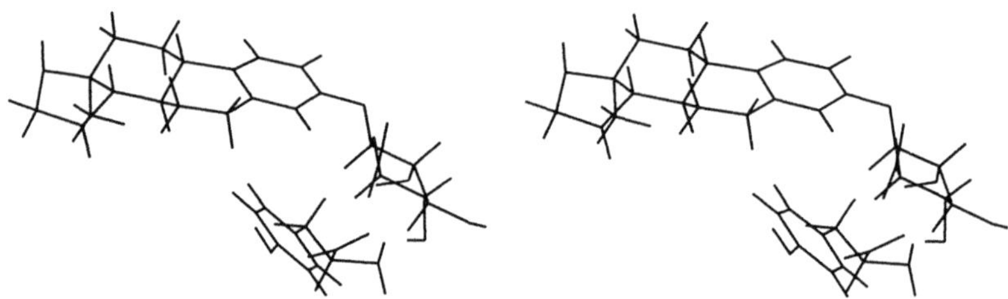

Figure 4. Stereo diagram of predicted bound-state conformation of E3G, illustrating the stacking interaction with an aromatic ring from the Fv binding site.

In the context of rational drug design, the derivation of the bound-state conformation of a glycan in the absence of structural data on the protein receptor is potentially of great value. In the above study the use of a ^{13}C-enriched glycan in combination with isotope-editing methods was quite successful in realising this goal, as evidenced by the good agreement between the predicted conformation of E3G in complex with anti-E3G Fv in comparison with the crystal structure of a closely related complex (Low et. al., 1996). Certain details of the architecture of the binding-site could also be delineated in the solution study. However, this was highly dependent upon the fact that significant ring-current shifts were measurable in the bound-state conformation of the ligand. Since carbohydrate-protein interactions examined to date always involve a stacking interaction with an aromatic ring (Cambillau, 1995), ring current shift measurements are potentially a very important tool for probing carbohydrate-protein interactions, particularly in view of the fact that these shifts can be detected in systems in the fast-exchange regime (as shown in the first example) as well as in the slow-exchange regime observed for the E3G-Fv interaction.

REFERENCES

Andrews, J. S., Weimar, T., Frandsen, T. B., Svensson, B., Pinto, B. M., 1995,
   Novel disaccharides containing sulfur in the ring and nitrogen in the interglycosidic linkage - conformation of methyl 5' - thio - 4 - N- alpha- inhibitor. *J. Am. Chem. Soc.* 117: 10799.

Arepalli, S. R., Glaudemans, C. P. J., Daves, G. D., Kovac, P., and Bax, A., 1995, Identification of protein-mediated indirect nOe effects in a disaccharide-Fab' complex by transferred ROESY, *J. Magn. Reson. B.* 106: 195.

Asensio, J. L., Cañada, F. J. and Jimenez-Barbero, J., 1995, Studies of the bound conformations of methyl alpha-lactoside and methyl beta-allolactoside to ricin b-chain using transferred NOE experiments in the laboratory and rotating frames, assisted by molecular mechanics and dynamics calculations, *Eur. J. Biochem.* 233: 618.

Bax, A., Clore, G. M., Driscoll, P. C., Gronenborn, A. M., Ikura, M. and Kay, L. E., 1990, Practical aspects of proton-carbon-carbon-proton 3-dimensional correlation spectroscopy of C-13-labelled proteins, *J. Magn. Reson.* 87: 620-627.

Bevilacqua, V. L., Thomson, D. S. and Prestegard, J. H., 1992, Conformation of methyl beta-lactoside bound to the ricin-b-chain - interpretation of transferred nuclear Overhauser effects facilitated by spin simulation and selective deuteration, *Biochemistry* 29: 5529.

Bundle, D. R., Baumann, H., Brisson, J. R., Gagne, S. M., Zdanov, A. and Cygler, M., 1994, Solution structure of a trisaccharide-antibody complex - comparison of NMR measurements with a crystal structure, *Biochemistry* 33: 5183.

Cambillau, C.,1995, The structural features of carbohydrate-protein interactions revealed by x-ray crystallography, in "New Comprehensive Biochemistry " eds. Neuberger A. and van Deenen, L. L. M. Vol 29a: pp 29 - 65.

Clore, G. M. and Gronenborn, A. M., 1982, Theory and applications of the transferred Overhauser effect to the study of the conformations of small ligands bound to proteins, *J. Magn. Reson.* 48: 402.

Clore, G. M. and Gronenborn, A. M., 1983, Theory of the time-dependent transferred nuclear Overhauser effect: applications to structural analysis of ligand-protein complexes in solution.,*J. Magn. Reson.* 53: 423.

Glaudemans, C. P. J., Lerner, L., Daves, G. D., Kovac, P., Venable, R. and Bax, A., 1990, Significant conformational changes in an antigenic carbohydrate epitope upon binding to a monoclonal antibody, *Biochemistry* 29: 10906.

Homans, S. W. and Forster, M., 1992, Application of restrained minimization, simulated annealing and molecular dynamics simulations for the conformational analysis of oligosaccharides, *Glycobiology* 2: 143.

Ikura, M., Kay, L. E. and Bax, A., 1990, A novel approach for sequential assignment of ^{1}H, ^{13}C and ^{15}N spectra of larger proteins: heteronuclear triple-resonance three-dimensional NMR spectroscopy. Application to calmodulin, *Biochemistry* 29: 4659.

London, R. E., Perlman, M. E., and Davis, D. G., 1992, Relaxation matrix analysis of the transferred Overhauser effect for finite exchange rates, *J. Magn. Reson.* 97: 79.

Low, D. G., Probert, M. A., Embleton, G., Seshadri, K., Field, R. A., Homans, S.W., Windust, J. and Davis, P. J., 1996, Structure of a glycoconjugate in solution and in complex with an antibody Fv fragment, *Glycobiology* in press.

Ni, F., 1994, Recent developments in transferred NOE methods. *Progr. NMR Spectr.* 26: 517.

Nyholm, P-G., Magnusson, G., Zheng, Z., Norel, R., Binnington-Boyd, B. and Lingwood, C. A., 1996, Two distinct binding sites for globotriaosyl ceramide on verotoxins: identification by molecular modelling and confirmation using deoxy analogues and a new glycolipid receptor for all verotoxins, *Chemistry and Biology* 3: 263.

Perkins, S. J., 1982, Application of ring current calculations to the protein and transfer RNA, in "Biological Magnetic Resonance" (eds. Berliner, L., and Reuben, J.) Plenum Press, New York. Vol. 4, Chapter 4, pp 193-336.

Petros, A. M., Kawai, M., Luly, J. R. and Fesik, S. W., 1992, Conformation of two immunosuppresive FK506 analogues when bound to FKBP by isotope-filtered NMR, *FEBS Letts.* 308: 309.

Poppe, L., Dabrowski, J., von der Lieth, C-W., Koike, K. and Ogawa, T., 1990, Three-dimensional structure of the oligosaccharide terminus of globotriaosylceramide and isoglobotriaosylceramide in solution, *Eur. J. Biochem.* 189: 313.

Richardson, J. M., Milton, M. J. and Homans, S. W., 1995, Solution dynamics of the oligosaccharide moiety of ganglioside Gm1: comparison of solution conformations with the bound state conformation in association with cholera toxin b-pentamer, *J. Mol. Recog.* 8: 358.

Scheffler, K., Ernst, B., Katapodis, A., Magnani, J. L., Wang, W. T., Weiseman, R. and Peters, T., 1995, Determination of the bioactive conformation of the carbohydrate ligand in the e-selectin sialyl lewis(x) complex, *Angew. Chem. Int. Ed. Engl.* 34: 1841.

Stein, P. E., Boodhoo, A., Tyrrell, G. J., Brunton, J. L., and Read, R. J., 1992, Crystal structure of the cell-binding B oligomer of verotoxin-1 from *E. coli*, *Nature* 355: 748.

Weber, C., Wider, G., von Freyberg, B., Traber, R., Braun, W., Widmer, H. and Wüthrich, K., 1991, The NMR structure of cyclosporine-A bound to cyclophilin in aqueous solution, *Biochemistry* 30: 6563.

Weimar, T. and Peters, T., 1994, Aleuria aurantia agglutinin recognizes multiple conformations of α-L-Fuc(1-6)-β-D-GlcNAc-OMe, *Angew. Chem. Int. Ed. Engl.* 33: 88.

OLIGOSACCHARIDES AND PROTEIN RECOGNITION

THE STRUCTURE OF A HUMAN RHEUMATOID FACTOR BOUND TO IgG Fc

Brian J. Sutton[1], Adam L. Corper[1], Maninder K. Sohi[1],
Roy Jefferis[2], Dennis Beale[3] and Michael J. Taussig[3]

[1] The Randall Institute, King's College London,
26-29 Drury Lane, London, WC2B 5RL, UK
[2] Department of Immunology, The Medical School,
University of Birmingham, Birmingham, B15 2TT, UK
[3] Department of Immunology, The Babraham Institute,
Babraham, Cambridge, CB2 4AT, UK

INTRODUCTION

Rheumatoid factors (RF) are autoantibodies with reactivity towards the Fc regions of IgG molecules (1). They are found in the sera and synovia of most patients with rheumatoid arthritis (RA), and high levels of RF are associated with severe disease and poor prognosis (2,3) suggesting that they have a causative role in the pathology of this disease. The IgM or IgG RFs form immune complexes with IgG molecules in the joints, which can then activate complement and cause inflammation. However, not all RFs cause disease, and the distinction between "pathological" and "physiological" RFs is not understood (4).

The mechanism by which these auto-reactive antibodies are produced is also a mystery. The sequences and germline gene origins of a large number of RFs have now been determined, and it appears that there is a considerable diversity in germline gene usage. While many RFs have sequences that are close to the germline configuration, somatic mutations are also found, indicating that the process of RF formation may be "antigen-driven" (5-7), as in a conventional immune response to antigen.

The specificities of RFs have also been defined with reference to their reactivity with the different subclasses of human IgG, and a common pattern is reactivity with subclasses 1, 2 and 4, but not 3 (*e.g.* 8). This mirrors the reactivity pattern of the *Staphylococcal aureus* protein A, which competes for the binding of many RFs, suggesting that they may share a common binding region in the IgG Fc. Other studies, involving mutagenesis (8,9) or peptide inhibition (10), have also mapped the epitopes of RFs, and they too point to a similar consensus binding region. If the majority of RFs do indeed share epitopes in IgG Fc, then perhaps it may be possible to interfere with immune complex formation and the subsequent inflammatory reactions by inhibiting the binding of RFs to IgG.

The work described below was undertaken in order to determine directly the nature of the interaction between a RF and its autoantigen, IgG Fc, in atomic detail. Such an analysis may provide an initial basis for designing molecules to interfere with the interaction, by revealing the residues of the antibody which recognise the antigen, and determining the precise epitope that is bound by the RF. However, knowledge of the structure of the complex at this level of detail also provide some insight into the process of RF generation.

RHEUMATOID FACTOR RF-AN AND IgG4

The RF chosen for this study was derived from the peripheral blood lymphocytes of an RA patient (11,12). It is an IgM with a λ light chain, and the sequences of its heavy and light chains reveal that while they are close to the germline gene sequences, both contain a few somatic mutations: seven amino acid differences in V_H and only three in V_L (13,14). The specificity of RF-AN is as described above, namely reactivity with human IgG1, 2 and 4 but not 3, and *Staphylococcal* protein A will compete for RF-AN binding to IgG. Its specificity is thus typical of many RFs.

The Fc of human IgG4 was selected for complex formation, and in particular a myeloma protein (Rea) for which the carbohydrate composition analysis revealed a relatively homogeneous population of glycoforms (15). The N-linked complex carbohydrate chains attached at Asn297 in each heavy chain of IgG4 Fc Rea, are predominantly (~80%) of a composition in which the terminal galactose residues are missing. This is in effect the "G0-IgG" which has been found to be associated with RA (16,17). Polyclonal IgG from rheumatoid sera has higher levels of G0-IgG than age-matched controls, and levels of G0-IgG have been related to disease activity (18). The use of IgG4 Fc Rea may therefore reflect the nature of complex formation in RA.

CRYSTAL STRUCTURE ANALYSIS OF THE RF-AN/IgG Fc COMPLEX

The Fab and Fc fragments of RF-AN IgM and IgG4 respectively were generated by proteolysis (14). Crystals of the complex grew from a 1:1 molar ratio mixture of the two components, although the stoichiometry of the complex formed in the crystals was found to be 2:1 Fab:Fc (14). The crystals diffracted to a resolution of 3Å, and diffraction data (11,600 independent measurements, 86% complete) were collected from a single specimen. The structure was determined by molecular replacement (19) using search models derived from the known crystal structures of Fab and human IgG Fc. In these search models, the complementarity determining regions (CDR) of the V domains and the carbohydrate chains of the Fc were excluded in order to eliminate the possibility of bias. While both V domains and the Cγ2 and Cγ3 domains of the Fc were readily identified, the two C domains of the Fab could not be located at first. This was found subsequently to be due to a degree of mobility or static disorder of these domains relative to the V domains. Flexibility at the "elbow" between V and C domains in Fab fragments has been seen in other antibodies. Crystallographic refinement of the final structure yielded an R-factor of 0.22 (R_{free} = 0.29).

THE STRUCTURE OF THE COMPLEX

The structure is shown in Figure 1, and the two Fabs may be seen binding symmetrically, one to each heavy chain of the Fc. In the crystal, an exact two-fold axis of symmetry relates the two halves of the Fc, and the two binding interactions are thus also identical. The epitope spans both the Cγ2 and Cγ3 domains, and involves residues in loop regions of both domains that line the cleft between them. A list of the contact residues is given in Table 1, and the buried surface area of each Fab:Fc interface is 1280Å^2. The epitope involves a total of 15 residues, which are contacted by 8 residues of RF-AN. It is very striking that the light chain only contributes two residues, Asp50 and Pro56, both in CDR-L2. CDR-L1 and CDR-L3 are not involved in binding at all. In the heavy chain, four consecutive residues at the start of CDR-H3 make numerous contacts including two hydrogen bonds, but both CDR-H1 and CDR-H2 contribute only one residue each. Asp31 in CDR-H1 makes a further hydrogen bond, and also a salt bridge to Arg255. The paucity of antibody contact residues in the RF-AN/Fc complex may account in part for a characteristic feature of this binding interaction, namely its low affinity. This has been estimated by surface plasmon resonance as $<10^5$ M^{-1} for the binding between isolated Fab and Fc fragments (B.J.Sutton, unpublished data), although the effective affinity (avidity) is higher for the polyvalent IgM (2 x 10^6 M^{-1}).

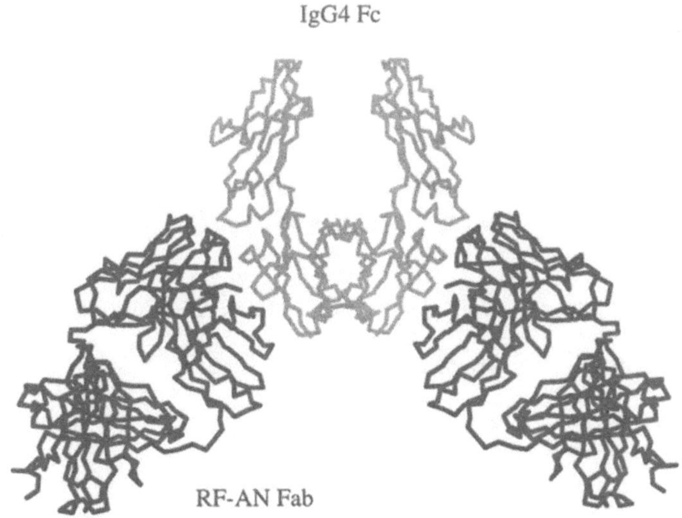

Figure 1. Structure of the complex between RF-AN Fab and human IgG4 Fc. The course of the polypeptide chain in each molecule is represented by a trace of the Cα atomic positions. N-linked carbohydrate chains in the Fc are not shown since no electron density was found for them.

The involvement of so few antibody residues is unusual, but more surprising still is the topology of the interaction, in which the Fc epitope makes contact with only one side of the potential combining site surface of the Fab. In other complexes between antibodies and protein antigens, the number of antibody contact residues can be as high as 20, with all six CDRs involved (20). Even in the complexes between Fab fragments and their anti-idiotypic antibody Fabs which have been studied crystallographically, the interaction involves contact between most if not all CDRs (*e.g.* 21) in a "head-to-head" fashion. In the RF-AN complex with Fc, the antigen is bound so far to one side that the central region of the combining site is unoccupied. With so many CDR residues apparently uninvolved in antigen binding, it is possible that the autoantibody may in fact be directed against another antigen as yet unknown, and that the binding reactivity towards IgG Fc is a chance cross-reactivity.

The contact residues in the first two heavy chain CDRs are germline encoded, and the four residues in CDR-H3 are from the D region. However in the light chain, one of the only two contact residues, and the more important of the two, Pro56 (Table 1), is a somatic mutation. In the germline sequence this residue is a serine, which could not make the several Van der Waals interactions of the proline residue. Indeed, this substitution may also affect the main-chain conformation, and it is therefore expected that with serine at this position the affinity for IgG Fc would be even lower. The structure thus provides the first evidence derived from three-dimensional structure for an antigen-driven process for rheumatoid factor generation.

Table 1. Contact residues between RF-AN Fab and IgG4 Fc

	RF-AN	IgG4 Fc
CDR L1	-	-
CDR L2	Asp 50	His 433
	Pro 56	Ser 424, Gln 438, Ser 440
CDR L3	-	-
CDR H1	Asp 31	Met 252, Ser 254, Arg 255
CDR H2	Trp 52a	Ile 253, Ser 254
CDR H3	Arg 96	Tyr 436
	Ser 97	Asn 434
	Tyr 98	Leu 251, Met 252, Ile 253, Met 428, Asn 434, Tyr 436
	Val 99	Asn 434, His 435

Contact residues were identified using CONTACSYM (22).
Residue numbering is according to Kabat (23).

THE IgG Fc OLIGOSACCHARIDE

In the crystal structure analysis, no electron density was present to correspond to either of the two N-linked oligosaccharide chains attached at residue Asn297 in each heavy chain. This may be due either to mobility of the carbohydrate chains relative to the polypeptide domains, or to static disorder, *i.e.* the carbohydrate chains adopting several different conformations within the crystal. In either case, this observation contrasts with all other crystal structures of IgG Fc, either free or complexed, in which the N-linked carbohydrate is relatively well-ordered and visible. In these structures, each $\alpha(1\text{-}6)$ linked branch from the central core mannose residue makes contact with the surface of its respective Cγ2 domain, and the $\alpha(1\text{-}3)$ linked branches occupy the space between the two domains (24,25). In all of these structures the carbohydrate chains display the characteristic micro-heterogeneity in composition typical of IgG Fc oligosaccharides, but the essential difference between these and the IgG Fc Rea protein of the present study is the virtual absence of galactose in the latter. It thus appears that a lack of galactose does affect the conformation of the oligosaccharide chains, such that they no longer adopt an ordered conformation or make contact with the Cγ2 domains. It is also striking that several hydrophobic residues of the Cγ2 domains which are normally covered by carbohydrate residues of the $\alpha(1\text{-}6)$ linked arms, are apparently exposed in the RF-AN/IgG Fc complex.

One possible caveat to the above conclusions is that the alteration in oligosaccharide conformation might be a result of the binding of the RF-AN Fab. This is very unlikely however, since the binding of RF-AN has no apparent effect upon the polypeptide structure of the IgG Fc as judged by comparison with the known uncomplexed structures, although the definitive answer will only come from X-ray analysis of uncomplexed IgG Fc Rea. The altered conformation, and apparent mobility of carbohydrate chains lacking terminal galactose residues, is consistent however with evidence from NMR analysis of G0-IgG (26). An earlier NMR study with ^{13}C-labelled galactose residues in IgG Fc had indicated that those on the $\alpha(1\text{-}6)$ linked arm were indeed immobile relative to the Cγ2 domain (27).

One functional significance of this altered oligosaccharide conformation was suggested by the discovery that G0-IgG could activate complement *via* the mannose binding protein, presumably recognising terminal N-acetyl glucosamine residues of mobile and accessible carbohydrate chains (26). However, it is also clear from the structure of the complex that a predominantly hydrophobic "patch" on the surface of the Cγ2 domain is uncovered when the terminal galactose is absent, and this is accessible to interaction with either a protein, such as a RF, or indeed with the carbohydrate component of a glycoprotein. Interaction between Fab oligosaccharides and the Fc regions of G0-IgG molecules has been proposed as a possible mechanism for IgG self-aggregation in RA (16,28).

RECOGNITION OF IgG Fc BY RF-AN

The epitope recognised by RF-AN consists entirely of polypeptide residues, as depicted in Figure 1 and summarised in Table 1. However, if the carbohydrate structure taken from that of the crystal structure of human IgG1 Fc (24) is modelled onto the Fc structure of IgG4 Fc Rea in the complex with RF-AN, the relationship between the epitope and the region normally occupied by the carbohydrate may be seen. Figure 2 shows the result of this modelling exercise. The V domains of RF-AN and one Fc heavy chain are represented in approximately the same orientation as in Figure 1, and the carbohydrate, superimposed upon the Fc in the same orientation as in the uncomplexed IgG1 Fc, is shown in lighter grey

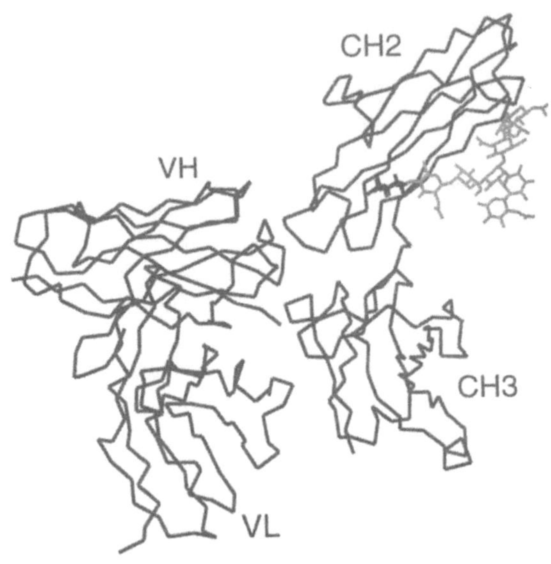

Figure 2. The interaction between RF-AN V domains and the IgG4 Fc C domains of one heavy chain, together with the branched carbohydrate chain modelled in a location structurally homologous to that found in IgG1 Fc. The terminal galactose residue of the α(1-6) linked arm, absent in almost all of the IgG4 Fc molecules in this study, is shown in black.

with the exception of the terminal galactose residue on the α(1-6) linked arm which is shown in black. This residue, viewed in the equatorial plane, is of course absent in the IgG4 Fc Rea protein used in this study. Clearly, neither this residue even if it were present, nor the Cγ2 domain residues with which it normally interacts, could make contact with the RF-AN antibody. However, while this particular RF does not apparently recognise the carbohydrate, many others are known to sense the presence or absence of the terminal galactose residue (29-31). It may be seen in Figure 2 that only a very slight shift of the epitope identified in this study would be required for either the galactose or preceding N-acetyl glucosamine residue, or polypeptide residues of the exposed hydrophobic "patch", to be encompassed within the epitope. Thus other RFs with the same subclass specificity and perhaps very similar epitopes may indeed interact directly with either the agalactosyl α(1-6) linked oligosaccharide chain, or the Cγ2 domain residues with which it normally interacts. RFs with enhanced affinity and avidity towards agalactosyl IgG are thought to have pathogenic potential (32).

All of the Fc residues that constitute the epitope are conserved in the human subclasses IgG1, 2 and 4 with which RF-AN reacts. A histidine to arginine substitution in IgG3 at position 435 has been identified as an important determinant of the non-reactivity of this subclass for many RFs (8,33), but it lies at the very edge of the epitope, and may not account for the non-reactivity of RF-AN; this may be due to other differences between the conformations of the IgG3 and IgG4 Fc regions (19). However, one very striking aspect of the epitope is its similarity to the binding sites for the bacterial proteins A and G. This had been predicted for protein A at least, on the basis of its identical IgG subclass specificity profile, as mentioned above. The crystal structures of the complexes between Fc and both protein A and G are known (24,34), and both proteins bind, as does RF-AN, to the cleft region between the Cγ2 and Cγ3 domains. The contact residues in all three complexes are compared in Table 2, and a more detailed structural comparison is presented elsewhere (19). (Intriguingly, the region of contact in the only other crystal structure of a complex with IgG Fc, that with the neonatal Fc receptor, is again the cleft region, although this was only defined in a very low resolution X-ray analysis (35)). It can be seen that Fc contact residues common to all three complexes are found particularly in regions 251-254 in Cγ2 and 428-438 in Cγ3. It has been suggested that RFs may originate as anti-idiotypic mimics of the bacterial immunoglobulin-binding proteins (36), and the striking coincidence of the RF-AN epitope with the protein A and G binding sites may appear to support this. However, at the atomic level, there is no evidence for mimicry, *i.e.* there is no correspondence between the interactions made by RF-AN and the bacterial proteins, with Fc.

Table 2. Comparison between the RF-AN epitope and the protein A and protein G binding sites in IgG Fc

RF-AN	Protein A	Protein G
251	251	251
252	252	252
253	253	253
254	254	254
255		
	309	
	310	
	311	311
	314	
		380
		382
384		
385		
424		
		426
428		428
	432	
433	433	433
434	434	434
435	435	
436		436
438		438
440		

SUMMARY

This is the first crystal structure analysis of a complex between an autoantibody and its autoantigen, and it reveals a mode of interaction never before seen in an antibody-antigen complex. Not only are there relatively few antibody contact residues, contributing perhaps to its very low affinity, but these residues are to be found on only one side of the potential combining site surface. Indeed, so many CDR residues are *not* involved in Fc binding, including those in the central region of the combining site, that it is easy to envisage that this RF may have another, entirely different, specificity. The antibody may therefore have originated in response to another, as yet unidentified, antigen, and the reactivity with IgG Fc may be an unfortunate cross-reactivity.

Certainly some of the CDR residues which *do* interact with IgG Fc are germline encoded, but significantly one of only two residues in the light chain, Pro56, which makes many contacts with Fc, is a somatic mutation. Since this mutation would appear to make a significant contribution to the binding affinity, it is therefore evidence for an antigen driven response to the IgG Fc in the generation of this autoantibody.

The Fc epitope recognised by RF-AN is strikingly similar to the binding sites for the bacterial binding proteins A and G, but the significance of this is not clear. What is clear however is that the epitope does not include any part of the Fc carbohydrate residues, although the structure of the complex does reveal that there is an alteration in the carbohydrate conformation when the galactose residues are absent. Loss of the interaction between the terminal galactose residue on the $\alpha(1\text{-}6)$ linked branch and the Cγ2 domain appears to allow the carbohydrate chains to become mobile, at the same time exposing a predominantly hydrophobic patch on the Cγ2 surface. Accessibility to either the agalactosyl carbohydrate chains or the newly exposed residues may account for the enhanced reactivity for G0-IgG that has been reported for certain RFs, and such an epitope need not be very different to that recognised by RF-AN.

In order to understand more completely the effect of the presence or absence of the terminal galactose residue, the fully galactosylated glycoform of Fc must be studied for comparison; this work is underway. It is also important now to study a RF which is known to sense this difference in oligosaccharide composition, and also to study RFs of higher affinity, of the IgG class, and from the synovium. RF-AN was the first RF to be immortalised as a cell line, and in many ways it is a typical RF (in terms of specificity, relationship to germline sequence and affinity), but we must now establish whether the novel structural features revealed in this analysis are indeed typical of other RFs. Only when comparisons can be made between RFs of different origin and with contrasting functional properties will we begin to understand what constitutes a pathogenic RF, and the mechanism by which such auto-reactive antibodies are generated.

ACKNOWLEGDEMENTS
This work is supported by the Arthritis and Rheumatism Council (U.K.) and a BBSRC studentship to A.L.C.

REFERENCES

1. Tighe, H. & Carson, D.A. Rheumatoid Factor. In *Textbook of Rheumatology* (Kelley, W.N., Harris, E.D., Ruddy, S. & Sledge, C.B. eds.) W.B. Saunders, Philadelphia, pp 241-249, 1997.

2. Zvaifler, N.J. The immunopathology of joint inflammation in rheumatoid arthritis. *Adv. Immunol.*, **16**, 265-336, 1973.

3. Vaughan, J.H. Pathogenetic concepts and origins of rheumatoid factor in rheumatoid arthritis. *Arthritis Rheum.*, **36**, 1-6, 1993.

4. Chen, P.P. & Carson, D.A. New insights on the physiological and pathological rheumatoid factors in humans. In *Autoimmunity: Physiology and Disease* (Coutinho, A. & Kazatchkine, M., eds.) Wiley-Liss, New York, pp 247-266, 1994.

5. Olee, T., Lu, E.W., Huang, D-F. *et al.* Genetic analysis of self-associating IgG rheumatoid factors from two rheumatoid synovia implicates an antigen-driven response. *J. Exp. Med.*, **175**, 831-842, 1992.

6. Randen, I., Thompson, K.M., Pascual, V. *et al.* Rheumatoid factor V genes from patients with rheumatoid arthritis are diverse and show evidence of an antigen-driven response. *Immunol. Rev.*, **128**, 49-71, 1992.

7. Ermel, R.W., Kenny, T.P., Chen, P.P. & Robbins, D.L. Molecular analysis of rheumatoid factors derived from rheumatoid synovium suggests an antigen-driven response in inflamed joints. *Arthritis Rheum.*, **36**, 380-388, 1993.

8. Bonagura, V.R., Artandi, S.E., Davidson, A. *et al.* Mapping studies reveal unique epitopes on IgG recognised by rheumatoid arthritis-derived monoclonal rheumatoid factors. *J. Immunol.*, **151**, 3840-3852, 1993.

9. Artandi, S.E., Calame, K.L., Morrison, S.L. & Bonagura, V.R. Monoclonal IgM rheumatoid factors bind IgG at a discontinuous epitope comprised of amino acid loops from heavy-chain constant-region domains 2 and 3. *Proc. Natl. Acad, Sci. USA*, **89**, 94-98, 1992.

10. Peterson, C., Malone, C.C. & Williams, R.C. Rheumatoid-factor-reactive sites on CH_3 established by overlapping 7-mer peptide epitope analysis. *Mol. Immunol.*, **32**, 57-75, 1995.

11. Steinitz, M., Izak, G., Cohen S., Ehrenfeld, M. & Flechner, J. Continuous production of monoclonal rheumatoid factor by EBV-transformed lymphocytes. *Nature*, **287**, 443-445, 1980.

12. Steinitz, M. & Tamir, S. Human monoclonal autoimmune antibody produced *in vitro*: rheumatoid factor generated by Epstein-Barr virus-transformed cell line. *Eur. J. Immunol.*, **12**, 126-133, 1982.

13. Pascual, V., Victor, K., Randen, I., Thompson, K., Steinitz, M., Forre, O., Fu, S-M., Natvig, J.B. & Capra, J.D. Nucleotide sequence analysis of rheumatoid factors and polyreactive antibodies derived from patients with rheumatoid arthritis reveals diverse use of V_H and V_L gene segments and extensive variability in CDR-3. *Scand. J. Immunol.*, **36**, 349-362, 1992.

14. Sohi, M.K., Corper, A.L., Wan, T., Steinitz, M., Jefferis, R., Beale, D., He, M., Feinstein, A., Sutton, B.J. & Taussig, M.J. Crystallization of a complex between the Fab fragment of a human immunoglobulin M (IgM) rheumatoid factor (RF-AN) and the Fc fragment of human IgG4 Fc. *Immunology*, **88**, 636-641, 1996.

15. Jefferis, R., Lund, J., Mizutani, H., Nakagawa, H., Kawazoe, Y., Arata, Y. & Takahashi, N. A comparative study of the N-linked oligosaccharide structures of human IgG subclass proteins. *Biochem. J.* **268**, 529-537, 1990.

16. Parekh, R.B., Dwek, R.A., Sutton, B.J. *et al.* Association of rheumatoid arthritis and primary osteoarthritis with changes in the glycosylation pattern of total serum IgG. *Nature*, **316**, 452-457, 1985.

17. Rahman, M.A.A. & Isenberg, D.A. Glycosylation of IgG in rheumatic disease. In *Abnormalities of IgG Glycosylation and Immunological Disorders* (Isenberg, D.A. & Rademacher, T.W. eds.) John Wiley & Sons Ltd., pp 101-118, 1996.

18. Parekh, R.B., Roitt, I.M., Isenberg, D.A., Dwek, R.A., Ansell, B.M. & Rademacher, T.W. Galactosylation of IgG associated oligosaccharides: reduction in patients with adult and juvenile onset rheumatoid arthritis and relation to disease activity. *Lancet*, i, 966-969, 1988.

19. Corper, A.L., Sohi, M.K., Bonagura, V.R., Steinitz, M., Jefferis, R., Feinstein, A., Beale, D., Taussig, M.J. & Sutton, B.J. Structure of a human IgM rheumatoid factor in complex with its autoantigen IgG Fc. *Nature Struct. Biol.*, in press, 1997.

20. Davies, D.R. & Cohen, G.H. Interactions of protein antigens with antibodies. *Proc. Natl. Acad. Sci. USA*, **93**, 7-12, 1996.

21. Fields, B.A., Goldbaum, F.A., Ysern, X., Poljac, R.J. & Mariuzza, R.A. Molecular basis of antigen mimicry by an anti-idiotope. *Nature*, **374**, 739-742, 1995.

22. Sheriff, S. Some methods for examining the interactions between two molecules. *Immunomethods*, **3**, 191-196, 1993.

23. Kabat, E.A., Wu, T.T., Perry, H.M., Gottesman, K.S. & Foeler, C. Sequences of Proteins of Immunological Interest. US Department of Health and Human Services, NIH, Bethesda, MD, USA, 1991.

24. Deisenhofer, J. Crystallographic refinement and atomic models of a human Fc fragment and its complex with fragment B of protein A from *Staphylococcus aureus* at 2.9- and 2.8-Å resolution. *Biochemistry*, **20**, 2361-2370, 1981.

25. Sutton, B.J. & Phillips, D.C. The three-dimensional structure of the carbohydrate within the Fc fragment of IgG. *Biochem. Soc. Trans.*, **11**, 130-132, 1983.

26. Malhotra, R., Wormald, M.R., Rudd, P.M., Fischer P.B., Dwek, R.A. & Sim, R.B. Glycosylation changes of IgG associated with rheumatoid arthritis can activate complement via the mannose-binding protein. *Nature Medicine*, **1**, 237-243, 1995.

27. Gilhespy-Muskett, A.M., Partridge, J., Jefferis, R. & Homans, S.W. A novel ^{13}C isotopic labelling strategy for probing the structure and dynamics of glycan chains *in situ* on glycoproteins. *Glycobiology*, **4**, 485-489, 1994.

28. Roitt, I.M., Dwek, R.A., Parekh, R.B. *et al.* Changes in carbohydrate structure of IgG in rheumatoid arthritis. *Recenti Prog. Medicina*, **79**, 314-317, 1988.

29. Soltys, A.J., Hay, F.C., Bond, A., Axford, J.A., Jones, M.G., Randen, I., Thompson, K.M. & Natvig, J.B. The binding of synovial tissue-derived human monoclonal immunoglobulin M rheumatoid factor to immunoglobulin G preparations of differing galactose content. *Scand. J. Immunol.* **40**, 135-143, 1994.

30. Soltys, A.J., Bond, A., Westwood, O.M.R. & Hay, F.C. The effects of altered glycosylation of IgG on rheumatoid factor-binding and immune complex formation. In *Glycoimmunology* (Alavi, A & Axford J.S., eds.), Plenum Press, New York, *Adv. Exp. Med. Biol.*, **376**, 155-160, 1995.

31. Newkirk, M. Fc glycosylation and rheumatoid factors. In *Abnormalities of IgG Glycosylation and Immunological Disorders* (Isenberg, D.A. & Rademacher, T.W. eds.) John Wiley & Sons Ltd., pp119-130, 1996.

32. Newkirk, M.M., Fournier, M-J. & Shiroky, J. Rheumatoid factor avidity in patients with rheumatoid arthritis: identification of pathogenic RFs which correlate with disease parameters and with the Gal(0) glycoform of IgG. *J. Clin. Immunol.*, **15**, 250-257, 1995.

33. Jefferis, R., Nik Jaafer, M.I. & Steinitz, M. Immunogenic and antigenic epitopes of immunoglobulins. VIII. A human monoclonal rheumatoid factor having specificity for a discontinuous epitope determined by histidine/arginine interchange at residue 435 of immunoglobulin G. *Immunol. Lett.*, 7, 191-194, 1984.

34. Sauer-Eriksson, A.E., Kleywegt, G.J., Uhlen, M. & Jones, T.A. Crystal structure of the C2 fragment of streptococcal protein G in complex with the Fc domain of human IgG. *Structure*, **3**, 265-278, 1995.

35. Burmeister, W.P., Huber, A.H. & Bjorkman, P.J. Crystal structure of the complex of rat neonatal Fc receptor with Fc. *Nature*, **372**, 379-383, 1994.

36. Oppliger, I.R., Nardella, F.A., Stone, G.C. & Mannik, M. Human rheumatoid factors bear the internal image of the Fc binding region of Staphylococcal protein A. *J. Exp. Med.*, **166**, 702-710, 1987.

50

CARBOHYDRATE RECOGNITION SYSTEMS IN INNATE IMMUNITY

Ten Feizi

The Glycosciences Laboratory
Imperial College School of Medicine
Northwick Park Hospital
Harrow
Middlesex HA1 3UJ, U.K.

INTRODUCTION

Carbohydrate-protein interactions have emerged as important recognition systems in the orchestration of mechanisms of innate immunity[1]. Unlike the usual situation of receptor-ligand pairs[2], a picture is emerging of recognition systems that operate as triads: receptors, ligands and carriers. Here the receptors are lectins[3], the ligands are oligosaccharides, and when they are optimally assembled on carriers (proteins or lipids) functional counter-receptors are formed, as depicted in Figure 1.

Such glyco recognition systems are now established as being among essential components of the cell-cell interaction cascades in leucocyte extravasation in inflammation where the receptors are the selectins[4-6], and in the first line of defence against bacterial pathogens the receptors are soluble proteins known as the collectins[7]. A striking aspect of the carbohydrate ligands thus far identified for the selectins and collectins is that they are not unique structures, rather they are sequences that occur on glycoconjugates in various compartments of the body[8,9]. Of course, the search continues for novel structures at the same time as examining ways in which biological specificities may be mediated by oligosaccharides that are not unique. Observations with the P-selectin indicate that on the glycoprotein counter-receptor for this selectin, PSGL-1, there is a co-operation between carbohydrate ligands and sulphated tyrosines for high affinity binding reviewed in refs:[1,5,6].

APPLICATIONS OF THE NEOGLYCOLIPID TECHNOLOGY IN THE MOLECULAR DISSECTION OF CARBOHYDRATE-PROTEIN INTERACTIONS IN INNATE IMMUNITY

In this lecture I discussed ways in which the neoglycolipid technology[10,11] has been exploited, not only for ligand discovery and assignments of binding specificities of the

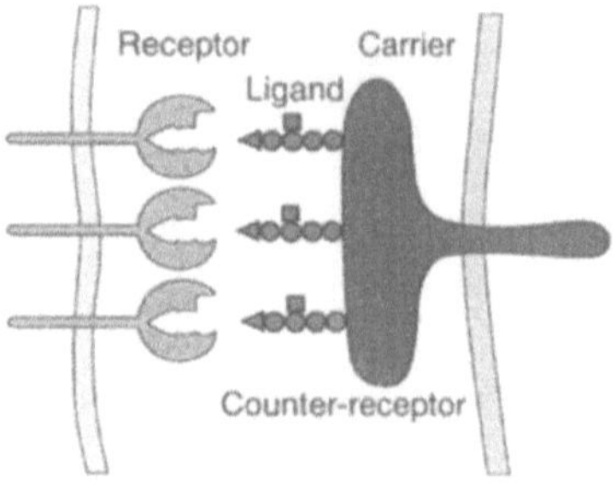

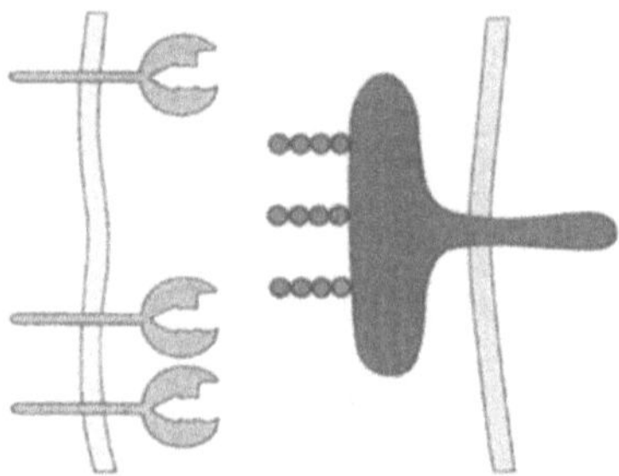

Figure 1. Schematic representation of a functional and a non-functional receptor-ligand assembly. The key ligating elements (*ligands*) are particular oligosaccharides that must be effectively presented on protein or lipid carriers, which together serve as functional *counter-receptors*. The *receptor* protein is the product of a single or a relatively small number of genes; by contrast, the counter-receptor is, typically, the product of many genes, the majority of which are involved in the post-translational and post-biosynthetic modifications. There are so many factors involved in determining whether a glycoprotein (or a glycolipid) does indeed function as a counter-receptor that in such glyco-recognition systems it seems appropriate to reserve the term ligand to the oligosaccharide and the term counter-receptor to the functional saccharide-carrier assembly. (taken from ref. 1 with permission).

selectins and the collectins, but also for providing insights into ways in which the collectin, conglutinin (a classical recognition system of glycoimmunology) specifically binds to an N-glycan (a high-mannose chain) as displayed on only certain of the proteolytic fragments of the complement glycoprotein C3[12,13].

The neoglycolipid technology was designed[10] and is being perfected as a means of revealing recognition codes in oligosaccharides[11,14,15]. It enables oligosaccharide probes (neoglycolipids) to be generated from structurally defined or chemically synthesized oligosaccharides, or mixtures of oligosaccharides released from glycoproteins[11] and glycolipids, or the surface of whole cells[16]. These are used as solid phase probes in binding studies with soluble proteins or with whole cells that express carbohydrate-recognizing proteins. The technology incorporates a microsequencing strategy as the neoglycolipids have unique ionization properties in mass spectrometry.

Using the oligosaccharide ligand sialyl-Le[a] in the form of a neoglycolipid, and sialyl-Le[x] as a glycosphingolipid, it was observed that these lipid-linked oligosaccharides can support the cell rolling and tethering interactions mediated by the E- and L-selectins[17]. Discovery of sulphated Le[a] and sulphated Le[x] as ligands for E- selectin was made using the neoglycolipid technology[18]. The marked influence of E-selectin density at the cell surface on avidity of binding to clustered, immobilized oligosaccharide ligands has been documented[19].

Recent observations with human L-selectin IgG chimera in paucivalent and multivalent form[20] have revealed pronounced differences in binding avidities to the high and low affinity oligosaccharide ligands sulphated Le^a/Le^x and sialyl-Le^a/Le^x, respectively. Moreover, the inhibitabilities of the binding of the paucivalent and multivalent L-selectin forms differ markedly. The binding of the paucivalent L-selectin is far less inhibitable than that of the multivalent L-selectin. This latter finding has therapeutic implications, and it highlights the need to reconsider the traditional assay methods which are biased toward inhibiting the high avidity interactions of selectins, and design assay systems that distinguish clearly compounds with desired broad spectrum anti-adhesive activities toward both the high and low affinity interactions of L-selectin.

Selectin-carbohydrate interactions are key initial events in inflammatory processes, and are important targets for therapeutic intervention in disorders of inflammation[4-6]. The choice of experimental systems is crucial not only for the initial selection of 'lead' compounds for developing therapeutic analogues, but also for the subsequent stages of improvement and fine tuning of inhibitor designs.

REFERENCES

1. P.R. Crocker and T. Feizi, Carbohydrate recognition systems: functional triads in cell-cell interactions, *Curr. Opin. Struct. Biol.* 6:679-691 (1996).
2. C. Liébecq, International Union of Biochemistry and Molecular Biology. *Biochemical Nomenclature and Related Documents*, Portland Press, London and Chapel Hill, (1992).
3. K. Drickamer, Increasing diversity of animal lectin structures, *Curr. Opin. Struct. Biol.* 5:612-616 (1995).
4. M.P. Bevilacqua and R.M. Nelson, Selectins. *J. Clin. Invest.* 91:379-387 (1993).
5. R.P. McEver, K.L. Moore, and R.D. Cummings, Leukocyte trafficking mediated by selectin-carbohydrate interactions, *J. Biol. Chem.* 270:11025-11028 (1995).
6. S.D. Rosen and C.R. Bertozzi, Leukocyte adhesion: Two selectins converge on sulphate, *Curr. Biol.* 6:261-264 (1996).
7. U. Holmskov, R. Malhotra, R.B. Sim, and J.C. Jensenius, Collectins: collagenous C-type lectins of the innate immune defense system, *Immunol. Today* 15:67-74 (1994).
8. T. Feizi, Oligosaccharides that mediate mammalian cell-cell adhesion, *Curr. Opin. Struct. Biol.* 3:701-710 (1993).
9. T. Feizi, Demonstration by monoclonal antibodies that carbohydrate structures of glycoproteins and glycolipids are onco-developmental antigens, *Nature* 314:53-57 (1985).
10. P.W. Tang, H.C. Gooi, M. Hardy, Y.C. Lee, and T. Feizi, Novel approach to the study of the antigenicities and receptor functions of carbohydrate chains of glycoproteins. *Biochem. Biophys. Res. Commun.* 132:474-480 (1985).
11. T. Feizi, M.S. Stoll, Yuen C-T, W. Chai, and A.M. Lawson, Neoglycolipids: probes of oligosaccharide structure, antigenicity and function, *Methods Enzymol.* 230:484-519 (1994).
12. T. Mizuochi, R.W. Loveless, A.M. Lawson, W. Chai, P.J. Lachmann, R.A. Childs, S. Thiel, and T. Feizi, A library of oligosaccharide probes (neoglycolipids) from N-glycosylated proteins reveals that conglutinin binds to certain complex type as well as high-mannose type oligosaccharide chains, *J. Biol. Chem.* 264:13834-13839 (1989).

13. D. Solis, T. Feizi, C.T. Yuen, A.M. Lawson, R.A. Harrison, and R.W. Loveless, Differential recognition by conglutinin and mannan-binding protein of *N*-glycans presented on neoglycolipids and glycoproteins with special reference to complement glycoprotein C3 and ribonuclease B, *J. Biol. Chem.* 269:11555-11562 (1994).

14. T. Feizi and R.A. Childs, Neoglycolipids: probes in structure/function assignments to oligosaccharides, *Methods Enzymol.* 242:205-217 (1994).

15. M.S. Stoll and T. Feizi, Preparation of neoglycolipids for structure/function assignments of oligosaccharides, in: "A laboratory guide to glycoconjugate analysis", P. Jackson et al., Birkhauser Verlag AG, Basel, Switzerland, pp. In press(1996).

16. T. Osanai, T. Feizi, W. Chai, A.M. Lawson, M.L. Gustavsson, K. Sudo, M. Araki, K. Araki, and C.-T. Yuen, Two families of murine carbohydrate ligands for E-selectin, *Biochem. Biophys. Res. Commun.* 218:610-615 (1996).

17. R. Alon, T. Feizi, C.-T. Yuen, R.C. Fuhlbrigge, and T.A. Springer, Glycolipid ligands for selectins support leukocyte tethering and rolling under physiologic flow conditions, *J. Immunol.* 154:5356-5366 (1995).

18. Yuen C-T, A.M. Lawson, W. Chai, M. Larkin, M.S. Stoll, A.C. Stuart, F.X. Sullivan, T.J. Ahern, and T. Feizi, Novel sulfated ligands for the cell adhesion molecule E-selectin revealed by the neoglycolipid technology among O-linked oligosaccharides on an ovarian cystadenoma glycoprotein, *Biochemistry* 31:9126-9131 (1992).

19. M. Larkin, T.J. Ahern, M.S. Stoll, M. Shaffer, D. Sako, J. O'Brien, Yuen C-T, A.M. Lawson, R.A. Childs, K.M. Barone, P.R. Langer-Safer, A. Hasegawa, M. Kiso, G.R. Larsen, and T. Feizi, Spectrum of sialylated and non-sialylated fuco-oligosaccharides bound by the endothelial-leukocyte adhesion molecule E-selectin. Dependence of the carbohydrate binding activity on E-selectin density, *J. Biol. Chem.* 267:13661-13668 (1992).

20. C. Galustian, R.A. Childs, C. Yuen, A. Hasegawa, M. Kiso, A. Lubineau, G. Shaw, and T. Feizi, Valency dependent patterns of reactivity of human L-selectin towards sialyl and sulfated oligosaccharides of Le^a and Le^x types: Relevance to anti-adhesion therapeutics, *Submitted for publication* (1996).

BIOSYNTHESIS OF SULFATED L-SELECTIN LIGANDS IN HUMAN HIGH ENDOTHELIAL VENULES (HEV)

Jean-Philippe Girard and François Amalric

Laboratoire de Biologie Moléculaire Eucaryote du CNRS
118 route de Narbonne, 31062 Toulouse, France

SUMMARY

High endothelial venules (HEVs) are specialized post-capillary venules found in lymphoid tissues, that support high levels of lymphocyte extravasation from the blood. Lymphocyte L-selectin plays a key role in the initial interaction of lymphocytes with HEVs by recognizing sulfated carbohydrate ligands on HEV mucin-like glycoproteins, GlyCAM-1, CD34 and MAdCAM-1. Sulfation is key to the uniqueness of the HEV ligands since 6 or 6'-sulfated-sLe^X isoforms have recently been identified as major capping groups of GlyCAM-1 and sulfation of both GlyCAM1 and CD34 has been shown to be required for high-affinity L-selectin binding and recognition by the HEV-specific monoclonal antibody MECA-79. To characterize the molecular mechanisms involved in the biosynthesis of sulfated L-selectin ligands in HEVs, we have started to isolate genes that play a role in sulfate metabolism in HEVs. Studies with chlorate, a selective inhibitor of the synthesis of the high energy donor of sulfate, PAPS (3'-phosphoadénosine 5'-phosphosulfate), had previously revealed that PAPS synthesis is required for sulfation of HEV ligands and recognition by L-selectin. Therefore, we screened an HEV cDNA library in order to isolate cDNAs encoding enzymes involved in PAPS synthesis. This strategy allowed us to isolate a novel cDNA encoding the PAPS synthetase from human HEVs. The molecular characteristics of PAPS synthetase and its role in biosynthesis of sulfated L-selectin ligands in HEVs are discussed.

LYMPHOCYTE MIGRATION THROUGH HEV

Patrolling the body in search of foreign antigen, lymphocytes continuously recirculate from blood, through lymphoid and other tissues, and back through the lymphatics to the blood (Butcher and Picker, 1996). This process, called lymphocyte recirculation, allows the dissemination of the immune response throughout the body, and thus provides an effective immune surveillance for foreign invaders and alterations in the body's own cells (Yednock and Rosen, 1989; Picker and Butcher, 1992). The first critical step in lymphocyte migration from circulation into tissue is the adhesion of lymphocytes to vascular endothelium. In lymphoid organs, lymphocyte adherence and transendothelial migration occur at specialized post-capillary vascular sites called high endothelial venules (HEVs) (for a review see Girard and Springer, 1995a). In humans, HEVs are found in all secondary lymphoid organs (with the exception of spleen, where lymphocyte emigration occurs via the blood sinusoids in the marginal zone), including hundreds of lymph nodes dispersed in the body, tonsils and adenoids in the pharynx, Peyer's patches in the small intestine, appendix, and small aggregates of lymphoid tissue in the stomach and large intestine. Moreover, HEV-like vessels are observed in chronically inflamed non lymphoid tissues and are believed to support lymphocyte recruitment into these sites (Freemont, 1988). In contrast to the endothelial cells from other vessels, the high endothelial cells of HEVs have a plump, almost cuboidal appearance, express specialized ligands for lymphocytes and are able to support high levels of lymphocyte extravasation (Girard and Springer, 1995a). The recruitment of lymphocytes by HEVs is a very specific and efficient process since about 25% of lymphocytes circulating in HEVs bind and emigrate (Bjerknes et al., 1986; Bargatze and Butcher, 1993). In the human body, it has been estimated that this results in more than 10^6 lymphocytes extravasating from the blood through HEV every second.

Lymphocyte sticking to HEVs is a multistep process that takes only a few seconds (Figure 1). This process requires four successive events: tethering, rolling, adhesion triggering and arrest (Bargatze and Butcher, 1993; Bargatze et al., 1995). First, free-flowing lymphocytes initiate contact with the HEV endothelium through microvillous receptors. This primary adhesion (tethering) is followed by loose rolling of lymphocytes along the vessel wall (rolling). Rolling slows the transit of lymphocytes in HEV, thus allowing sufficient time for lymphocyte activation through G-protein-linked receptors. Activation of lymphocyte integrins (adhesion triggering) leads to tight adhesion and lymphocyte arrest that is stable under shear forces (arrest). Sticking to the HEV endothelial cells is then followed by lymphocyte entry into the lymphoid tissue, a process that occurs in approximately 10 minutes and involves lymphocyte transendothelial migration and crossing of the HEV basal lamina.

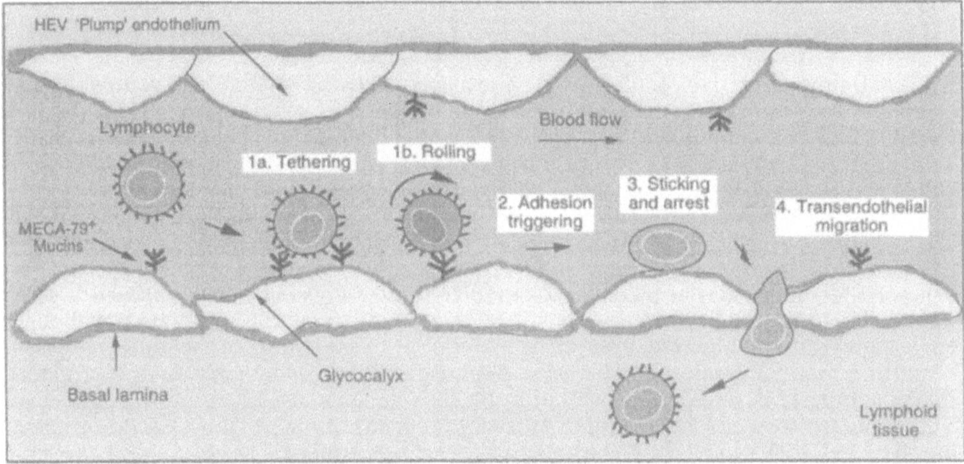

Figure 1. Lymphocyte migration through high endothelial venules (HEVs). Lymphocytes circulating in the blood initiate contact with the high endothelial cells via microvillous receptors (1a. Tethering). This primary adhesion is transient and is often manifested in rolling of the interacting cells along the HEV endothelium (1b. Rolling). Activation of lymphocyte adhesiveness through G-protein coupled receptors (2. Adhesion triggering) results in firm attachment, which becomes stable to physiological shear force (3. Sticking and arrest). Sticking is then followed by lymphocyte migration through endothelial cell junctions and the HEV basal lamina (4. Transendothelial migration).

SULFATED L-SELECTIN COUNTER-RECEPTORS IN HEV

At the molecular level, the initial interaction of lymphocytes with HEVs *in vivo* is mediated by lymphocyte L-selectin, a cell surface adhesion molecule containing a C-type lectin domain. In situ videomicroscopy has revealed that functional inactivation of L-selectin by blocking antibodies (Gallatin et al., 1983) or gene-knockout (Arbones et al., 1994) completely abolishes lymphocyte tethering and rolling in HEVs, and results in a 99% decrease of lymphocyte emigration in peripheral lymph nodes (PLNs) HEVs. L-selectin recognizes sulfated carbohydrates on the following HEV mucin-like glycoproteins, glycosylation-dependent cell adhesion molecule 1 (GlyCAM-1; Lasky et al., 1992), CD34 (Baumhueter et al., 1993) and mucosal addressin cell adhesion molecule 1 (MAdCAM-1;

Berg et al., 1993). These three glycoproteins contain highly rigid, extended, serine/threonine-rich, mucin-like domains that are densely substituted with O-linked oligosaccharides. HEV-specific O-glycosylation of these three mucin-like proteins is required for L-selectin binding. For example, although CD34 is expressed in many other vessels in addition to HEVs, it functions as a high affinity L-selectin counter-receptor only when appropriately decorated by HEV-specific, sulfated, sialylated, fucosylated, O-linked oligosaccharides (Baumhueter et al., 1995).

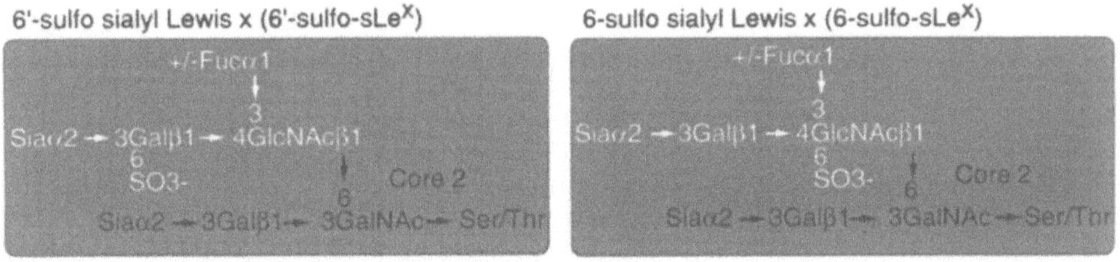

Figure 2. Structure of the sulfated O-linked oligosaccharides present on the HEV mucin-like glycoprotein GlyCAM-1. The oligosaccharides bear the sLe[X] tetrasaccharide (highlighted) and are sulfated on the 6 position of Gal (6'-sulfo-sLe[X]) or the 6 position of GlcNac (6'-sulfo-sLe[X]). These sulfated sLe[X] derivatives are incorporated into the core 2 structure (Galb1,3(GlcNacb1,6)GalNac), a common structural precursor for many O-linked carbohydrates. Abbreviations: Fuc, fucose; Gal, galactose; GalNac, N-acetylgalactosamine; GlcNac, N-acetylglucosamine; Sia, sialic acid; SO3-, sulfate.

Sialic acid and fucose are critical for recognition of HEV mucins by L-selectin since sialidase treatment (Rosen et al., 1985; Rosen et al., 1989) or knock-out of the $\alpha(1,3)$ fucosyltransferase FucTVII gene (Maly et al., 1996), both abolish L-selectin binding and lymphocyte adhesion to lymph node HEVs *in vitro* and *in vivo* . The tetrasaccharide sLe[X] - i.e. Siaα2-3Galβ1-4(Fucα1-3)GlcNAc - has been shown to function as a low affinity ligand for L-selectin. However, the detection of sLe[X] in endothelial cells that do not mediate lymphocyte recruitment *in vivo* indicates that the biological HEV ligands for L-selectin are more complex than this simple tetrasaccharide. The detailed structure of the O-linked oligosaccharides present on GlyCAM-1 has recently been elucidated (Figure 2). GlyCAM-1 O-glycans are extensively sulfated on the 6 position of galactose (Gal-6-sulfate) and N-acetylglucosamine (GlcNac-6-sulfate). Two major capping groups have been identified as sulfated derivatives of sLe[X] (Hemmerich and Rosen, 1994): 6'-sulfo-sLe[X] (Siaα2-3(SO4-6)Galβ1-4(Fucα1-3)GlcNAc) and 6-sulfo-sLe[X] (Siaα2-3Galβ1-4(Fucα1-3)(SO4-6) GlcNAc). In addition, although no direct evidence has been provided yet, a disulfated sLe[X] derivative 6',6-disulfo-sLe[X] (Siaα2-3(SO4-6)Galβ1-4(Fucα1-3)(SO4-6)GlcNAc) may also be a determinant on GlyCAM-1. The majority of GlyCAM-1 sulfated sLex derivatives are incorporated into the core 2 structure (Galβ1-3(GlcNAcβ1-6)GalNAc), a common structural precursor for O-linked carbohydrates on many glycoproteins (Hemmerich et al., 1995). Thus, the oligosaccharide ligands for L-selectin on GlyCAM-1 consist of a common underlying core sequence decorated with unique sulfated capping structures. Sulfation may thus be key to the uniqueness of the HEV ligands. In agreement with this possibility, sulfation of both GlyCAM1 and CD34 has been shown to be required for L-selectin recognition (Imai et al., 1993; Hemmerich et al., 1994). Undersulfated GlyCAM1 and CD34, produced in organ culture of lymph nodes treated with chlorate, a highly selective metabolic inhibitor of sulfation, do not bind to L-selectin. Interestingly, recognition of GlyCAM-1 and CD34 by MECA-79, the only HEV-specific monoclonal antibody currently available, is also abrogated after treatment with chlorate (Hemmerich et al., 1994). MECA-79 recognizes HEV counter-receptors for L-selectin and inhibits lymphocyte emigration through HEVs into lymph nodes *in vivo* and lymphocyte adhesion to lymph node and tonsil HEVs *in vitro* (Michie et al., 1993). The fact that MECA-79 recognition of GlyCAM-1 and CD34 is sulfation-dependent further emphasizes the critical role of sulfation for L-selectin binding. Moreover, the total absence of crossreactivity of MECA-79 with non-HEV cells of the human body (Michie et al., 1993) suggests that the HEV-specific sulfated oligosaccharides recognized by MECA-79 and L-selectin are unique structures that are probably even more complex than simple sulfated sLe^X derivatives. Alternatively, specificity may be explained by the clustering on mucin-like domains of oligosaccharides containing sulfated sLe^X. This would be in agreement with the fact that free oligosaccharides released from GlyCAM-1 do not bind with recognizable affinity to L-selectin, and that all HEV counter-receptors for L-selectin identified to date contain highly O-glycosylated mucin-like domains.

BIOSYNTHESIS OF SULFATED L-SELECTIN LIGANDS IN HEV

Sulfation of L-selectin counterreceptors in HEV is likely to involve at least four types of genes (Figure 3): a sulfate transporter, a PAPS synthetase (bifunctional ATP sulfurylase-APS kinase), a PAPS transporter and one or more sulfotransferases. A high affinity sulfate transporter is known to exist in HEV since, both in humans and rats, the HEV endothelium has been shown to be unique amongst vascular endothelium by virtue of its capacity to incorporate large amounts of $^{35}SO_4$ (Andrews et al., 1982). Although, the HEV sulfate transporter(s) is not yet molecularly characterized, two human sulfate transporters, DTD and DRA, have recently been cloned, that may play a role in sulfate incorporation in HEV (Hastbacka et al., 1994). Both DTD and DRA are members of the superfamily of membrane transporters exhibiting 12 transmembrane domains. DTD is ubiquitously expressed while DRA has a more restricted expression with high levels in intestine and colon. Both genes seem to play a role in human diseases. DRA is down-regulated in colon adenomas and adenocarcinomas while DTD is mutated in three different chondrodysplasias (diastrophic dysplasia, atelosteogenesis type II, achondrogenesis type IB). DTD mutations result in undersulfation of proteoglycans in cartilage. These sulfation defects in DTD genetic diseases clearly show that sulfate transporters, by allowing efficient sulfate incorporation, control the degree of sulfation of proteoglycans or glycoproteins. Similarly, sulfate transporters are likely to play a key role in the control of sulfation of L-selectin ligands in HEV. Therefore, it will be important, in the future, to determine whether DTD and DRA are expressed in HEV and/or to identify putative HEV-specific sulfate transporter(s).

To be used by sulfotransferases, inorganic sulfate entering the cell, needs first to be activated by a two steps process. First, ATP sulfurylase (ATP sulfate adenylyltransferase) catalyzes the production of adenosine-5'-phosphosulfate (APS) from sulfate and ATP. Subsequently, APS kinase (ATP adenosine-5'-phosphosulfate 3'-phosphotransferase) transfers a phosphate group from ATP to APS to yield PAPS (adenosine 3'-phosphate 5'-phosphosulfate), the activated sulfate donor. ATP sulfurylase and APS kinase have been cloned in prokaryotes, yeast, funghi and plants and found to be encoded by separate genes. In contrast, we have recently identified the human homologues of ATP sulfurylase and APS kinase and found that both enzymes activities are encoded by a single gene and reside on a single protein, the bifunctional ATP sulfurylase-APS kinase or PAPS synthetase (see below). A key role for PAPS synthetase in the biosynthesis of sulfated L-selectin ligands is supported by the fact that treatment with chlorate, a potent anion inhibitor of ATP sulfurylase, abrogates L-selectin binding to GlyCAM-1 and CD34. Moreover, the observation in a number of systems that PAPS synthesis is the rate-limiting step for sulfation suggests that regulation of PAPS synthetase levels may play an important role in the modulation of L-selectin ligands sulfation.

Sulfation of glycoproteins and proteoglycans is a late synthetic event that occurs within the lumen of the Golgi apparatus. Therefore, to serve as a substrate for sulfotransferases, PAPS must be transported from the cytosol, its site of synthesis, into the lumen of the Golgi apparatus. Although, the PAPS transporter within the Golgi membrane has recently been purified, its molecular characterisitics have not yet been reported. Similarly, little is known about the sulfotransferases involved in mucin sulfation. The large number of distinct sulfated structures and the presence of these structures on a limited number of glycoproteins in a given cell or tissue indicate that the sulfotransferases that transfer sulfate to glycoprotein oligosaccharides are both numerous and highly specific. The presence of distinct sulfated sLe^X structures on GlyCAM-1 (Figure 2) suggests the existence of at least two different HEV sulfotransferases involved in the transfer of sulfate to the 6 position of galactose and N-acetylglucosamine.

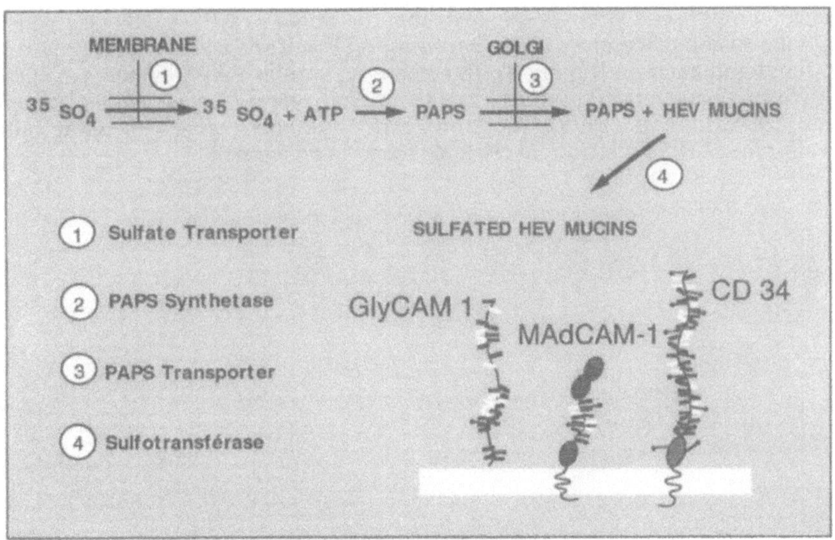

Figure 3. Sulfation of L-selectin counterreceptors in high endothelial venules (HEVs). GlyCAM-1, MAdCAM-1 and CD34, are three HEV glycoproteins containing mucin-like domains (with a high percentage of serine and threonine residues), that present dense clusters of O-linked sulfated carbohydrates to lymphocyte L-selectin. Sulfation of these three mucin-like proteins requires at least four types of genes: a sulfate transporter, to allow sulfate incorporation; a PAPS synthetase (bifunctional ATP-sulfurylase/APS kinase), to synthesize PAPS, the activated sulfate donor; a PAPS transporter to transport PAPS in the Golgi and one or more sulfotransferases to transfer sulfate from PAPS to the O-linked carbohydrates of HEV mucin-like glycoproteins.

Although the transfer of sulfate from PAPS to HEV mucins by sulfotransferases seems to be the most specific step in the pathway, it is important to keep in mind that sulfation of HEV mucins may also be controlled at earlier steps. For instance, the very high levels of sulfate incorporation in HEVs could explain the extensive sulfation of O-linked carbohydrates from GlyCAM-1 and other HEV mucins. Similarly, PAPS synthetase and the Golgi PAPS transporter, by modulating PAPS production and availability, could also play a major role in the control of HEV mucins sulfation.

MOLECULAR CLONING OF THE PAPS SYNTHETASE (ATP-SULFURYLASE/APS KINASE) FROM HUMAN HEV

Since treatment of lymph nodes with chlorate, a potent inhibitor of PAPS synthesis, has previously been shown to abrogate synthesis of sulfated L-selectin ligands in HEVs, we decided to isolate the gene encoding the target for chlorate inhibition, the HEV ATP sulfurylase. For this purpose, we used an HEV cDNA library, that we had previously generated with mRNA from purified HEV cells (Girard and Springer, 1995b). Screening of this HEV library with a human expressed sequence tag (EST) homologous to plants ATP·sulfurylase cDNAs allowed us to isolate 51 independent positive clones, corresponding to a unique cDNA. The 2.5 kb full length cDNA contains a single open reading frame encoding a putative 627 amino-acid protein with a predicted molecular weight of 71.1 kDa. Sequence comparisons with databases revealed that the carboxy-terminal part of the predicted protein exhibits extensive homologies with ATP sulfurylases from plants, yeast and fungi (Figure 4). Surprisingly, the amino-terminal part of the protein is closely related to APS kinases from bacteria, yeast and plants. This indicates that the human cDNA, we have isolated, encodes a bifunctional ATP sulfurylase/APS kinase or PAPS synthetase.

Sequence comparisons strongly suggests that the human PAPS synthetase gene evolved through the fusion of separate genes encoding ATP sulfurylase and APS kinase in plants, yeast, fungi and bacteria (Figure 4). Interestingly, similar PAPS synthetase cDNAs have recently been characterized in mouse and the marine worm *Urechis caupo* (Rosenthal and Leustek, 1995; Li et al., 1995). Therefore, PAPS synthetase genes encoding bifunctional ATP sulfurylase/APS kinase are likely to be found in all animals.

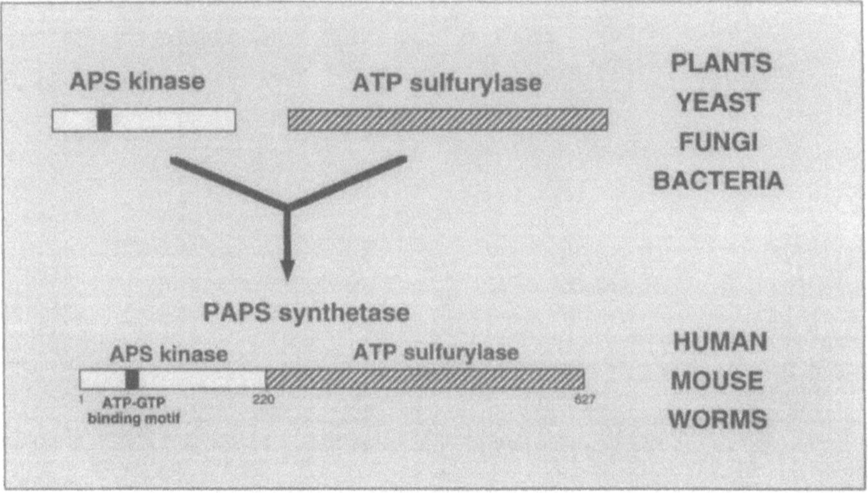

Figure 4. The human PAPS synthetase gene evolved through the fusion of the ATP sulfurylase- and APS kinase-encoding genes from lower organisms. The amino-terminal part of human PAPS synthetase (aa 1-220) shows extensive homologies to APS kinases from bacteria, yeast and plants, while the carboxy-terminal part is very similar to ATP sulfurylases from yeast, fungi and plants.

To confirm that human PAPS synthetase has both ATP sulfurylase and APS kinase activities, we tested the capacity of the cloned PAPS synthetase to synthesize PAPS *in vivo*. Functional expression of the human PAPS synthetase in CHO cells resulted in high levels of PAPS production, as revealed by high-voltage paper electrophoresis after metabolic labeling with $^{35}SO_4$. These experiments also showed that PAPS synthetase is sensitive to chlorate inhibition since treatment of transfected CHO cells with chlorate prevented PAPS synthesis. Northern blot analysis revealed that the 2.5 kb PAPS synthetase mRNA is expressed in many other human tissues in addition to HEVs. This later result suggests that PAPS synthetase may play a key role in the sulfation not only of HEV mucins but also of many other molecules, including leukocyte PSGL-1, heparan sulfate proteoglycans, glycolipids, glycoprotein hormones and toxic compounds.

CONCLUSION

Sulfation of L-selectin ligands is a major biosynthetic activity of HEVs, that contributes significantly to the uniqueness of the HEV ligands. Therefore, characterization of the genes and molecular mechanisms involved in the sulfation of HEV mucins will provide a better understanding of the mechanisms controlling the recruitment and migration of lymphocytes through HEVs. We have started to isolate these genes and succeeded in cloning a novel human cDNA encoding PAPS synthetase, a bifunctional ATP sulfurylase/APS kinase. PAPS synthetase appears to play a key role in the synthesis of sulfated L-selectin ligands in HEVs since treatment of lymph nodes with chlorate, a potent inhibitor of PAPS synthesis, has previously been shown to abrogate L-selectin recognition of HEV mucins GlyCAM-1 and CD34. PAPS synthetase is widely expressed in the human body and may also play a role in the biosynthesis of sulfated L-selectin ligands in chronically inflamed non-lymphoid tissues. In the future, it will be important to characterize the other genes (sulfate transporter, PAPS transporter, sulfotransferases) involved in the pathway of mucin sulfation in HEVs, in order to determine how modulation of the activities of all these genes regulates the biosynthesis of sulfated L-selectin ligands and the recruitment of lymphocytes in HEVs.

REFERENCES

Andrews, P., Milsom, D.W., and Ford, W.L., 1982, Migration of lymphocytes across specialized vascular endothelium V. Production of a sulphated macromolecule by high endothelial cells in lymph nodes, *J. Cell. Sci.*, 57: 277.

Arbones, M.L., Ord, D.C., Ley, K., Ratech, H., Maynard-Curry, C., Otten, G., Capon, D.J., and Tedder, T.F., 1994, Lymphocyte homing and leukocyte rolling and migration are impaired in L-selectin-deficient mice. *Immunity*, 1:247.

Bargatze, R.F. and Butcher, E.C., 1993, Rapid G protein-regulated activation event involved in lymphocyte binding to high endothelial venules, *J. Exp. Med.*, 178: 367.

Bargatze, R.F., Jutila, M.A., and Butcher, E.C., 1995, Distinct roles of L-selectin and integrins a4b7 and LFA-1 in lymphocyte homing to Peyer's patch-HEV in situ: the multistep model confirmed and refined, *Immunity*, 3:99.

Baumhueter, S., Singer, M.S., Henzel, W., Hemmerich, S., Renz, M., Rosen, S.D., and Lasky, L.A., 1993, Binding of L-selectin to the vascular sialomucin CD34, *Science*, 262: 436.

Baumhueter, S., Dybdal, N., Kyle, C., and Lasky, L.A., 1994, Global vascular expression of murine CD34, a sialomucin-like endothelial ligand for L-selectin, *Blood*, 84: 2554.

Berg, E.L., McEvoy, L.M., Berlin, C., Bargatze, R.F., and Butcher, E.C., 1993, L-selectin-mediated lymphocyte rolling on MAdCAM-1, *Nature*, 366: 695.

Bjerknes, M., Cheng, H., and Ottaway, C.A., 1986, Dynamics of lymphocyte-endothelial interactions in vivo, *Science*, 231:402.

Butcher, E.C. and Picker, L.J., 1996, Lymphocyte homing and homeostasis, *Science*, 272:60.

Freemont, A.J., 1988, Functional and biosynthetic changes in endothelial cells of vessels in chronically inflamed tissues: evidence for endothelial control of lymphocyte entry into diseased tissues, *J. Pathol.*, 155:225.

Gallatin, W.M., Weissman, I.L., and Butcher, E.C., 1983, A cell-surface molecule involved in organ-specific homing of lymphocytes, *Nature*, 304:30.

Girard, J-P. and Springer, T.A., 1995, Cloning from purified high endothelial venule cells of hevin, a close relative of the antiadhesive extracellular matrix protein SPARC, *Immunity*, 2:113.

Girard, J-P. and Springer, T.A., 1995, High endothelial venules (HEVs): specialized endothelium for lymphocyte migration, *Immunol. today*, 16:449.

Hastbacka, J., de la Chapelle, A., Mahtani, M.M., Clines, G., Reeve-Daly, M.P., Daly, M., Hamilton, B.A., Kusumi, K., Trivedi, B., Weaver, A., Coloma, A., Lovett, M., Buckler, A., Kaitila, I., Lander, E.S., 1994, The diastrophic dysplasia gene encodes a novel sulfate transporter: positional cloning by fine-structure linkage desequilibrium mapping, *Cell*, 78:1073.

Hemmerich, S., Butcher, E.C., and Rosen, S.D., 1994, Sulfation-dependent recognition of HEV-ligands by L-selectin and MECA 79, an adhesion-blocking mAb, *J. Exp. Med.*, 180:2219.

Hemmerich, S. and Rosen, S.D., 1994, 6'-Sulfated sialyl Lewis x is a major capping group of GlyCAM-1, *Biochemistry*, 33:4830.

Hemmerich, S., Leffler, H., and Rosen, S.D., 1995, Structure of the O-glycans in GlyCAM-1, an endothelial-derived ligand for L-selectin, *J. Biol. Chem.*, 270:12035.

Imai, Y., Lasky, L.A., and Rosen, S.D., 1993, Sulphation requirement for GlyCAM-1, an endothelial ligand for L-selectin, *Nature*, 361:555.

Lasky, L.A., Singer, M.S., Dowbenko, D., Imai, Y., Henzel, W.J., Grimley, C., Fennie, C., Gillett, N., Watson, S.R., and Rosen, S.D., 1992, An endothelial ligand for L-selectin is a novel mucin-like molecule, *Cell*, 69:927.

Li, H., Deyrup, A., Mensch, J. R., Domowicz, M., Konstantinidis, A., and Schwartz, N.B., 1995, The isolation and characterization of cDNA encoding the mouse bifunctional ATP sulfurylase-Adenosine 5'-phosphosulfate kinase, *J. Biol. Chem.*, 270:29453.

Maly, P., Thall, A., Petryniak, B., Rogers, C.E., Smith, P.L., Marks, R.M., Kelly, R.J., Gersten, K.M., Cheng, G., Saunders, T.L., Camper, S.A., Camphausen, R.T., Sullivan, F.X., Isogai, Y., Hindsgaul, O., von Andrian, U.H., and Lowe, J.B., 1996, The $\alpha(1,3)$ fucosyltransferase Fuc-TVII controls leukocyte trafficking through an essential role in L-, E-, and P-selectin ligand biosynthesis, *Cell*, 86:643.

Michie, S.A., Streeter, P.R., Bolt, P.A., Butcher, E.C., and Picker, L.J., 1993, The human peripheral lymph node vascular addressin: An inducible endothelial antigen involved in lymphocyte homing, *Am. J. Pathol.*, 143:1688.

Picker, L.J. and Butcher, E.C., 1992, Physiological and molecular mechanisms of lymphocyte homing, *Annu. Rev. Immunol.*, 10:561.

Rosen, S.D., Singer, M.S., Yednock, T.A., and Stoolman, L.M., 1985, Involvement of sialic acid on endothelial cells in organ-specific lymphocyte recirculation, *Science*, 228:1005.

Rosen, S.D., Chi, S.I., True, D.D., Singer, M.S., and Yednock, T.A., 1989, Intraveneously injected sialidase inactivates attachment sites for lymphocytes on high endothelial venules, *J. Immunol.*, 142:1895.

Rosenthal, E., and Leustek, T., 1995, A multifunctional *Urechis caupo* protein, PAPS synthetase, has both ATP sulfurylase and APS kinase activities, *Gene*, 165:243.

Yednock, T.A. and Rosen, S.D., 1989, Lymphocyte homing, *Adv. Immunol.*, 44:313.

ENDOTHELIAL SIALYL LEWIS X AS A CRUCIAL GLYCAN DECORATION ON L-SELECTIN LIGANDS

Risto Renkonen

Department of Bacteriology and Immunology
Haartman Institute
University of Helsinki
Helsinki, Finland

INTRODUCTION

Inflammatory reactions, such as organ transplant rejections, are characterized by lymphocyte infiltration into the tissue[1]. This extravasation of lymphocytes is initiated by the interaction of members of the selectin family and their ligands, which leads to a vascular shear flow-dependent rolling on the endothelial surfaces[2-5]. Of the three identified selectins, L-selectin is expressed on leukocyte surfaces and it recognizes glycoprotein ligands on endothelium[6-8]. Three characterized mucin-like heavily O-glycosylated proteins; GlyCAM-1, CD34 and MAdCAM-1 are endothelial L-selectin ligands[9-11]. These ligands recognize L-selectin only when posttranslationally glycosylated in a proper manner. So far only the crucial glycoforms of murine GlyCAM-1 have been characterized in great detail and they have been shown to be $\alpha2,3$ sialylated, $\alpha1,3$ fucosylated and sulfated, i.e. carry sialyl Lewis x (sLex) and/or sulfated sLex, respectively[12-16]. L-selectin was first characterized to guide lymphocyte traffic to lymph nodes and to sites of inflammation[17]. Today it is also known to participate in the rolling of leukocytes on vascular endothelium[6-8]. The two other members of selectin adhesion molecule family, E- and P-selectin, are expressed on activated endothelium[3, 18] and their glycoprotein ligands on leukocytes are active only when properly decorated with fucosylated oligosaccharides.

STRUCTURE OF SLEX AND RELATED MOLELCULES

All selectin ligand glycoproteins are decorated by sLex-type of glycans, but clearly this is not enough[19]. sLex is a tetrasaccharide NeuNAcα2-3Galβ1-4(Fucα1-3)GlcNAc. Already very early observations pointed out that also another closely realted structure sialyl Lewis a (sLea, NeuNAcα2-3Galβ1-3(Fucα1-4)GlcNAc) might also act as a selectin ligand[20-22]. The sialic acid can be replaced within a sLex or sLea glycans by a sulfo group and the affinity towards E-selectin increases[23, 24]. The glycans of murine GlyCAM-1 recognizing L-selectin carry sulfated sLex[16].

GLYCOSYLTRANSFERASES IN SLEX SYNTHESIS

If sLex type of glycans play an essential role in the regulation of leukocyte extravasation, how is their expression regulated on endothelial cells? Starting from the very abundant distal sequence, N-acetyllactosamine, two enzymatic steps which are required for the synthesis of sLex, first one being the $\alpha(2,3)$sialylation[19]. To date several $\alpha(2,3)$ sialyltrasferases have been cloned and two of them are involved in the sLex synthesis adding sialic acid with a $\alpha(2,3)$ linkage to the galactose of a N-acetyllactosamine[25-28]. After the N-acetyllactosamine has become sialylated the $\alpha(1,3)$fucosyltransferase will act and transfer fucose from GDP-fucose to GlcNAc. Five human $\alpha(1,3)$fucosyltransferases have been cloned and characterized in some detail. They seem to have slightly different, but mainly overlapping acceptor specificities; four of them fucosylate efficiently sialylated acceptors, while Fuc-TIV prefers neutral acceptors [20, 29-35]. Fuc-TIII is shown to have two different activities ($\alpha(1,3)$ and $\alpha(1,4)$fucosylation) leading to the generation of sLex and sLea, respectively.

The sLex synthesis in a cell is under several regulatory processes. If the N-acetyllactos-amine would be $\alpha(2,6)$sialylated no further $\alpha(1,3)$fucosylation could take place [36]. The primary acceptor N-acetyllactosamine can also be $\alpha(1,2)$fucosylated to the galactose, which leads to a H-type of blood group O-structure. Furthermore, if the N-acetyllactosamine becomes first $\alpha(1,3)$fucosylated to the GlcNAc, $\alpha(2,3)$sialyltransferases are unable to act anymore on the newly formed Lewis x (Lex) structures. These examples show how complicated the synthesis of an even relatively small oligosaccharide can be. The essential components of this cascade are thereby the availability of oligonucleotide donors (such as CMP-sialic acid or GDP-fucose) and the amount and activity of the corresponding glycosyltransferases within a given cell.

EXPRESSION, SYNTHESIS AND DEGRADATION OF SLEX IN ENDOTHELIAL CELLS

sLex glycans have been characterized both on normal peripheral blood granulocytes as well as on certain carcinoma cells in the early 1980's[37], but it was not before the 1990's the sLex expression on endothelial cells was resolved. While endothelium normally did not express sLex glycans there was one exception. Lymph node high endothelial venules (HEV) lined by a special plump high endothelium is the site of extravasation for lymphocytes. These HEVs were shown to express sLex strongly and thereby facilitate the L-selectin-dependent lymphocyte homing to lymph nodes[38-40].

After demonstrating the sLex expression in lymph node high endothelium we wanted to analyse would the endothelial cells have the required machinery to synthesize sLex, ie. possess the appropriate glycosyltransferases and would this expression be under inflammatory regulation. The expression of sLex oligosaccharides was documented on cultured human umbilical endothelial cells (HUVECs) with flow cytometry and a panel of anti sLex mAbs[41]. Further evidence of the presence of sLex on HUVECs came from flow cytometry experiments where the cells were treated with a broad spectrum sialidase in order to remove sialic acid from the sLex glycans and the expression of Lex antigens was thereafter analysed. As expected the expression of Lex enhanced significantly after sialidase treatment with a concomitant abolisment of reactivity with anti sLex antibodies [41]. Taken together these data suggest that *in vitro* cultured HUVECs express sLex related glycans.

Cultured endothelial cells expressed at least transcripts for one of the essential $\alpha(2,3)$ sialyltransferases (ST3N/ST3Gal II) and for several $\alpha(1,3)$fucosyltransferases (Fuc-TIII, Fuc-TIV and Fuc-TVII) as analysed by Northern blots[41]. The presence of a given glycosyltransferase mRNA does not directly provide evidence for the presence of the active corresponding enzyme. Thus assays for enzyme activities were carried out, where N-acetyllactosamine served as the acceptor, CMP-[^{14}C]sialic acid as a donor and endothelial lysates as the source on enzyme and the structure of newly synthesized glycans was analysed by anion exchange and paper chromatography. HUVECs were shown to preferentially add sialic acid into a $\alpha(2,6)$ linkage, but the $\alpha(2,3)$ sialylated product was obtained in a 10% yield[41]. Its synthesis could not be enhanced by *in vitro* proinflammatory cytokine stimulations.

Resting endothelial cells could transfer fucose to both neutral N-acetyllactosamine and sialylated N-acetyllactosamine with a $\alpha(1,3)$ linkage to GlcNAc. Interestingly tumor necrosis factor (TNF) -stimulated endothelial cells could synthesize over 4-times more sLex from sialylated N-acetyllactosamine than resting endothelial cells, indicating that proinflammatory signals can induce endothelial sLex synthesis[41]. Taken together these data suggested that cultured endothelial cells have the capacity to synthesize sLex-type of glycans and the rate of synthesis can be further induced by inflammatory stimuli.

We were able to show a feedback loop in the L-selectin dependent adhesion cascade as cultured endothelial cells could also degrade the sLex motifs primarily by $\alpha(2,3)$ sialidase function followed to a minor extent with $\alpha(1,3)$fucosidase function. Besides the *de novo* expression of upregulated sLex expression which occur in the begining of an inflammation reaction it is necessary to also analyse the events leading to the termination of inflammation. *In vitro* evidence with purified enzymes had already suggested that sLex is not degraded by $\alpha(1,3)$fucosidases, but rather by $\alpha(2,3)$sialidase activity. A radiolabelled sLex glycan was synthesized, which was resistant to spontaneus β-elimination of the fucose residue. Endothelial cell lysates desialylated most of the sLex glycan was within a few hours[42] and only a small proportion of the molecule experienced also further defucosylation. Concomitantly no defucosylation occurs with the sialylated original molecule, indicating that the removal of sialic acid is the primary event in the degradation of sLex glycan by endothelial cells[42]. These data suggest that the synthesis and degradation of sLex motifs is under strict control as they are key regulators of extravasation and the consecutive lymphocyte homing and generation of inflammation.

ROLE OF ENDOTHELIAL SLEX IN THE INDUCTION OF INFLAMMATION

To be able to interpret the role of endothelial sLex synthesis in the context of inflammatory reactions we characterized two organ transplant models in rats[1, 43]. Allogeneic kidney of heart grafts between different inbred rat strains are rejected within 5 days. The hallmark of allograft rejection is lymphocyte infiltration into the transplant. One of to the key questions in the interventions of this lymphocytic infiltrate is to understand how and where do the lymphocytes get into the graft. We analyzed this lymphocyte traffic with the Stamper-Woodruff *in vitro* lymphocyte-endothelial binding assay. We could show that the adhesion of lymphocytes into the kidney and heart allografts increased already on the second postoperative day[43, 44]. This enhanced lymphocyte adhesion was solely due to an increased binding to capillary endothelium as practially no other endothelial or other structures in the grafts possessed capacity to adhere lymphocytes. Concomitantly with the increased binding of lymphocytes to capillary endothelium, these cells obtained several morphological features common to high endothelium of lymphoid organs as analysed by electron microscopy.

These observations indicate that capillary endothelium is the site of lymphocyte entry into the rejecting allografts. Of the endothelial adhesion molecules, ICAM-1 was already expressed on the endothelium of normal grafts, and its expression was strongly enhanced during rejection without site-specific restriction[45]. VCAM-1 was not expressed on the endothelium of normal or syngeneic transplants, but its expression was induced during allograft rejection not only in capillaries, but occasionally also on the endothelium of larger vessels. On the other hand sLex showed a very restricted pattern of expression; endothelium was sLex negative both in control and syngeneic grafts whereas capillary endothelium reacted strongly only in allografts, but not in control organs[45]. Only these capillaries in the allografts, but not in syngeneic grafts or normal tissues bound an L-selectin-IgG fusion protein indicating that ligands for L-selectin were induced during rejection[45].

All the above mentioned data provided circumstantial evidence for the crucial role of sLex glycans in the generation of inflammatory infiltrates during transplant rejection. Still direct evidence for the role of fucose containing sLex glycans participating in lymphocyte extravasation to lymph nodes and to sites of inflammation was needed. Five mammalial $\alpha(1,3)$fucosyltransferases have been cloned and recently the first mice deficient in one of these genes was constructed and charaterized[20, 29-35] The Fuc-TVII knock-out mice resemble to some extent patients suffering of the leukocyte adhesion deficiency II (LAD II) syndrome. The leukocytes in the knock-out mice do not express active E- or P-selectin ligands, which leads to marked leukocytosis and defective leukocyte extravasation to sites on inflammation. Concomitantly no functional endothelial ligands for L-selectin are expressed in these animals, leading to severe attenuation of lymphocyte homing to lymph nodes[25]. These Fuc-TVII knock-out mice provide the first direct evidence that fucose is an essential component of the L-selectin ligands together with sialic acid and sulfate.

There is some controversy whether rodent leukocytes synthetize sLex type of oligosaccharides as they do not express sLex as analysed by anti sLex immunoreactivity[46]. However, murine GlyCAM-1 secreted from lymph node endothelium bears sulfated sLex epitopes[13, 14, 16] and we have shown that capillary endothelium on rat kidney and heart allografts start to express sLex *de novo* during rejection episodes[45, 47]. The sLex-type structures on endothelium of rat lymph nodes or inflammed tissues react with the anti sLex mAbs. Furthermore cultured rat endothelial cells have been shown to synthetize sLex glycans [42]. Yet again the crucial role of $\alpha(1,3)$fucosylated glycans on L, E- and P-selectin ligands have been shown by gene knock-out mice[25], suggesting that sLex-type glycans still would be expressed also in the murine system. So, it may well be that sLex-type oligosaccharides are not expressed on rodent leukocytes, or that they are modifications not reacting with the presently available anti sLex mAbs.

OLIGOSACCHRIDE ANTAGONISTS FOR SELECTINS

Immediately after the selectin ligands had been shown to contain sLex-type of oligosaccharide decorations a number of laboratories initiated projects aiming at characterizing putative selectin antagonists. Most of these approaches used the synthesis of oligosaccharides and the preliminary animal experiments have been very promising[48].

As the *in vitro* synthesis of sLex glycans is tedious and expensive so far these glycans have been tested only in short-term inflammation models involving a rapid influx of granulocytes to the site of inflammation. The first one was a model where a snake venom caused P-selectin-dependent granulocytic inflammation into the lungs. By intravenous infusion of sLex, a ligand for P-selectin, the number of granulocytes accumulated into the lungs of these rats could be significantly decreased with a concomitant reduction of the lung injury[49].

Reperfusion of an ischemic coronary or brain vascular bed enhances the tissue injury despite the reestablishment of blood flow to these tissues. This reperfusion injury, is characterized by extravasating granulocytes. As in the snake venom-induced granulocytosis also here in most cases the exogenous sLex injected into blood circulation just before the initiation of reperfusion could dramatically reduce the granulocyte accumulation into ischemic tissues and consecuently decrease the reperfusion injury[50-57]. Sialylated N-actelylactosamine (sLN) was a crucial control in most of these studies. It lacks only the fucose compared to the sLex structure and it was always without of any effects in these inflammation models.

Data from *in vivo* experiments with animals demonstrated the efficacy of very low concentrations of sLex in inhibiting short-term P-selectin mediated inflammation[49, 50, 54]. Even long-term inflammatory responses occurring within a few days could be inhibited by continuous infusion of anti L-selectin mAb to animals. L-selectin knockout mice or animals treated with anti L-selectin Mabs had normal differential counts of peripheral blood leukocytes [58, 59]. These data suggest that it might be possible to treat long term inflammations by inhibiting selectin-mediated leukocyte traffic with oligosaccharides antagonists.

All the above mentioned work has been done with monovalent sLex glycans. This is mainly due to the fact that multivalent sLex glycans were not readily available. Chemical synthesis cannot be used easily to generate very large oligosaccharides and even the enzymatic synthesis is tedious due to the need to purify appropriate glycosyltransferases, optimize the reaction conditions and purify the synthesized glycans. Previously it was known that the affinity of otherwise very low oligosaccharide-lectin interactions could be enhanced by the multivalency of the oligosaccharide component[60-64]. Therefore a program was initiated to synthesize enzymatically large complex multivalent sLex glycans in order to analyse their potential capacity in inhibiting selectin-dependent inflammation[64-68].

A series of sLex and sLN glycans on a branched polylactosmine backbone was synthesized. The definitive structures of the glycans in Figure 1 were established by a number of various techniques such as anion and high pH anion exchange (HPAEC) chromatography, nuclear magnetic resonance (NMR)-spectroscopy and matrix assisted laser desorption ionisation (MALDI)-mass spectrometry[64-68].

The ability of the family of branched sLex- and sLN-glycans to inhibit L-selectin mediated adhesion of lymphocytes to endothelium of rat kidney transplants undergoing acute rejection was tested. All exogenous $\alpha(2,3)$sialic acid- and $\alpha(1,3)$fucose-containing polylactosamines (i.e. mono-, branched di- and branched tetravalent sLex respectively) inhibited the L-selectin-dependent lymphocyte binding to graft capillary endothelium both in the kidney and heart transplantion models significantly[47, 66]. The potency of these molecules increased as the number of sLex-determinant in them increased the branched tetravalent sLex being superior to other branched glycans. *In vitro* data by others also suggest that oligosaccharide constructs bearing two sLex groups (divalent sLex structures) are 5 times better inhibitors of E-selectin dependent adhesion compared to monovalent sLex[62, 63]. None of the fucose-free structures, *i.e.* sLNs, whether mono-, di- or tetravalent inhibited the lymphocyte adhesion[47, 66]. These results support the concept of the crucial role of sLex and $\alpha(1,3)$fucosyltransferases, the final enzymes synthesizing it, in the L-selectin-dependent extravasation of lymphocytes.

We synthesized also a tetravalent sLex on a linear backbone. This glycan has the benefit that its synthesis is much simpler, requiring a smaller number of enzymatic reaction steps than the branched tetravalent sLex[68]. The linear tetravalent sLex was shown to be as effective as the branched one, IC_{50} values were in the low nanomolar range. Once again the control glycan lacking fucoses, but being otherwise similar to linear tetravalent sLex, was without any effect. These tetravalent sLex glycans, either branched or linear, consisting of 22

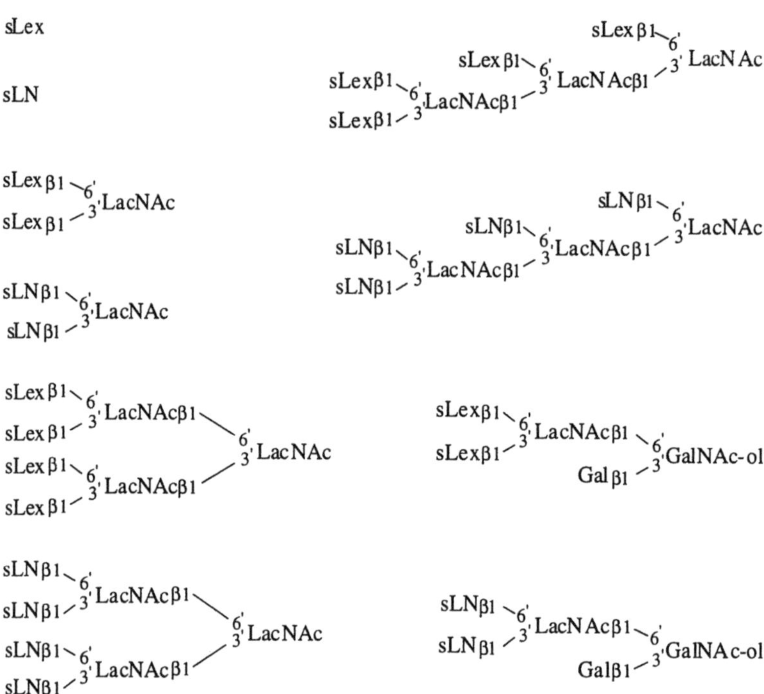

Figure 1. Schematic representation of enzymatically synthesized glycans discussed in this review. sLex represents NeuNAcα2-3Galβ1-4(Fucα1-3)GlcNAc, sLN represents NeuNAcα2-3Galβ1-4GlcNAc and LacNAc is Galβ1-4GlcNAc.

monosaccharides in a precisely defined array, probably represent the largest oligosaccharides constructed so far enzymatically *de novo* [68]. We have also synthesized an O-linked divalent sLex, which inhibit the L-selectin-dependent lymphocyte adhesion to capillary endothelium 3-fold better compared to a truncated divalent sLex molecule[67]. This data suggests that also other parts of these glycans than just the terminal sLex-determinant contribute to the selectin-dependent adhesion.

The high inhibitory potency of tetravalent glycans for L-selectin is probably generated by the multiplicity of sLex epitopes, implying that one glycan may bind to several L-selectin molecules on lymphocyte surface. Even monomeric receptors may become crosslinked on cell surface, like individual hemagglutinin trimers on intact influenza virus by bivalent sialosides[69]. Of potential importance for the high affinity is also the length of the saccharide chains linking the sLex determinants in tetravalent glycans. These chains contain GlcNAcβ1-6Gal bonds, lending them extra length and flexibility. Such spacers between the binding epitopes are linked to enhance the possibility of multisite binding of these glycans. These data suggest that multiply branched polylactosamines may prove themselves valuable inhibitors of several different saccharide-dependent recognition processes[47, 66, 68]. It is possible that multiply branched or linear polylactosamines act as natural ligands in some of the processes.

A decrease in the inhibitory efficacy of glycans was observed at the highest concentration used[47, 66, 68]. The explanation for this migth be that too few L-selectin molecules may be available for the glycan to bind in a multivalent manner in the high micromolar concentration. The decrease of binding efficiency at higher concentrations can also be due to increasing glycan-glycan interactions[37].

Taken together sLex type of glycans participate in the of extravasation of lymphocytes. Furthermore enzymatically synthesized oligosaccharides might prove to be good candidates for anti-inflammatory, -metastatic and -infection agents.

REFERENCES

1. R. Renkonen, A. Soots, E. von Willebrand, P. Häyry, Lymphoid cell subclasses in rejecting renal allograft in the rat. *Cell Immunol.* 77:188-195 (1983).
2. K. Ley, T. Tedder, Leukocyte interactions with vascular endothelium. *J Immunol.* 155:525-528 (1995).
3. R.P. McEver, K.L. Moore, R.D. Cummings, Leukocyte trafficking mediated by selectin-carbohydrate interactions. *J Biol Chem.* 270:11025-11028 (1995).
4. E.C. Butcher, L.J. Picker, Lymphocyte homing and homeostasis. *Science.* 272:60-66 (1996).
5. T.A. Springer, Traffic signals on endothelium for lymphocyte recirculation and leukocyte migration. *Annu Rev Physiol.* 57:827-872 (1995).
6. L.A. Lasky, Selectin-carbohydrate interactions and the initiation of the inflammatory response. *Annu Rev Biochem.* 64:113-139 (1995).
7. S.D. Rosen, C.R. Bertozzi, The selectins and their ligands. *Current opinion in cell biology.* 6:663-673 (1994).
8. D. Vestweber, Ligand-specificity of the selectins. *J Cell Biochem.* 61:585-91 (1996).
9. L.A. Lasky, M.S. Singer, D. Dowbenko, Y. Imai, W.J. Henzel, C. Grimley, C. Fennie, N. Gillett, S.R. Watson, S.D. Rosen, An endothelial ligand for L-selectin is a novel mucin-like molecule. *Cell.* 69:927-938 (1992).
10. S. Baumhueter, M.S. Singer, W. Henzel, S. Hemmerich, M.R. Renz, S.D. Rosen, L.A. Lasky, Binding of L-selectin to vascular sialomucin CD34. *Science.* 262:436-438 (1993).
11. M.J. Briskin, L.M. McEvoy, E.C. Butcher, MAdCAM-1 has homology to immunoglobulin and mucin-like adhesion receptors and to IgA1. *Nature.* 363:461-464 (1993).
12. Y. Imai, L.A. Lasky, R.D. Rosen, Sulphation requirements for GlyCAM-1, an endothelial ligand for L-selectin. *Nature.* 361:555-557 (1993).
13. S. Hemmerich, C.R. Bertozzi, H. Leffler, S.D. Rosen, Identification of the sulfated monosaccharides of GlyCAM-1, an endothelial-derived ligand for L-selectin. *Biochemistry.* 33:4820-4829 (1994).
14. S. Hemmerich, S.D. Rosen, 6'-sulfated sialyl Lewis x is a major capping group of GlyCAM-1. *Biochemistry.* 33:4830-4835 (1994).
15. D. Crommie, S.D. Rosen, Biosynthesis of GlyCAM-1, a mucin-like ligand for L-selectin. *Journal of Biological Chemistry.* 270:22614-24 (1995).
16. S. Hemmerich, H. Leffler, S.D. Rosen, Structure of the O-glycans in GlyCAM-1, an endothelial-derived ligand for L-selectin. *Journal of Biological Chemistry.* 270:12035-47 (1995).
17. W.M. Gallatin, I.L. Weisman, E.C. Butcher, A cell surface molecule involved in organ-specific homing of lymphocytes. *Nature.* 303:30-34 (1983).
18. M.P. Bevilacqua, R.M. Nelson, G. Mannori, O. Cecconi, Endothelial-leukocyte adhesion molecules in human disease. *Annual Review of Medicine.* 45:361-78 (1994).
19. C. Bertozzi, Cracking the carbohydrate code for selectin recognition. *Chemistry and Biology.* 2:703-708 (1995).

20. J.F. Kukowska-Latallo, R.D. Larsen, R.P. Nair, J.B. Lowe, A cloned human cDNA determines expression of a mouse stage-specific embryonic antigen and the Lewis blood group $\alpha(1,3/1,4)$fucosyltransferase. *Genes&Development.* 4:1288-1303 (1990).

21. E.L. Berg, M.K. Robinson, O. Mansson, E.C. Butcher, J.L. Magnani, A carbohydrate domain common to both sialyl Lea and sialyl Lex is recognized by the endothelial cell leukocyte adhesion molecule ELAM-1. *J Biol Chem.* 266:14869-14872 (1991).

22. A. Takada, K. Ohmori, N. Takahashi, K. Tsoyuoka, A. Yago, K. Zenita, A. Hasegawa, R. Kannagi, Adhesion of human cancer cells to vascular endothelium mediated by carbohydrate antigens, sialyl Lewis A. *Biochem Biophys Res Comm.* 179:713-719 (1991).

23. C.-T. Yuen, K. Bezouska, J. O'Brien, M. Stoll, R. Lemoine, A. Lubineau, M. Kiso, A. Hasekawa, N. Bockovich, K.C. Nicolaou, T. Feizi, Sulfated blood group Lewis a. A superior oligosaccharide ligand for human E-selectin. *J Biol Chem.* 269:1596-1598 (1994).

24. C.-T. Yuen, A.M. Lawson, W. Chai, M. Larkin, M.S. Stoll, A.C. Stuart, F.X. Sullivan, T.J. Ahern, T. Feizi, Novel sulfated ligands for cell adhesion molecule E-selectin revealed by the neoglycolipid technology among O-linked oligosaccharides on an ovarian cystadenoma glycoprotein. *Biochemistry.* 31:9126-9131 (1992).

25. P. Maly, A.D. Thall, B.R. Petryniak, C.E., P.L. Smith, R.M. Marks, R.J. Kelly, K.M. Gersten, G. Cheng, T.L. Saunders, S.A. Camper, R.T. Camphausen, F.X. Sullivan, Y. Isogai, O. Hindsgaul, U.H. von Andrian, J.B. Lowe, The $\alpha(1,3)$fucosyltransferase Fuc-TVII controls leukocyte trafficing through an essential role in L-, E-, and P-selectin ligand biosynthesis. *Cell.* 86:643-653 (1996).

26. D.X. Wen, B.D. Livigston, K.F. Medzihradszky, S. Kelm, A.L. Burlingame, J.C. Paulson, Primary structure of Galβ1,3(4)GlcNAc α2,3-sialyltransferase determined by mass spectrometry sequence analysis and molecular cloning. *JBiolChem.* 267:21011-21019 (1992).

27. K. Sasaki, E. Watanabe, K. Kawashima, S. Sekine, T. Dohi, M. Oshima, N. Hanai, T. Nishi, M. Hasegawa, Expression cloning of a novel Gal beta (1-3/1-4) GlcNAc alpha 2,3-sialyltransferase using lectin resistance selection. *Journal of Biological Chemistry.* 268:22782-7 (1993).

28. K. Sasaki, Molecular cloning and characterization of sialyltransferases. *Trends in Glycoscience and Glycotechnology.* 8:195-215 (1996).

29. S.E. Goelz, C. Hession, D. Goff, B. Griffiths, R. Tizard, B. Newman, G. Chi-Rosso, R. Lobb, ELFT: a gene that directs the expression of an ELAM-1 ligand. *Cell.* 63:1349-1356 (1990).

30. R. Kumar, B. Potvin, W.A. Muller, P. Stanley, Cloning of a human $\alpha(1,3)$-fucosyltransferase gene that encodes ELFT but does not confer ELAM-1 recognition on chinese hamster ovary cell tranfectants. *JBiolChem.* 266:21777-21783 (1991).

31. J.B. Lowe, J.F. Kukowska-Latallo, R.P. Nair, R.D. Larsen, R.M. Marks, B.A. Macher, R.J. Kelly, L.K. Ernst, Molecular cloning of a human fucosyltransferase gene that determines expression of the Lewis[X] and VIM-2 epitopes but not ELAM-1-dependent cell adhesion. *J Biol Chem.* 266:17467-17477 (1991).

32. B.W. Weston, R.P. Nair, R.D. Larsen, J. Lowe, B., Isolation of a novel human $\alpha(1,3)$fucosyltransferase gene and molecular comparison of the human Lewis blood group $\alpha(1,3/1,4)$ fucosyltransferase gene. *J Biol Chem.* 267:4152-4260 (1992).

33. B.W. Weston, P.L. Smith, R.J. Kelly, J.B. Lowe, Molecular cloning of a fourth member of a human $\alpha(1,3)$fucosyltransferase gene family. *JBiolChem.* 267:24575-24584 (1992).

34. K. Sasaki, K. Kurata, K. Funayama, M. Nagata, E. Watanabe, S. Ohta, N. Hanai, T. Nishi, Expression cloning of a novel alpha 1,3-fucosyltransferase that is involved in biosynthesis of the sialyl Lewis x carbohydrate determinants in leukocytes. *Journal of Biological Chemistry.* 269:14730-7 (1994).

35. S. Natsuka, K.M. Gersten, K. Zenita, R. Kannagi, J.B. Lowe, Molecular cloning of a cDNA encoding a novel human leukocyte α-1,3-fucosyltransferase capable of synthesizinga the sialyl Lewis x determinant. *J BIol Chem.* 269:16789-16794 (1994).

36. J.C. Paulson, J.P. Prieels, L.R. Glasgow, R.L. Hill, Sialyl- and fucosyltransferases in the biosynthesis of asparaginyl-linked oligosaccharides in glycoproteins. *Journal of Biological Chemistry.* 253:5617-24 (1978).

37. S.-I. Hakomori, Possible functions of tumor-associated carbohydrate antigens. *Current Opinion in Immunol.* 3:646-653 (1991).

38. T. Paavonen, R. Renkonen, Selective expression of sialyl-Lewis[x] and sialyl Lewis[a], putative ligands for L-selectin, on peripheral lymph node high endothelial venules. *Am J Pathol.* 141:1259-1264 (1992).

39. J.M. Munro, S.K. Lo, C. Corless, M.J. Robertson, N.C. Lee, R.L. Barhill, D.S. Weinberg, M.P. Bevilacqua, Expression of sialyl-Lewis x, an E-selectin ligand, in inflammation, immune processes and lymphoid tissues. *Am J Pathol.* 141:1397-1408 (1992).

40. M. Sawada, A. Takada, I. Ohwaki, N. Takahashi, H. Tatene, J. Sakamoto, R. Kannagi, Specific expression of a complex sialyl Lewis x antigen of high endothelial venules in human lymph nodes: Possible candidate for L-selectin ligand. *Biochem Biophys Res Comm.* 193:337-347 (1993).

41. M. Majuri, M. Pinola, R. Niemelä, S. Tiisala, O. Renkonen, R. Renkonen, α2,3 sialyl- and α1,3-fucosyltransferase-dependent synthesis of sialyl Lewis x, an essential oligosaccharide present in L-selectin counterreceptor in cultured endothelial cells. *Eur J Immunol.* 24:3205-3210 (1994).

42. M. Majuri, J. Räbinä, S. Tiisala, E. Aavik, P. Mattila, M. Miyasaka, O. Renkonen, R. Renkonen, High endothelial cells synthetize and degrade sLex via different pathways. Putative impications for the L-selectin-dependent extravasation. *JBiol Chem submitted for publication.* (1997).

43. J.P. Turunen, P. Mattila, J. Halttunen, P. Häyry, R. Renkonen, Evidence that lymphocyte traffic into rejecting cardiac allograft is CD11a- and CD49d-dependent. *Transplantation.* 54:1053-1058 (1992).

44. R. Renkonen, J.P. Turunen, P. Häyry, Site of influx of inflammatory white cells into a rejecting rat renal allograft. *Transplantation.* 47:577-579 (1989).

45. J. Turunen, T. Paavonen, M. Majuri, S. Tiisala, P. Mattila, A. Mennander, C. Gahmberg, P. Häyry, T. Tamatani, M. Miyasaka, R. Renkonen, Sialyl-Lewis[x] and L-selectin-dependent site-specific lymphocyte extravasation into the renal transplants during acute rejection. *Eur J Immunol.* 24:1130-1136 (1994).

46. K. Ito, K. Handa, S. Hakomori, Species-specific expression of sialosyl-Le(x) on polymorphonuclear leukocytes (PMN), in relation to selectin-dependent PMN responses. *Glycoconjugate J.* 11:232-237 (1994).

47. J. Turunen, M. Majuri, A. Seppo, S. Tiisala, T. Paavonen, M. Miyasaka, K. Lemström, L. Penttilä, O. Renkonen, R. Renkonen, De novo expression of endothelial sialyl Lewis a and sialyl Lewis x during cardiac transplant rejection. Superior capacity of a tetravalent sLex-oligosaccharide in inhibiting L-selectin-dependent lymphocyte adhesion. *J Exp Med.* 182:1133-1142 (1995).

48. P.A. Ward, M.S. Mulligan, Blocking of adhesion molecules *in vivo* as anti-inflammatory therapy. *Therapeutic Immunology.* 1:165-71 (1994).

49. M.S. Mulligan, J.C. Paulson, S. deFrees, Z.-L. Zheng, J.B. Lowe, P.A. Ward, Protective effects of oligosaccharides in P-selectin-dependent lung injury. *Nature*. 364:149-151 (1993).

50. M.S. Mulligan, J.B. Lowe, R.D. Larsen, J. Paulson, Z.-I. Zheng, S. DeFrees, K. Maemura, M. Fukuda, P.A. Ward, Protective effects of sialylated oligosaccharides in immune complex-induced acute lung injury. *J Exp Med.* 178:623-631 (1993).

51. D.J. Lefer, D.M. Flynn, M.L. Phillips, M. Ratcliffe, A.J. Buda, A novel sialyl LewisX analog attenuates neutrophil accumulation and myocardial necrosis after ischemia and reperfusion. *Circulation*. 90:2390-401 (1994).

52. C. Skurk, M. Buerke, J.P. Guo, J. Paulson, A.M. Lefer, Sialyl Lewisx-containing oligosaccharide exerts beneficial effects in murine traumatic shock. *American Journal of Physiology*. 267:H2124-31 (1994).

53. A. Seekamp, G.O. Till, M.S. Mulligan, J.C. Paulson, D.C. Anderson, M. Miyasaka, P.A. Ward, Role of selectins in local and remote tissue injury following ischemia and reperfusion. *American Journal of Pathology*. 144:592-8 (1994).

54. M. Buerke, A.S. Weyrich, Z. Zheng, F.C.A. Gaeta, M.J. Forrest, A.M. Lefer, Sialyl-Lewis x-containing oligosaccharide attenuates myocardial reperfusion injury in cats. *J Clin Invest*. 93:1140-1148 (1994).

55. M.J. Silver, J.M. Sutton, S. Hook, P. Lee, J.L. Malycky, M.L. Phillips, S.G. Ellis, E.J. Topol, F.A. Nicolini, Adjunctive selectin blockade successfully reduces infarct size beyond thrombolysis in the electrolytic canine coronary artery model. *Circulation*. 92:492-9 (1995).

56. E.A. Gill, Y.N. Kong, L.D. Horwitz, An oligosaccharide, sialyl Lewis x analoque does not reduce myocardial infarct size after ischemia and reperfusion in dogs. *Circulation*. 94:(1996).

57. K.T. Han, S.R. Sharar, M.L. Phillips, J.M. Harlan, R.K. Winn, Sialyl Lewis(x) oligosaccharide reduces ischemia-reperfusion injury in the rabbit ear. *Journal of Immunology*. 155:4011-5 (1995).

58. M.L. Arbones, D.C. Ord, K. Ley, H. Ratech, C. Maynard-Curry, G. Otten, D.J. Capon, T.F. Tedder, Lymphocyte homing and leukocyte rolling and migration are impaired in L-selectin-deficient mice. *Immunity*. 1:247-260 (1994).

59. P. Pizcueta, F.W. Luscinskas, Monoclonal antibody blockade of L-selectin inhibits mononuclear leukocyte recruitment to inflammatory sites *in vivo*. *Am J Pathol*. 145:461-469 (1994).

60. R.T. Lee, P. Lin, Y.C. Lee, New synthetic cluster ligands for galactose/N-acetylgalactosamine-specific lectin of mammalian liver. *Biochemistry*. 23:4255-4261 (1984).

61. R.T. Lee, K.G. Rice, N. Rao, Y. Ichikawa, T. Barthel, Binding characteriztics opf N-acetylglucosamine-specific lectin of the isolated chicken hepatocytes: similarities to mammalial hepatic galactose/N-acetylgalactosamine-specific lectin. *Biochemistry*. 28:(1989).

62. S.A. DeFrees, C.A. Gaeta, Y.-C. Lin, Y. Ichikawa, C.-H. Wong, Ligand recognition by E-selectin: Analysis of conformation and activity of synthetic monomeric and bivalent sialyl Lewis x analogs. *J Am Chem Soc*. 115:7549-7550 (1993).

63. S.A. DeFrees, W. Kosch, W. Way, J.C. Paulson, S. Sabesan, R.L. Halcomb, D.-H. Huang, Y. Ichikawa, C.H. Wong, Ligand recognition by E-selectin: Synthesis, inhibitory activity and conformational analysis of bivalent sialyl Lewis x analogs. *J Am Chem Soc*. 117:66-79 (1995).

64. R. Niemelä, L. Penttilä, A. Seppo, J. Helin, A. Leppänen, J. Räbinä, L. Uusitalo, H. Maaheimo, J. Taskinen, C. Costello, R. O., Enzyme-assisted synthesis of a divalent

high-affinity oligosaccharide inhibitor of mouse gamete adhesion. *Febs Lett.* 367:67-72. (1995).

65. R. Niemelä, J. Natunen, E. Brotherus, A. Saarikangas, O. Renkonen, α1,3-fucosylation of branched I-type oligo-(N-acetyllactosamino)-glycans by human milk transferases is restricted to distal N-acetyllactosamine units: Resulting isomeres are separated by WGA-agarose chromatography. *Glycoconjugate J.* 12:36-44 (1994).

66. A. Seppo, J.P. Turunen, L. Penttilä, A. Keane, O. Renkonen, R. R., Enzymatically synthesized oligovalent sialyl Lewis x glycans are high-affinity inhibitors of L-selectin-mediated lymphocyte binding to endothelium. *Glycobiology.* 6:65-71 (1996).

67. H. Maaheimo, R. Renkonen, J. Turunen, L. Penttilä, O. Renkonen, Synthesis of a divalent sialyl Lewis x O-glycan, a potent inhibitor of lymphocyte-endothelium adhesion. Evidence that multivalency enhances the saccharide binding to L-selectin. *Eur J Biochem.* 234:616-625 (1995).

68. O. Renkonen, S. Toppila, L. Penttilä, J. Helin, H. Maaheimo, C. Costello, J. Turunen, R. Renkonen, Enzyme-assisted synthesis of a tetravalent sialyl Lewis x glycan, derived from a linear polylactosamine. *Glycobiology accepted for publication.* (1997).

69. G.D. Glick, P.L. Toogood, D.C. Wiley, J.J. Skehel, J.R. Knowles, Ligand recognition by influenza virus. The binding of bivalent sialosides. *J Biol Chem.* 266:23660-23669 (1991).

ROLE OF LECTIN-GLYCOCONJUGATE RECOGNITIONS IN CELL-CELL INTERACTIONS LEADING TO TISSUE INVASION

Claudine Kieda

Glycobiologie, Centre de Biophysique Moléculaire
Centre National de la Recherche Scientifique
Rue Charles Sadron, 45071, Orléans CEDEX 2, France

SUMMARY

Endogenous lectins are cell receptors, expressed in normal and transformed cells both circulating as well as in organized tissues.   Their biological significance was shown in developmentally regulated processes of cell migration, embryonic maturation, differentiation and during various other normal and pathological processes.

This work will focus on the role of endogenous lectins and their glycoconjugate ligands in homing of circulating normal and cancer cells.

During the normal immune process of lymphocyte recirculation and their journey among the whole body through the secondary lymphoid organs in the search for antigen, lectins are decisive molecules that allow the very first interaction of arrest onto the endothelial cell layer.

It has been demonstrated that dual lectin-glycoconjugate interactions were taking part in the initiation of the whole adhesion cascade between adhering and endothelial cell. This indicates the role of the endothelium which will be described here. Indeed, using high endothelial cell lines immortalized by us, we could demonstrate that endothelium of microcapillaries is characterized by its tissue-specific properties although with a high microenvironment dependency. Both are decisive for selecting cells to stop and undergo further invasion.

Such normal properties of endothelial cells and homing cells could be taken as example to understand pathologies like specific establishment of metastases in the case of cancer cells.

Consequently, we shall take into account the potential offered by lectins and the knowledge of the structure of their ligand to design efficient adhesion blockers or enhancers as invasion inhibitors or immunomodulators.

INTRODUCTION

Homing of lymphocytes is a fundamental process by which immunological responders are made available to face the foreign antigens and neutralize them. It means entry of lymphocytes from the blood into the lymphoid organs and the further active recirculation from lymph to blood. This process is a massive flow of very large numbers of lymphocytes the turnover of which is roughly 30 minutes in the blood of an adult human, as shown in Figure 1. This shows the efficacity of the process which goes along with a high degree of selectivity. The selection of cells occurs at the level of the endothelium lining the post capillary veinules of the secondary lymphoid organs and lymphoid formations within the various tissues. Such a selection was exemplified by Bradley and Watson (1996) who described the phenotype of CD4+ T cells homing into distinct lymph nodes based on the adhesion molecules they express. This could not be separated from the adhesion molecule phenotype of endothelial cells lining the venules of the tissues. This results in the distinction of naive CD4+ T cells (IL2Rec-) homing into lymph nodes from memory CD4+ Tcells able to adhere on normal inflamed endothelium as well as on the mucosal lymph node endothelium. While, effector Il2Rec+ T cells mainly recognize the inflamed sites endothelium.

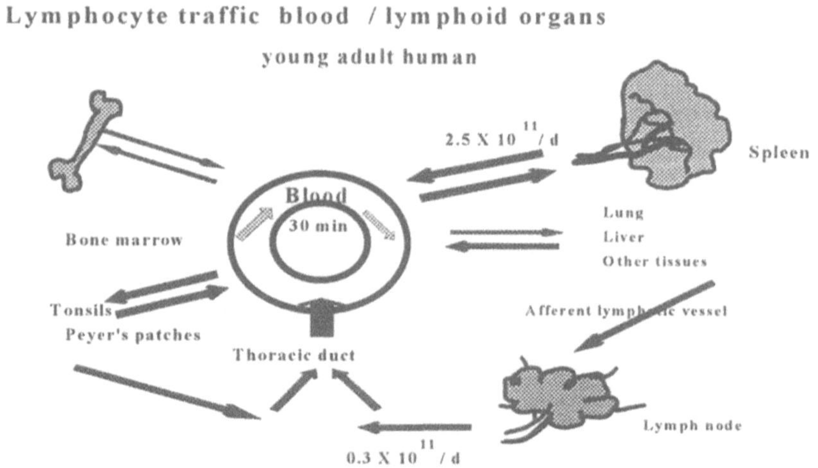

Figure 1. Lymphoid cells traffic among blood and lymphoid organs in a young adult human.

This report will focus on adhesion molecules of lectin-type of which have been intensely studied during the past ten years. It is now kown that molecules first described for their involvement in cell-cell interactions are frequently primarily acting as lectins. This is true for selectins which were first found as leukocyte adhesion molecules -1,-2,-3 and which are presently called selectins L- (leukocyte), E- (endothelial), P- (platelet) (see Bevilacqua *et al.*, 1991).

This lectin-like property gets to be a more general feature of adhesion molecules. Indeed, CR3, (Thornton *et al.*, 1996), ICAM-1, VCAM-1 and CD44 (Bennet *et al.*, 1996) have also lectin activities.

We have discovered the presence of lectins on lymphocytes and their distinct roles among subpopulations (Kieda *et al.*, 1978, 1979) as well as the biological modulation of their expression (Grillon *et al.*, 1990,1991) We have further evidenced the double lectin/glycoconjugate recognitions in cell adhesion as summarized on Figure 2 and as shown by Kieda and Monsigny (1986) in the case of mouse LL2 lung carcinoma metastasizing cells.

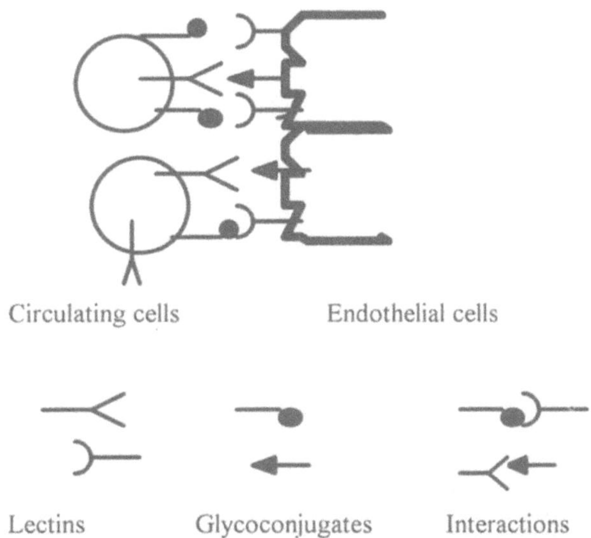

Figure 2. Double lectin/glycoconjugate recognition (Kieda and Monsigny, 1986).

The part played by endothelial cells endogeneous lectins in the lymphocyte homing could be characterized by pointing out their general features:
- recalling the main properties of selectins and their ligands ;
- underline the unique property of endothelial cells according to their specific endogenous lectins and,
- presenting an example of such a new lectin evidenced on peripheral lymph node high endothelial venules cells.

A- Lymphocyte homing : selectins and ligands

It has been shown by the famous pioneer work of Stamper and Woodruff (1977) that lymphocytes (naive) are selected at the level of the high endothelium of the lymph nodes to enter into the lymphoid organs and there, initiate an immune response or go further towards other lymphoid organs and / or reach the blood to follow on the journey. Lymphocytes which have previously reacted to an antigen display a preferential recruitment into the peripheral sites of antigen entry and inflammation and thus, can respond rapidly. Butcher and Weismann (1980) have introduced the concept of homing receptors by raising adhesion blocking monoclonal antibodies which happened to react with L-selectin (see the review by Butcher and Picker, 1996).

It was further demonstrated that two other lectin-like molecules are acting early in the adhesion cascade which were also designated as selectins due to their structural similarities (Fig.3) between one another and with L selectin. They are expressed on the endothelium surface in response to exogenous stimuli and react with distinct ligands (Fig.3, see the review by Whelan, 1996).

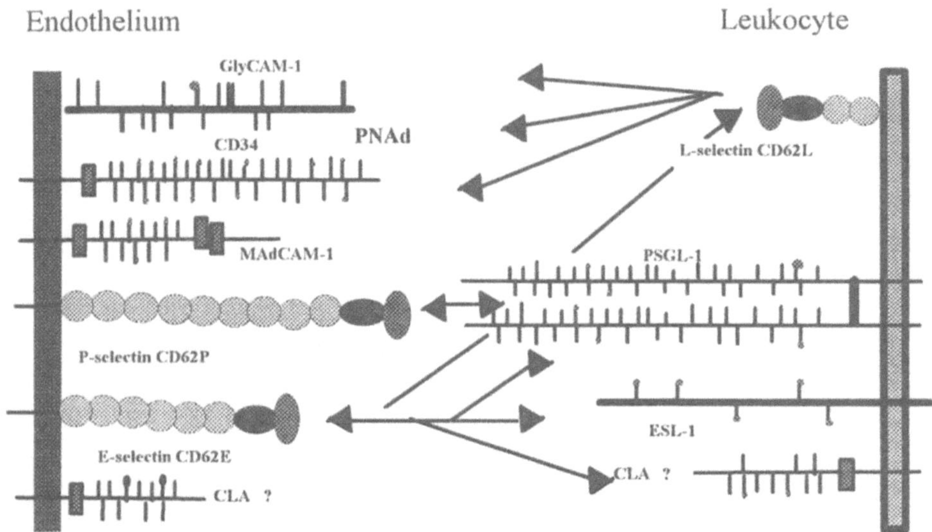

Figure 3.   Structure and distribution of selectins and their ligands (adapted from Whelan, J. 1996).

Endothelial cells express : - Selectins : E-selectin and P-selectin.
- Ligands for L-selectin :
      GlyCAM-1 : glycosylation dependent cell adhesion molecule-1,
      PNAd : peripheral lymph node addressin ;
      MadCAM-1 : mucosal vascular addressin cell adhesion molecule-1,
      CD 34 : sialo glycoprotein 90,
      CLA : labelled by HECA452.

Leukocytes express : - L-selectin : "homing receptor".
- Ligands for E- and P-selectins :
      PSGL-1 : P-selectin glycoprotein ligand-1,
      ESL-1 : E-selectin-ligand-1,
      CLA : cutaneous lymphocyte-associated antigen.

We have accumulated data showing that the whole lymphocyte homing process together with the inflammatory reaction as well as other various pathological invasion reactions cannot be satisfactorily fully explained by the expression of those three selectins and their ligands.

B- Endothelial cells endogenous lectins

We have found that lymphocyte homing is a lectin-dependent process which is occuring at the endothelium/lymphocyte interaction step. Because of the double lectin/glycoconjugate recognition evidences, it was likely that glycoconjugates, provided they were properly presented could be recognized by endothelial cells endogenous lectins and thus could be selectively targeted towards these cells and further to underneath tissues. This could be fodemonstrated by an *in vivo* homing assay of beads (fluorescent covaspheres, Polysciences) covered by neoglycoproteins (Monsigny *et al.*, 1983). The specific localisation of glycosylated beads among lymphoid organs after a short time *in vivo* homing assay is presented in figure 4.

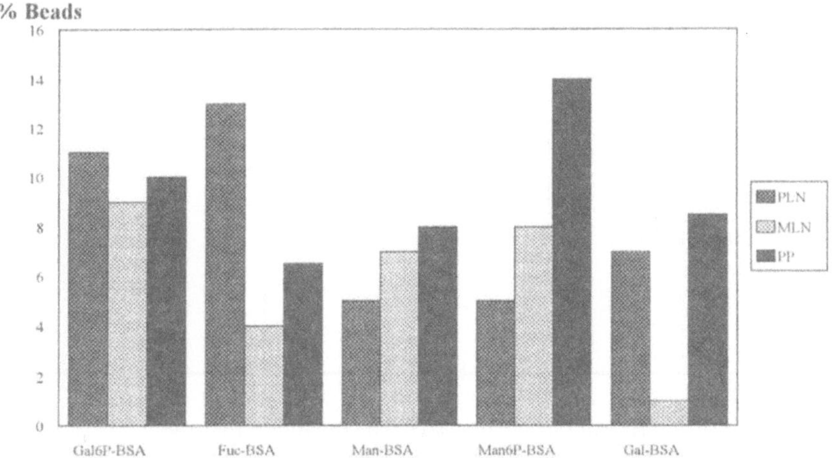

Figure 4. *in vivo* homing of neoglycoprotein-substituted beads, as detected by fluorescence activated cell sorter counting.
Percentages are estimated from the quantity of injected beads.

Fluorescent 6-phosphomannosylated beads injected intraveneously into mice were detected two hours after, maily in Peyer's patches while fucosylated beads were preferentially found in peripheral lymph nodes.

These data suggested that it is possible to target glycoconjugates towards lymphoid organs.

In order to find out the way glycosylated beads get to reach secondary lymphoid organs in a distinct manner : i.e. according to the sugar moiety they carry, it was necessary to study the capacity of endothelial cells to bind sugars *in vitro*.

Endothelial cells were isolated from mice peripheral lymph nodes and Peyer's patches according to Bizouarne *et al.* (1993b), cultured and labelled using the same neoglycoprotein-substituted fluorescent beads then studied by fluorescence microscopy.

Data given on table 1, indicate the specificity of endothelial cells towards sugars as a function of their tissue origin, pointing out the possibility that glycoconjugates can reach secondary lymphoid organs by being recognized by sugar specific receptors present on the endothelium lining the high endothelial venules.

Table 1. *in vitro* binding of neoglycoprotein-substituted fluorescent beads to high endothelial cells from PLN and Peyer's patches, in primary cultures.

NGP-Beads	HEC from PLN	HEC from Peyer's patches
Gal-6P-BSA	++	++
GalNAc-BSA	0	0
Fuc-BSA	++++	0
Gal-BSA	0	++
Man-6P-BSA	+	++++
Rha-BSA	0	0
BSA	0	0

From both, *in vivo* and *in vitro* assays we concluded that mouse high endothelial cells from peripheral lymph nodes carry endogenous sugar-specific receptors binding fucosylated glycoconjugates while high endothelial cells from Peyer's patches carry endogenous sugar-specific receptors binding 6-phosphomannosylated glycoconjugates. These receptors are lectins which are different from selectins as shownby us from the studies on the fucosyl-binding lectin (Flectin) from PLN (Bizouarne *et al.*, 1993a,b).

C- Endothelial cells specific membrane lectins differ from selectins

Our hypothesis relies on previous data (Kieda and Monsigny, 1986) and on the limited variety of recognitions provided by selectins/ligands expression, not sufficient to explain the high diversity of lymphocyte homing specificity. In figure 5, selectins are indicated part as playing a role in the early adhesion/recognition between endothelial and circulating cells. The added question mark indicates the place where endogenous lectins can take part. The arguments to support our hypothesis are as follows :

i) in conditions where cultured HECa10 endothelial cells, immortalized from peripheral lymph nodes (Bizouarne *et al.*, 1993a) high endothelial cells were not stimulated, fluorescent neoglycoproteins detected fucose-specific receptors -Flectin - which were not enhanced by stimulation with LPS as opposed to E-selectin ;

ii) in conditions where P-selectin was induced to be expressed on the HECa10 surface, the Flectin was not. In other words Flectin was not expressed by the action of classical P-selectin inducers (i.e. histamin) ;

iii) upon Flectin-specific ligand binding E-selectin and P-selectin are more efficiently and more rapidly expressed by HECa10 cells than upon action of either LPS, TNF or IL-1 (unpublished data).

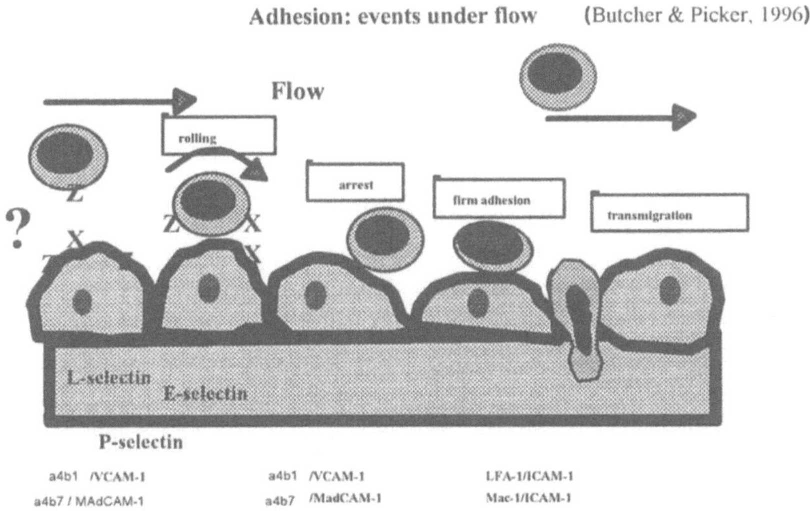

Figure 5. Endothelial cell endogenous lectin part in the adhesion cascade (adapted from Butcher and Picker, 1996).

CONCLUSION

Selectins have opened a new area for the understanding the steps of the adhesion cascade. The studies demonstrating their high sensitivity to modulation by environmental factors have brought a lot to the general knowledge of the adhesion/recognition mechanisms. In the same line, the demonstration of lectin molecules expressed in a constitutive manner on the high endothelial cells surface and the description of their modulation as well as their role in other adhesion molecules induction might be of very high interest for further studies of the immunological reaction mechanism. The design of molecules able to modify the recruitment of lymphoid cells in an inflammatory site, either by reducing the flux of lymphocytes, thus reducing the inflammatory damages or, by increasing the adhesion and thus the invasion of the underlining tissues by competent lymphocytes is of highly valuable potential. This may be reached by the knowledge of endogenous lectins which, as we have mentionned in this report, are : constitutively expressed on endothelium, highly tissue specific and can induce the expression of other adhesion molecules namely selectins on the endothelial cell surface. We think that at this point of the research advancement and the number of companies working on the adhesion-based therapy, knowledge of endogenous lectins of leukocytes and endothelial cells can bring a breakthrough to overcome the actual limitations of their developments.

Acknowledgements
C.K. wishes to thank M. Mitterrand and M.L.Welter for their valuable technical help. Dr Drs D. Dus for critically reading the manuscript and M. Paprocka for her scientific interest. This work was supported by ARC grants 6515 and 1117.

References

Bennet,K.L, Modrell,B., Greenfield,B., Bartolazzi,A., Stamenkovic,I., Peach,R., Jackson,D.G., Spring,F. and Aruffo, A., 1996, Regulation of CD44 binding by glycosylation of variably sliced exons. *J. Cell Biol.*,131:1623-1633

Bevilacqua, M.P. , Butcher, E.C., Furie,B., Gallatin, M., Gimbrone, M., Harlan, J., Kishimoto,K., Lasky,L., Mc Ever,P., Paulson,J., Seed,B., Siegelman, M., Springer, T., Stoolman,L., Tedder,T., Varki,A.,Wagner, D., Weissman, I. and Zimmerman, G., 1991, Selectins : a familly of adhesion receptors. *Cell,*67:233

Bizouarne , N., Denis, V., Legrand, A., Monsigny, M. and Kieda, C., 1993a, A SV-40 immortalized murine endothelial cell line from peripheral lymph node high endothelium expresses a   new α-L-fucose binding protein. *Biol. Cell,*79:209-218

Bizouarne , N., Mitterrand, M., Monsigny, M. and Kieda, C., 1993b, Characterization of membrane sugar-specificreceptor in hhigh endothelial cells from mouse peripheral lymph nodes. *Biol. Cell,* 79:27-35

Bradley, L.M. and Watson, S. A., 1996, Lymphocyte migration into tissue : the paradigm derived from CD4 subsets. *Curr. Opin. Immunol.*, 8:312-316.

Butcher, E.C. and Picker, L.J., 1996, Lymphocyte homing and homeostasis, *Science*, 22:60-66

Butcher, E.C., Scollay, R.G.and Weissman, I.L., 1980, Organ specificity of lymphocyte migration : mediation by highly selective lymphocyte interaction with organ-specific determinants on high endothelial venules. *Eur.J. Immunol.*, 10:556-561

Grillon, C., Monsigny, M. and Kieda, C., 1990, Cell surface lectins of human granulocytes: their expression is modulated by mononuclear cells and granulocyte/macrophage colony stimulating factor. *Glycobiology,*1:33-38

Grillon, C., Monsigny, M. and Kieda, C., 1991, Changes in the expression of lectins in human T lymphocyte membrane upon mitogenic stimulation. *Carbohyd. Res.,*213:283-292

Kieda, C. Bowles, D.J., Ravid, A. and Sharon, N., 1978, Lectins in lymphocyte membranes. *FEBS Lett.,* 94:391-396

Kieda, C. and Monsigny, M., 1986, Involvement of membrane sugar receptors and membrane glycoconjugates in the adhesion of 3LL cell subpopulations to cultured primary cells. *Invasion metastasis,* 6:347-366

Kieda, C., Roche, A.C., Delmotte, F. and Monsigny, M., 1979, Lymphocyte membrane lectins. Direct visualization by the use of fluoresceinyl-glycosylated cytochemical markers. *FEBS Lett.,* 99:329-332

Monsigny, M. Kieda, C. and Roche, A.C., 1983, Membrane glycoproteins, glycolipids and membrane lectins as recognition signals in normal and malignant cells. *Biol. Cell,* 47:95-110

Stamper, H. B. and Woodruff, J.J., 1977, An *in vitro* model for lymphocyte homing. *J. Immunol.,*119:772-780

Thornton, B.P., Vetvicka, V., Pitman,M., Goldman,R.C.,Ross,G.D.,1996, Analysis of the sugar specificity and molecular location of the beta-glucan binding lectin site of complement receptor type 3 (CD11b/CD18).*J. Immunol.*, 156:1235-1246

Whelan, J., 1996, Selectin synthesis and inflammation, *TIBS*, 21:65-69

OLIGOSACCHARIDES AND BIOLOGICAL FUNCTION

PROTEIN O-GlcNAcylation: POTENTIAL MECHANISMS FOR THE REGULATION OF PROTEIN FUNCTION

Bradley K. Hayes and Gerald W. Hart

Department of Biochemistry and Molecular Genetics
University of Alabama at Birmingham
Birmingham, AL 35294-0005

INTRODUCTION

Protein O-GlcNAcylation is the process whereby single N-acetylglucosamine residues are glycosidically linked to the hydroxyl side chains of specific serine and threonine residues. O-GlcNAc was originally identified while probing the surfaces of lymphocytes using UDP-[^{3}H]galactose and highly purified galactosyltransferase (1). O-GlcNAc was not a substrate for galactosyltransferase unless the cell membrane was first disrupted with detergents indicating that it is an intracellular glycosylation. Subcellular fractionation further demonstrated that O-GlcNAc is found exclusively on nuclear and cytosolic proteins (2,3). Galactosyltransferase labeling of mouse liver nuclei with subsequent analysis by 2-dimensional gel electrophoresis and fluorography indicates that a large number of nuclear proteins are modified with O-GlcNAc residues and suggests that O-GlcNAc is as abundant as phosphorylation (4).

O-GlcNAc is an evolutionarily conserved post-translational modification that has been found in all eukaryotes studied to date, but which is apparently absent in prokaryotes. For instance, several nuclear proteins from the filamentous fungi *Aspergillus orzyae* are O-GlcNAcylated as are intracellular proteins from *Giardi lamblia* (5,6). Interestingly, some cell surface O-linked saccharides of the intracellular parasites *Plasmodium falciparum* and *Trypanosoma cruzi* have GlcNAc instead of GalNAc at their reducing terminus (7,8). In the case of *P. falciparum*, an endogenous O-GlcNAc transferase has been identified that differs from the host enzyme in its substrate specificity and its stimulation by metal ions (7). Whether these intracellular parasites additionally use the host O-GlcNAc transferase to initiate the synthesis of their surface saccharides remains a possibility, but they likely use their own compliment of other glycosyltransferases to extend the reducing O-GlcNAc. Many viruses likewise have made use of the O-GlcNAc modification with the rotavirus protein NS26 (9), the cytomegalovirus basic phosphoprotein (10), the baculovirus gp41 (11), the adenovirus fibre protein (12), and the SV-40 large T antigen (13) all being O-GlcNAcylated. For additional recent reviews containing an expanded list of O-GlcNAcylated proteins see (14).

O-GlcNAcylated PROTEINS

Dozens of proteins have been identified as O-GlcNAcylated and one way to classify them is according to function. For instance, many proteins involved in transcription such as Sp1, Jun, Fos, serum response factor, estrogen receptor, Oct 1, HNF 1 and HNF 6, c-Myc, v-*erbA*, and p53 have all been identified as bearing O-GlcNAc residues (15-23). Cytoskeletal proteins such as the cytokeratins 8, 13, and 18, neurofilaments L-, M-, and H and microtubule associated proteins Tau, Map 1, 2 and 4 are similarly modified (24-29). Another group of O-GlcNAcylated protein connect cytoskeletal elements to other structures such as membranes and includes, Band 4.1, talin, vinculin, synapsin I and ankyrin (30-34). The list of enzymes known to be O-GlcNAcylated is surprisingly short and includes a nuclear protein tyrosine phosphatase, casein kinase II and RNA polymerase II (35-37).

Most O-GlcNAcylated proteins identified so far are soluble within the cell. The only exceptions that we are aware of are the β-amyloid precursor protein, a 92 kDa protein from the Golgi apparatus, proteins of 46, 56, 72 and 110 kDa from the endoplasmic reticulum (38-40). These proteins were identified as intrinsic membrane proteins based either on the inability to solubilize them with sodium carbonate pH 11.5 or on the ability to biotinylate their extracellular domains on intact cells. That so few O-GlcNAcylated integral membrane proteins have been identified suggests that either they are not good substrates for the O-GlcNAc transferase; do not have access to the O-GlcNAc transferase; are more susceptible to O-GlcNAcase; or have simply been selected against during cell fractionation. Therefore, describing the nature, number and diversity of membrane bound O-GlcNAcylated proteins remains an area for further study.

It is important to remember, however, that the list of O-GlcNAcylated proteins was compiled largely by asking "does this protein have O-GlcNAc on it?" or "can I purify this protein on a Wheat Germ agglutinin affinity column?". Thus, many classes of cytosolic and nuclear proteins may be under-represented. We have recently devised an approach to search for O-GlcNAcylated proteins using a combination of lectin affinity chromatography and liquid chromatography electrospray mass spectrometry (LC-ESMS) that does not introduce the biases of protein purification (41).

In this approach, a crude subcellular fraction devoid of extracellular glycans, such as whole cytosol or a nuclear preparation, is digested with a protease of defined specificity such as trypsin. The resulting mixture of peptides is then labeled with UDP-[^{3}H]galactose and galactosyltransferase. The resulting galactosylated O-GlcNAcylated peptides (O-LacNAc peptides) are then purified using a galactose specific lectin affinity column (42). Peptides not glycosylated pass through the column unimpeded, whereas those with a single O-LacNAc disaccharide are retarded in their elution and those with multiple O-LacNAc disaccharides bind strongly and require lactose for their elution. The resulting glycopeptide mixtures are then subjected to LC-ESMS. Glycopeptides are identified by the partial fragmentation of the O-LacNAc saccharide from the peptide backbone that occurs during ionization at even modest orifice potentials (43). This results in the formation of three characteristic ions for a given O-LacNAc peptide: the 366 mass ion representing the released glycan, the corresponding peptide that lost the Gal-GlcNAc and the intact glycopeptide that did not undergo the fragmentation. Once identified, the glycopeptides are subjected to collision induced dissociation. From the fragments generated, the sequence of the peptide can often be deduced. Then Using the Genbank, it should be possible to determine which protein the glycopeptide was originally derived from. We have applied this approach to a crude nuclear envelope fraction and identified a new glycopeptide from the amino terminal GFSFG repeat region of the nuclear pore protein, p62 (41). Several other peptides from this preparation are currently under investigation and it is hoped that we will be able to identify additional O-GlcNAcylated proteins as well as additional sites of glycosylation on the nuclear pore proteins.

O-GlcNAc AS A REGULATORY MODIFICATION

Protein O-GlcNAcylation has all of the hallmarks of a regulatory modification. First, a high energy pyrophosphate (UDP-GlcNAc) is used as the donor. Second, O-GlcNAc is a reversible modification. A UDP-GlcNAc:polypeptide N-Acetylglucosaminyl-transferase (O-GlcNAc transferase) responsible for its addition and an O-GlcNAc selective β-D-N-Acetylglucosaminidase (O-GlcNAcase) responsible for its removal have been described (44, 45). Third, O-GlcNAc turns over faster than the polypeptide to which it is attached. In the cases of the cytokeratins and αB-crystallin the half life of the O-GlcNAc residue is over ten times shorter than the half life of the polypeptide backbone (24, 46). Fourth, O-GlcNAc fluctuates across the cell cycle. Arresting human colonic HT29 cells at the G2/M phase with colcemid results in increased phosphorylation of cytokeratin 8 on the one hand but increased O-GlcNAcylation of cytokeratin 18 on the other (47). Likewise, glycosylation of the nuclear pore protein, p62, changes during the cell cycle (Kelly, Roquemore and Hart, unpublished observations). Protein O-GlcNAcylation also varies in response to specific physiologic stimuli. Mitogenic stimulation of T lymphocytes leads to a rapid and transient increase in the O-GlcNAcylation of nuclear proteins and similar decrease in the O-GlcNAcylation of a subset of cytosolic proteins (48). Additionally, the DNA binding activity of the tumor suppressor p53 is strongly correlated with its being O-GlcNAcylated (23). Thus, all evidence indicates that O-GlcNAc is a regulatory modification similar to phosphorylation in many respects.

MECHANISMS OF ACTION

Many interesting proteins have been identified as O-GlcNAcylated but few reports of the impact of O-GlcNAc on protein function are available. The identification of new O-GlcNAcylated proteins and the identification of their sites of glycosylation can give important clues as to which biological processes may be modulated by O-GlcNAcylation. However, for the current discussion we will focus on the mechanisms by which O-GlcNAc may exert its effects.

Regulation of Protein Phosphorylation

Most, if not all, identified O-GlcNAcylated proteins are also phosphoproteins, suggesting a possible inter-relationship between the two modifications. In many cases, such as the C-terminal domain (CTD) of the large subunit of RNA polymerase II glycosylation and phosphorylation are mutually exclusive (37). This suggests that protein kinases and O-GlcNAc transferase(s) may directly compete for the hydroxyl group of a given amino acid. Alternatively, the presence of an O-GlcNAc residue may either prevent or enhance the addition of a phosphate to nearby amino acids and vice versa.

Direct competition. Sites of O-GlcNAcylation are often similar to the consensus phosphorylation sites used by MAP kinase, casein kinase I or glycogen synthase 3 kinase suggesting that the same amino acid may be either phosphorylated or glycosylated under different physiologic conditions. One example of this is the onco-protein c-Myc. Myc acts as a transcription factor when dimerized with Max and is important in cellular proliferation, differentiation and transformation. The transactivation domain of Myc is phosphorylated on both Thr58 and Ser62 and the phosphorylation of these amino acids regulates the transactivation by Myc. Myc has recently be shown to be O-GlcNAcylated in the trans-activation domain and the site of glycosylation has been identified as Thr58 (21, 49). Since there is no known additional modification of O-GlcNAc itself, phosphorylation and glycosylation of Thr58 are necessarily mutually exclusive. Thus, it is predicted that if phosphorylation is essential for increased transactivation, then glycosylation may either suppress any basal transactivation by Myc or block the stimulation resulting from phosphorylation.

A second example comes from the microtubule associated protein Tau which has recently been shown to be heavily O-GlcNAcylated with perhaps as many as a dozen O-GlcNAcylation sites and an apparent stoichiometry of four GlcNAc residues per molecule of Tau (28). Hyperphosphorylated Tau forms the paired helical filaments observed in Alzheimer's disease brains. Phosphorylation of Ser262, inhibits the binding of Tau to microtubules leading to their instability (50). Ser262 is also uniquely phosphorylated in Alzheimer's disease and preliminary evidence suggests that Ser262 is perhaps one of the major sites of glycosylation on Tau (28). Thus, it is postulated that the abnormal phosphorylation of Tau seen in Alzheimer's disease may actually arise from a defect in glycosylation that would otherwise block its phosphorylation (28).

Indirect interference. There is no precedence in the literature for the effect of O-GlcNAc on nearby sites of phosphorylation. However, the α-toxin produced by *Clostridium novyio* has recently been shown to be an O-GlcNAc transferase that specifically glycosylates members of the Rho subfamily of small G proteins (51). Rho is also a target for the ADP-ribosylation at Asn41 catalyzed by the *Clostridium botulinum* C3 exoenzyme (51). Prior O-GlcNAcylation of Thr37 by α-toxin completely inhibits subsequent ADP-ribosylation.

We are currently exploring the possible influence of glycosylation on nearby phosphorylation and vice versa, again using Myc as the model protein. As noted above, Myc is glycosylated on Thr58 and phosphorylated on both Thr58 and Ser62. It is also known that the phosphorylation of Thr58 is facilitated by phosphorylation of Ser62. The possible inter-relationships between the post-translational modifications Myc are depicted in Figure 1. If one starts with unmodified peptide, then the first addition could either be O-GlcNAc at Thr58 or phosphate at Ser62. If the phosphorylation is the first reaction, then the facilitated addition of phosphate to Thr58 is the next likely step, although the addition of GlcNAc cannot be ruled out. If, on the other hand, the addition of the GlcNAc occurs first, then the addition of phosphate at Ser62 could be either facilitated or blocked. If it is facilitated, then the GlcNAc residue may ultimately lead to stimulated transactivation by Myc. Also, the GlcNAc residue may be an unexpected part of the kinase recognition sequence. If, the GlcNAc blocks subsequent phosphorylation of Ser62, then the GlcNAc would lead to decreased transactivation by Myc. Finally, if the presence of GlcNAc stimulates the addition of phosphate to Ser62, the presence of that phosphate may stimulate the removal of the GlcNAc by O-GlcNAcase paving the way for the addition of the second phosphate at Thr58. Dissecting these possibilities using defined, synthetic glyco- and phospho-peptides and purified kinases and O-GlcNAc transferase should clarify the relationship between glycosylation and phosphorylation.

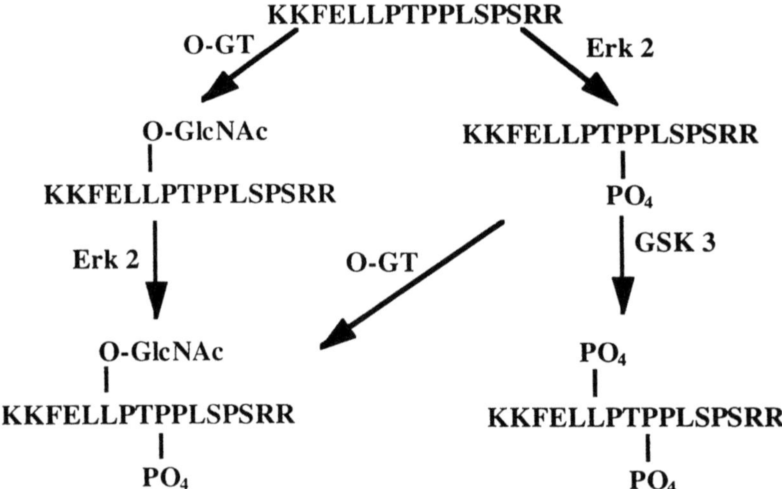

Figure 1. Possible inter-relationships between the phosphorylation and O-GlcNAcylation of the c-Myc transactivation domain. O-GT: O-GlcNAc transferase, GSK 3: Glycogen synthase 3 kinase.

Effects on Protein Folding

Coupled *in vitro* transcription and translation systems derived from reticulocyte lysates are capable of O-GlcNAcylating newly synthesized proteins (52). This indicates that the sugar nucleotide donor and O-GlcNAc transferase are present and active in these preparations. This also suggests that the addition of O-GlcNAc may occur either co-translationally or shortly after the completion of protein synthesis. Indeed, pulse chase analysis of the tumor suppressor p53 indicates that it is O-GlcNAcylated within 5 minutes of being synthesized indicating that glycosylation occurs either co-translationally or at about the same time the proteins would adopt their native conformation (23).

The isomerization of Xaa-Pro peptide bonds is often a limiting step in the adoption of native protein conformation and is catalyzed by the enzyme proline *cis/trans* isomerase. Additionally, the sequences surrounding the proline residue can influence the rate of protein folding *in vitro* (53). Since many sites of O-GlcNAc addition are near proline residues it is possible that the presence of a GlcNAc added before isomerization may interfere with this reaction. The placement of an O-GlcNAcylated Ser residue on the C-terminal side of proline does not effect its ability to undergo the non-enzymatic *cis/trans* isomerization in solution (54). However, O-GlcNAc residues are perhaps more commonly found on the C-terminal side of Pro residues and the effect of O-GlcNAcylation on the enzymatic isomerization of nearby proline residues remains an open question.

Alterations of Protein Conformation

The ability of a post-translational modification to regulate protein function is perhaps most fundamentally realized by its ability to alter the target protein's conformation. There is a growing body of evidence that glycosylation can effect protein conformation. For instance, mucins adopt a more extended or stiffened random coil conformation due to the presence of many small O-linked saccharide chains (55). The effect of single GalNAc residues on peptide conformation is more ambiguous, and appears to depend on peptide sequence and length. NMR studies of the small peptide FFWKTF indicate that attachment of a single GalNAc to the Thr residue alters the average conformation of the peptide. This is believed to be caused by the GalNAc residue sterically excluding the adoption of several possible conformations (56). On the other hand, the addition of GalNAc residues to Thr3 and Ser4 of the peptide GVTSAPDTRPAPGSTA from the protein MUC-1 had no apparent effect on peptide conformation (57). However, a direct comparison of these two studies is complicated because they were performed in different solvents.

We are currently trying to compare the effects on peptide conformation caused by both phosphorylation and O-GlcNAcylation using the synthetic Myc-derived peptides described above. Myc may present a special case because its sequence (see Figure 1) differs from other sites of O-GlcNAc addition in that there are several closely spaced proline residues flanking the site of glycosylation and relatively few hydroxy amino acids nearby. The results obtained using Myc will therefore need to be compared to other glycopeptides before generalizations can be made.

The presence of an O-GlcNAc residue may also alter protein structure through long range interactions. In the case of the tumor suppressor p53, O-GlcNAcylation is correlated with increased DNA binding (23). Charged synthetic peptides from the carboxy terminus of p53 also are capable of de-repressing DNA binding suggesting that p53 forms intra-molecular associations which inhibit DNA binding. It is therefore postulated that the addition of bulky group such as O-GlcNAc may disrupt these intra-molecular associations leading to increased DNA binding (23).

Regulation of Subunit Associations

In addition to being phosphoproteins, all known O-GlcNAc proteins are also members of reversible multimeric complexes suggesting that the O-GlcNAc may play a role complex assembly, maintenance or disassembly. For instance, O-GlcNAc may mediate the subunit associations required for the formation of the transcription pre-initiation complex. *In vitro* transcription from the adenovirus major late promoter by nuclear extracts is inhibited by synthetic O-GlcNAc modified peptides representing the consensus heptapeptide repeat from the RNA polymerase II CTD (58). The non-glycosylated and α-O-GalNAc peptides were without effect suggesting a sugar-specific disruption of the transcription cycle prior to elongation, possibly during the formation of the pre-initiation complex. PUGNAC, a potent hexosaminidase inhibitor, similarly blocks the formation of mature transcript indicating that the removal of O-GlcNAc may be important in allowing the transition from the pre-initiation to elongation complex (58).

Additionally, many O-GlcNAcylated proteins require denaturation or proteolytic fragmentation in order to expose the O-GlcNAc residues for labeling with galactosyltransferase and UDP-[^{3}H]galactose. A notable example is the fibre protein from adenovirus. Fibre is a capsid protein that is important in virus attachment and which trimerizes into stabile structures. Fibre from serotype Ad2 and Ad5 are modified with 2-4 O-GlcNAc residues per protein molecule as determined by composition analysis but which are particularly resistant to galactosyltransferase labeling or lectin binding unless first denatured. Because of the relative inaccessibility, it has been suggested that fibre glycosylation is important in assembly or stabilization of the fibre trimers (12). All three neurofilament proteins (N-, M- and L-) likewise form oligomers and all are O-GlcNAcylated (26, 27). These proteins also required prior proteolytic fragmentation for efficient galactosyltransferase labeling suggesting that the sugar residues are buried within the native protein.

Regulation of Enzyme Activity

As noted above, glycosylation of the C-terminal domain large subunit of RNA polymerase II may play a role in the formation of the pre-initiation complex. However, O-GlcNAcylation of the CTD may also directly effect its catalytic activity. Indeed, an alternative explanation of the data described above is that the O-GlcNAc residues block the catalytic incorporation of nucleotides into the nascent mRNA and removal of the O-GlcNAc residues may therefore be necessary to release the inhibition of catalysis and allow the transition to message elongation.

As noted above, the α-toxin from *Clostridium novyio* O-GlcNAcylates the small G protein Rho. Other members of the rho family of small G proteins that are also targets for α-toxin include Rac and cdc42. Rac is glycosylated at Thr35 which is located in the GTP-binding and hydrolyzing domain and is likely involved in the coordination of the Mg^{2+} ions required for the binding of guanine nucleotides. Thus, the addition of an O-GlcNAc to Thr35 leads to inactivation of the GTPase activity and ultimately the redistribution of the actin cytoskeleton (51).

REGULATION OF THE REGULATOR

If O-GlcNAc is a regulatory modification that alters protein function, then it must be responsive to changing physiologic conditions as was seen during lymphocyte activation (48). A proteins state of O-GlcNAcylation will likely be determined by the relative activities of O-GlcNAc transferase and O-GlcNAcase together with substrate availability.

O-GlcNAc Transferase

An enzyme activity capable of transferring GlcNAc from UDP-GlcNAc to Ser/Thr residues of synthetic peptides has been identified, characterized and purified from rat liver (44). The enzyme is soluble, does not require metal ions for activity and has a Km of 0.5 µM and 10 mM for UDP-GlcNAc and the peptide acceptor, YSDSPSTST, respectively. The enzyme fractionates during sedimentation and gel filtration with an apparent molecular weight of 340 kDa and seems to be composed of two 110 kDa subunits and one 78 kDa subunit. Photoaffinity labeling with 4-[β-^{32}P]thio-UDP indicate that the 110 kDa subunit contains the catalytic domain. The 110 kDa subunit has recently been cloned form rat liver (59). The identity of cloned gene product was confirmed either by overexpressing the protein or by direct immunoprecipitation of enzyme using polyclonal antibodies raised against the recombinant protein expressed in *E. coli*. The transferase is unlike other glycosyltransferases in that it has no signal sequence or transmembrane domain. Additionally, the amino terminal portion of the enzyme contains eleven consensus tetratricopeptide repeats (TPR) which have been implicated in mediating protein-protein interactions in a variety of systems (60). In the case of the O-GlcNAc transferase, these TPRs may interact with other factors that may modulate substrate specificity; dictate intracellular location; or regulate activity. The presumed catalytic domain is therefore located in the C-terminal region of the molecule. Interesting, there is a single tyrosine phosphorylation consensus sequence in this region that may modulate enzyme activity or dictate intracellular location. These questions are currently under investigation.

Southern blot analyses suggest that there is only one gene encoding this polypeptide. Northern blot analysis suggests that the transcript may be alternatively spliced, possibly giving rise to different forms of the enzyme. However, the existence of additional O-GlcNAc transferases is likely. For instance, in the original description of the O-GlcNAc transferase, an activity that bound to membranes but which was eluted with salt was identified. A nuclear O-GlcNAc transferase activity from adipocytes has also been described (61). Finally, enzyme activities form different organisms such as *P. falciparum* and *Clostridium novyi* have different characteristics in terms of metal ion requirements and molecular weight (7, 51).

Recently, the SPINDLY protein from *Arabidopsis thaliana* (62) has been found to be homologous to O-GlcNAc transferase. Recessive mutations in this enzyme cause a phenotype characteristic of the constitutive activation of the gibberellins signal transduction pathway. If the SPINDLY protein is indeed an O-GlcNAc transferase, it further supports the hypothesis that O-GlcNAc is a regulatory modification. Additionally, the point mutations identified in the enzyme may give important clues for dissecting the mammalian O-GlcNAc transferase.

O-GlcNAcase

A β-D-N-Acetylglucosaminidase that preferentially removes O-GlcNAc residues from glycopeptides (O-GlcNAcase) has been identified, characterized and purified (45). The enzyme is specific for terminal β-linked GlcNAc residues but interestingly, does not efficiently cleave chitobiose. O-GlcNAcase is similar to Hexosaminidase C in that it has a neutral pH optima, does not utilize terminal β-GalNAc is not a substrate, and is soluble and cytosolic disposed. While there is very little known about the regulation of O-GlcNAcase the cloning of this enzyme should allow for rapid progress.

Acceptor Splice Variants

An alternative approach to regulating a protein's state of glycosylation is through the generation of splice variants that may add or remove sites of glycosylation. This mechanism is at work in the generation of the different forms of CD44, a cell surface hyaluronan-binding protein. There are several splice variants of CD44. One of these, CD44E contains variably spliced exons V8-V10 and binds to hyaluronan less well than does CD44 lacking these exons. Interestingly, the decrease in binding can be attributed, in part, to the multiple O-linked saccharide chains added to the Ser/Thr rich region encoded by exon V8-V10 (63). A single example of splice-specific O-GlcNAcylation is Protein 4.1   O-GlcNAc site mapping indicated that O-GlcNAc was added within a 34 amino acid peptide in the amino terminal portion of Protein 4.1 (30). This peptide contains the only identified site of glycosylation on Protein 4.1 and arises from alternative splicing. Thus, the availability of acceptor and ultimately the potential for regulation by O-GlcNAc can be governed at the level of the mRNA.

CONCLUSIONS

O-GlcNAc is a post-translational regulatory modification analogous to protein phosphorylation in many respects. The enzymes that both add and remove the O-GlcNAc residues have been identified and are currently under study. There are several model systems for the study of protein O-GlcNAcylation that have recently been reviewed (14). Our purpose here was not to recapitulate those arguments, but rather to consider O-GlcNAc function from a mechanistic viewpoint in the hopes of finding underlying principles from which to make predictions that can be tested experimentally. While there are some recurring themes, such as modulation as protein phosphorylation, it is too early to discount the possibility that the addition of a single monosaccharide may modulate protein function in several different ways.

REFERENCES

1. C.-R. Torres, and G.W. Hart, Topography and polypeptide distribution of terminal N-acetylglucosamine residues on the surfaces of intact lymphocytes. *J. Biol .Chem.* 259:3308 (1984).
2. G. D. Holt, and G. W. Hart, The subcellular distribution of terminal N-acetylglucosamine moieties. Localization of a novel protein-saccharide linkage, O-linked GlcNAc, *J. Biol. Chem.* 261:8049 (1986).
3. K. P. Kearse, and G. W. Hart, Topology of O-linked N-acetylglucosamine in murine lymphocytes, *Arch. Biochem. Biophys.* 290:543 (1991).
4. G. W. Hart, K. D. Greis, D. L.-Y. Dong, M. A. Blomberg, T.-Y. Chou, M.-S. Jaing, E. P. Roquemore, D. M. Snow, L. K. Kreppel, R. N. Cole, and B. K. Hayes, Ubiquitous and temporal glycosylation of nuclear and cytoplasmic proteins, *Pure and Appl. Chem.* 67:1637 (1995).
5. M. Machida, and Y. Jigami, Glycosylated DNA-binding proteins from filamentous fungus, *Aspergillus orzyae*: modification with N-acetylglucosamine monosaccharide through an O-glycosidic linkage *Biosci. Biotech. Biochem.* 58:344 (1994).
6. E. Ortega-Barria, H. D. Ward, J. E. Evans, and M. E. A. Pereira, N-acetylglucosamine is present in cysts and trophozoites of *Giardia lamblia* and serves as receptor for wheat germ agglutinin, *Mol. Biochem. Parasitol.* 43:151 (1990).
7. A. Dieckmann-Schuppert, E. Bause, and R. T. Schwarz, Studies on O-glycans of *Plasmodium falciparum*-infected human erythrocytes: evidence for O-GlcNAc and O-GlcNAc transferase in malaria parasites, *Eur. J. Biochem.* 216:779 (1993).
8. J. O. Previato, C. Jones, L. P. B. Goncalves, R. Wait, L. R. Travassos, and L. Mendonca-Previato, O-Glycosidically linked N-acetylglucosamine-bound oligosaccharides from glycoproteins of *Trypanosoma cruzi, Biochem. J.* 301:151 (1994).
9. S. A. Gonzalez, and O. R. Burrone, Rotavirus NS26 is modified by addition of single O-linked residues of N-acetylglucosamine, *Virology* 182:8 (1992).

10. K. D. Greis, W. Gibson, and G. W. Hart, Site-specific glycosylation of the human cytomegalovirus tegument basic phosphoprotein (UL32) at serine 921 and serine 952, *J. Virol.* 68:8339 (1994).

11. M. Whitford, and P. Faulkner, A structural polypeptide of the baculovirus *Autographa californica* nuclear polyhedris virus contains O-linked N-acetylglucosamine, *J. Virol.* 66:3324 (1992).

12. K. G. Mullis, R. S. Haltiwanger, G. W. Hart, R. B. Marchase, and J. A. Engler, Relative accessibility of N-acetylglucosamine in trimers of the Adenovirus types 2 and 5 fiber proteins *J. Virol.* 64:5317 (1990).

13. L. Medina-Vera, and R. S. Haltiwanger, SV-40 Large T antigen is modified with O-linked N-acetylglucosamine, *Mol. Biol. Cell* 5(S):340a (1994).

14. G. W. Hart, Dynamic O-GlcNAcylation of nuclear and cytoskeletal proteins, *Ann. Rev. Biochem.* In Press (1997).

15. S. P. Jackson, and R. Tjian, O-Glycosylation of eukaryotic transcription factors: implications for mechanisms of transcriptional regulation, *Cell* 55:125 (1988).

16. A. J. Reason, H. R. Morris, M. Panico, R. Marais, R. H. Treisman, R. S. Haltiwanger, G. W. Hart, W. G. Kelly, and A. Dell, Localization of O-GlcNAc modification on the serum response transcription factor, *J. Biol. Chem.* 267:16911 (1992).

17. M.-S. Jiang and G. W. Hart, A subpopulation of estrogen receptors are modified by O-linked N-acetylglucosamine, *J. Biol. Chem.* In Press (1997).

18. S. Murphy, A. Pierani, C. Scheidereit, M. Melli, and R. G. Roeder, Purified octamer binding transcription factors stimulate RNA polymerase III-mediated transcription of the 7SK RNA gene, *Cell* 59:1071 (1989).

19. S. Lichtsteiner, and U. Schibler, A Glycosylated liver-specific transcription factor stimulates transcription of the albumin gene, *Cell* 57:1179 (1989).

20. F. P. Lemaigre, S. M. Durviaux, O. Truong, V. J. Lannoy, J. J. Hsuan, and G. G. Rousseau, Hepatocyte nuclear factor 6, a transcription factor that contains a novel type of homeodomain and single cut domain, *Proc. Natl. Academ. Sci. U. S. A.* 93:9460 (1996).

21. T.-Y. Chou, C. V. Dang, and G. W. Hart, Glycosylation of c-Myc transactivation domain, *Proc. Natl. Academ. Sci. U. S. A.* 92:4417 (1995).

22. M. L. Privalsky, A subpopulation of the avian Erythroblastosis virus v-*erbA* protein, a member of the nuclear hormone receptor family, is glycosylated, *J. Virol.* 64:463 (1990).

23. P. Shaw, J. Freeman, R. Bovey, and R. Iggo, Regulation of specific DNA binding by p53: evidence for a role for O-glycosylation and charged residues at the carboxy terminus, *Oncogene* 12:921 (1996).

24. C.-F. Chou, A. J. Smith, and M. B. Omary, Characterization and dynamics of O-linked glycosylation of human cytokeratins 8 and 18, *J. Biol. Chem.* 267:3901 (1992).

25. I. A. King, and E. F. Hounsell, Cytokeratin 13 contains O-glycosidically linked N-acetylglucosamine, *J. Biol. Chem.* 264:14022 (1989).

26. D. L.-Y. Dong, Z.-S. Xu, M. R. Chevrier, R. J. Cotter, D. W. Cleveland, and G. W. Hart, Glycosylation of mammalian neurofilaments: localization of multiple O-linked N-acetylglucosamine moieties on neurofilament polypeptides L and M, *J. Biol. Chem.* 268:16679 (1993).

27. D. L.-Y. Dong, Z.-S. Xu, G. W. Hart, and D. W. Cleveland, Cytoplasmic O-GlcNAc modification of the head domain and the KSP repeat motif of the neuofilament protein neurofilament H, *J. Biol. Chem.* 271:20845 (1996).

28. C. S. Arnold, G. V. W. Johnson, R. N. Cole, D. L.-Y. Dong, M. Lee, and G. W. Hart, The microtubule-associated protein Tau is extensively modified with O-linked N-acetylglucosamine, *J. Biol. Chem.* 271:28741 (1996).

29. M. Ding, and D. D. Vandre, High molecular weight microtubule-associated proteins contain O-linked N-acetylglucosamine, *J. Biol. Chem.* 271:12555 (1996).

30. M. Inaba, and Y. Meade, O-N-Acetyl-D-glucosamine moiety on discrete peptide of multiple Protein 4.1 isoforms regulated by alternative pathways, *J. Biol. Chem.* 264:18149 (1989).

31. J. Hagmann, M. Grob, and M. M. Burger, The cytoskeletal protein Talin is O-glycosylated, *J. Biol. Chem.* 267:14424 (1992).

32. J. G. Vostal, and D. M. Krasnewich, Dynamic O-linked N-acetylglucosamine glycosylation of platelet vinculin, *Mol. Biol. Cell* 5(S):263a (1994).
33. T. Luthi, R. S. Haltiwanger, P. Greengard, and M. Bahler, Synapsins contain O-linked N-acetylglucosamine, *J. Neurochem.* 56:1493 (1991).
34. X. Zhang, and V. Bennett, Identification of O-linked N-acetylglucosamine modification of Ankyrin$_G$ isoforms targeted to Nodes of Ranvier, *J. Biol. Chem.* 271:31391 (1996).
35. W. Meikrantz, D. M. Smith, M. M. Sladicka, and R. A. Schlegel, Nuclear localization of an O-glycosylated protein phosphotyrosine phosphatase from human cells, *J. Cell. Sci.* 98:303 (1991).
36. T. Matsuoka, G. V. W. Johnson and G. W. Hart, In Preparation.
37. W. G. Kelly, M. E. Dahmus, and G. W. Hart, RNA polymerase II is a glycoprotein: modification of the COOH-terminal domain by O-GlcNAc, *J. Biol. Chem.* 268:10416 (1993).
38. L. S. Griffith, M. Mathes, and B. Schmitz, β-Amyloid precursor protein is modified with O-linked N-acetylglucosamine, *J. Neurosci. Res.* 41:270 (1995).
39. C. Abeijon, and C. B. Hirschberg, Intrinsic membrane glycoproteins with cytosol-oriented sugars in the endoplasmic reticulum, *Proc. Natl. Academ. Sci. U. S. A.* 85:1010 (1988).
40. J. M. Capasso, C. Abeijon, and c. B. Hirschberg, An intrinsic membrane glycoprotein of the Golgi apparatus with O-linked N-acetylglucosamine facing the cytosol, *J. Biol. Chem* 263:19778 (1988).
41. B. K. Hayes, and G. W. Hart, In search of O-GlcNAcylated proteins, *Glycobiology* 6:737 (1996).
42. B. K. Hayes, K. D. Greis, and G. W. Hart, Specific isolation of O-linked N-acetylglucosamine glycopeptides from complex mixtures, *Anal. Biochem.* 228:115 (1995).
43. K. D. Greis, B. K. Hayes, F. I. Comer, M. Kirk, S. Barnes, T. L. Lowary, and G. W. Hart, Selective detection and site-analysis of O-GlcNAc-modified glycopeptides by β-elimination and tandem electrospray mass spectrometry, *Anal. Biochem.* 234:38 (1996).
44. R. S. Haltiwanger, M. A. Blomberg, and G. W. Hart, Glycosylation of nuclear and cytoplasmic proteins: purification and characterization of a UDP-GlcNAc:polypeptide β-N-Acetylglucosaminyltransferase, *J. Biol. Chem.* 267:9005 (1992).
45. D. L.-Y. Dong, and G. W. Hart, Purification and characterization of an O-GlcNAc selective N-Acetyl-β-D-glucosaminidase from rat spleen cytosol, *J. Biol. Chem.* 269:19321 (1994).
46. E. P. Roquemore, M. R. Chevrier, R. J. Cotter, and G. W. Hart, Dynamic O-GlcNAcylation of the small heat shock protein αB-crystallin, *Biochemistry* 35:3578 (1996).
47. C.-F. Chou, and M. B. Omary, Mitotic arrest-associated enhancement of O-linked glycosylation and phosphorylation of human keratins 8 and 18, *J. Biol. Chem.* 268:4465 (1993).
48. K. P. Kearse, and G. W. Hart, Lymphocyte activation induces rapid changes in nuclear and cytoplasmic glycoproteins, *Proc. Natl. Academ .Sci. U. S. A.* 88:1701 (1991).
49. T.-Y. Chou, G. W. Hart, and C. V. Dang, c-Myc is glycosylated at threonine 58, a known phosphorylation site and a mutational hot spot in lymphomas, *J. Biol. Chem.* 270:18961 (1995).
50. E. M. Mandelkow, O. Schweers, G. Dewes, J. Biernat, N. Gustke, B. Trinczek, and E. Mandelkow, Structure, microtubule interactions, and phosphorylation of the Tau protein, *Annals N. Y. Academ Sci.* 777:96 (1996).
51. J. Selzer, F. Hofmann, G. Rex, M. Wilm, M. Mann, I. Just, and K. Aktories, *Clostridium novyi* α-toxin-catalyzed incorporation of GlcNAc into Rho subfamily proteins, *J. Biol. Chem.* 271:25173 (1996).
52. C, M. Starr, and J. A. Hanover, Glycosylation of nuclear pore protein p62. reticulocyte lysate catalyzes O-linked N-acetylglucosamine addition *in vitro, J. Biol. Chem.* 265:6868 (1990).
53. G. Fisher, and F. X. Schmid, The mechanism of protein folding: Implications *of in*

vitro refolding models for *de novo* protein folding and translocation in the cell, *Biochemistry* 29:2205 (1990).

54. Y.-L. Pan, M. R. Wormarld, R. A. Dwek, and A. C. Lellouch, Effect of serine O-glycosylation on *cis-trans* proline isomerization, *Biochem. Biophys. Res. Commun.* 219:157 (1996).

55. N. Jentoft, Why are proteins O-glycosylated?, *Trends Biochem. Sci.* 15:291 (1990).

56. A. H. Andreotti, and D. Kahne, Effects of glycosylation on peptide backbone conformation, *J. Amer. Chem. Soc.* 115:3352 (1993).

57. X. Liu, J. Sejbal, G. Kotovych, R. R. Koganty, M. A. Reddish, L. Jackson, S. S. Gandhi, A. J. Mendonca, and B. M. Longenecker, Structurally defined synthetic cancer vaccines: analysis of structure, glycosylation and recognition of cancer associated mucin, MUC-1 derived peptides, *Glyconjugate J.* 12:607 (1995).

58. F. I. Comer, and G. W. Hart, Investigating the role of O-GlcNAc on RNA polymerase II, *FASEB J.* 10:A1119 (1996). '

59. L. K. Kreppel, M. A. Blomberg, and G. W. Hart, Dynamic glycosylation of nuclear and cytosolic proteins: cloning and characterization of unique O-GlcNAc transferase with multiple tetratricopeptide repeats, *J. Biol. Chem.* In Press (1997).

60. J. R. Lamb, S. Tugendreich, and P. Hieter, Tetratricopeptide repeat interactions: to TPR or not to TPR? *Trends Biochem. Sci.* 20:257 (1995).

61. W. K. Gottschalk, J. Stuart, T. Wang, J. Weiel, and S. Marshall, Nuclear localization of a novel glycosyltransferase, *FASEB J.* 9:A1362 (1995).

62. S. E. Jackson, K. A. Binkowski, and N. E. Olszewski, SPINDLY, a tetratricopeptide repeat protein involved in gibberellin signal transduction in *Arabidopsis, Proc. Natl. Academ .Sci. U. S. A.*93:9292 (1996).

63. K. L. Bennett, B. Modrell, B. Greenfield, A. Bartolazzi, I. Stamenkovic, R. Peach, D. G. Jackson, F. Spring, and A. Aruffo, Regulation of CD44 binding to hyaluronan by glycosylation of variably spliced exons, *J. Cell Biol.* 131:1623 (1995).

A LONGITUDINAL STUDY OF GLYCOSYLATION OF A HUMAN IgG3 PARAPROTEIN IN A PATIENT WITH MULTIPLE MYELOMA

Muhammad Farooq[1], Noriko Takahashi[2], Mark Drayson[1], John Lund[1] and Royston Jefferis[1]

[1]Department of Immunology, The University of Birmingham Medical School, Edgbaston, Birmingham, B15 2TT
[2]Glycolab Nakano Vinegar Company, Handa City 475, Japan

INTRODUCTION

The IgG antibody molecule is a structural paradigm for members of the immuoglobulin super family[1]. Whilst the oligosaccharide moiety of the IgG molecule accounts for only 2-3% of its mass it has been shown to be essential for optimal activation of effector mechanisms leading to the clearance and destruction of pathogens[2,3,4]. Human antibody molecules of the IgG class have N- linked oligosaccharide attached at the amide side chain of Asn-297 in the heavy chain[5]. The oligosaccharide moiety is of the complex bianntennary type having a heptasaccharide "core" structure ($GlcNAc_2Man_3GlcNAc_2$) and variable outer arm "non-core" sugar residues, such as fucose, bisecting N-acetylglucosamine, galactose and sialic acid (Figure 1). The number of variant oligosaccharides that may be attached to heavy chains is 32 (Figure 2) and the total number of possible glycoforms >800[6,7]. This level of heterogeneity is evident for polyclonal IgG whilst a more restricted heterogeneity may be observed for monoclonal proteins[4]. In addition, ~30% of polyclonal IgG has been reported to bear a complex N- linked oligosaccharide in the Fab region[8]. It is apparent, therefore, that glycosylation is a post-translational modification that can introduce a very significant structural and, possibly, functional heterogeneity into the IgG molecule, such that glycoforms can alter the biological activity[9].

SA ($\alpha2{\to}6$) — Gal ($\beta1{\to}4$) — GlcNAc ($\beta1{\to}2$) — Man ($\alpha1{\to}6$) Fuc ($\alpha1{\to}6$)

GlcNAc ($\beta1{\to}4$) — Man ($\beta1{\to}4$) — GlcNAc ($\beta1{\to}4$) — GlcNAc

SA ($\alpha2{\to}6$) — Gal ($\beta1{\to}4$) — GlcNAc ($\beta1{\to}2$) — Man ($\alpha1{\to}3$)

Figure 1. Biantennary oligosaccharide structure present on IgG antibody molecules. The bold print represents the common heptasaccharide core, heterogeneity arises from the addition of variable terminal sugar residues represented in normal type.
Key: S.A, sialic acid; Gal, galactose; GlcNAc, N-acetylglucosamine, Man, mannose; Fuc, fucose.

Glycoimmunology 2
Edited by Axford, Plenum Press, New York, 1998

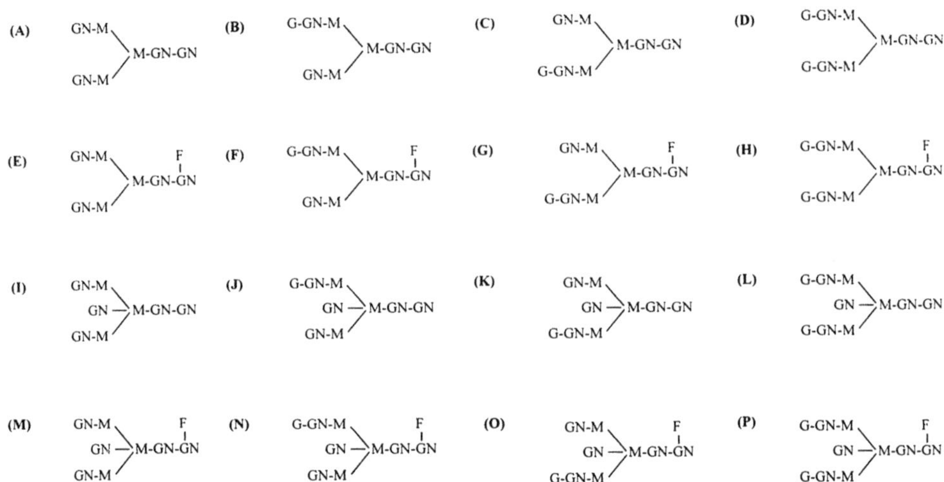

Figure 2. Profile of neutral sugars derived from human IgG antibody molecules, addition of sialic acid to terminal galactose residues results in 32 glycoforms. Key: G, galactose; GN, N-acetylglucosamine; M, mannose; F, fucose.

Changes in the profile of IgG oligosaccharides have been reported for several disease states[10,11,12] and for rheumatoid arthritis it has been claimed to be a prognostic indicator for disease progression[10,13]. It has been suggested that changes in glycosylation result from deficits in glycosyltransferase levels, however, exposure of plasma cells to altered micro-environments, particularly cytokines released in inflammatory reactions, could be the determining factor.

Since a majority of glycoprotein molecules currently undergoing development for *in vivo* therapeutic application are monoclonal antibodies or members of the immunoglobulin gene super family it is evident that glycosylation is a parameter over which it is important to exercise control. In this report we investigate the glycosylation of IgG paraproteins associated with the disease of multiple myeloma as a model system in which some parameters that influence *in vivo* glycosylation may be defined. At the time of diagnosis patients with multiple myeloma usually have progressive disease typically with rising paraprotein levels and symptoms including lethargy and bone pain. Administration of chemotherapy over a variable period (6-12 months) induces a stable disease state (plateau phase) in about two thirds of patients. In plateau phase paraprotein levels remain stable without further chemotherapy and patients do not suffer symptons attributable to multiple myeloma. In a cohort of patients in plateau phase, half will enter a further period of disease progression within 18 months from the start of observation. This second phase of disease progression may respond to chemotherapy and a second plateau phase be achieved. We have studied glycosylation of the paraprotein and the polyclonal IgG at different stages of individual patients disease course. The results suggest that while the glycosylation of the paraprotein is clone specific, the glycoform profile fluctuates over the course of the disease. This may reflect disease activity and consequent changes in microenvironments or may be indicative of heterogeneity within the neoplastic clone and variable dominance or the emergence of subclones. Thus, changes in the glycosylation of the IgG paraprotein may be indicative of similar changes that will take place for other glycoproteins, some of which, e.g. surface receptors, may be of direct relevance to disease activity and progression. Alternatively, the fluctuating glycoform profile may reflect typical clonal behaviour, due to fluctuations in the relative levels of glycosyltransferase and glycosidase enzymes.

MATERIALS AND METHODS

Serum Samples

Myeloma serum samples were selected from archival material available from the Medical Research Council Myeloma Trials VI - VIII. The longitudinal study was conducted with the serum of patient FL who had responded to cytotoxic therapy and achieved first plateau phase which was maintained for 69 months without need of treatment. FL then suffered disease progression but achieved a second plateau phase on re-introduction of the original therapeutic regimen. The paraprotein was of the IgG3 subclass. Control sera were obtained from three normal, age matched controls; all sera were preserved with azide (0.05% ^{W}/v) and stored frozen at -20°C

Purification of IgG

Sera were dialysed versus 10mM sodium phosphate buffer, pH 7.0 and passed over a DEAE(diethylaminoethyl)- cellulose column (10ml), equilibrated with the same buffer and the IgG collected as the breakthrough. The IgG fraction was passed over a SpA(staphylococcal protein A)-Sepharose affinity column which allowed the paraprotein to be collected in the flow-through whilst the polyclonal IgG was retained, and subsequently recovered on elution with 3M KCNS. Purity was assessed by SDS-PAGE on 12.5% gels, and haemagglutination assays that detected the presence of each IgG subclass protein.

Carbohydrate analyses

Oligosaccharide analyses were carried out as described previously[14]. Between 10 and 15 nmoles of each IgG was used for isolation of the oligosaccharides. Following digestion with chymotrypsin and trypsin (1% w/w), the glycopeptide fraction was digested with 100-200 μunits of glycoamidase (almond). The oligosaccharide was purified by passage over 1 ml columns of cation (Dowex 50WX8) and anion (Dowex 1) exchangers. The filtrate was evaporated to dryness, and reductively aminated with a fluorescent reagent, 2-aminopyridine, and sodium cyanoborohydride[15]. Pyridylaminated oligosaccharides were purified by gel filtration on a Sephadex G-15 column with 10 mM ammonium bicarbonate. The pyridylaminated oligosaccharide mixture was divided into two parts. The first part was analysed for sialic acid content by HPLC on a DEAE-5PW TSK gel column (7.5 by 75 mm, Tosoh Corporation, Japan)[14]. Each oligosaccharide fraction resolved by the DEAE column was evaporated *in vacuo* without desalting and subjected to further HPLC on an ODS-silica column (Shimpack CLC-ODS, 6 by 150 mm; Shimadzu, Japan). The second part of the pyridylaminated oligosaccharide mixture was treated with 20-40 milliunits of sialidase from *Arthrobacter ureafaciens* (Nacalai Tesque) and applied directly to the ODS-silica column. All subsequent analytical procedures including chromatographic conditions have been reported previously[16].

RESULTS

HPLC profiles of oligosaccharides derived from age matched controls (Figure 3 and Table 1) were similar to that obtained for poloyclonal IgG from a pooled population (Figure 4 and Table 1).

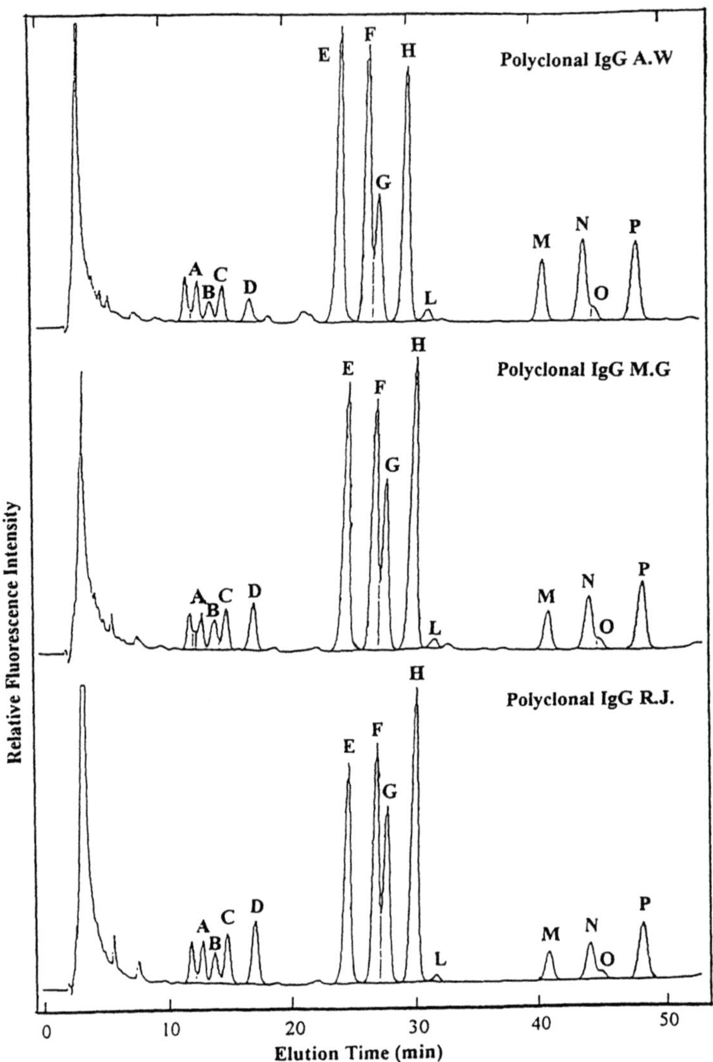

Figure 3. HPLC profiles of pyridylamino oligosaccharide derivatives obtained from polyclonal IgG of age matched controls. Peaks A-P correspond to the structures shown in Figure 2.

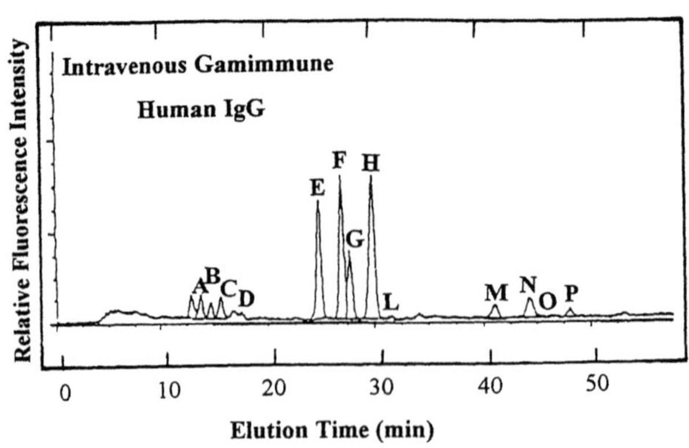

Figure 4. HPLC profiles of pyridylamino oligosaccharide derivatives obtained from intravenous gamimmune, that is polyclonal IgG from a pooled population of approximately 10,000 donors. Peaks A-P correspond to the structures shown in Figure 2.

Table 1. Percentage composition of pyridylamino derived neutral sugar structures from polyclonal IgG. The age matched controls are represented by samples A.W, M. G and R. J, whilst the pooled population is represented by the intravenous gamimmune. Structures A-P correspond to the structures shown in Figure 2.

Oligosaccharide structure	Polyclonal IgG A.W	Polyclonal IgG M.G	Polyclonal IgG R. J	Intravenous Gamimmune
A	2.3	0.5	2.6	3.9
B	1.3	2.2	2.1	3.9
C	2.2	2.7	3.4	2.5
D	1.7	3.4	4.8	3.6
E	21.8	20.3	17.3	20.2
F	20.0	18.6	18.6	23.6
G	9.3	13.1	14.1	10.9
H	19.3	22.6	23.6	24.4
L	0.9	0.7	0.5	-
M	5.4	3.6	2.8	2.3
N	7.7	5.2	3.8	} 3.2
O	0.8	0.6	0.6	}
P	7.3	6.5	5.7	1.5
TOTAL	100	100	100	100

Serum samples had been obtained and stored over a nine year period from a patient with multiple myeloma. Five of these samples were selected on the basis of disease stage at the time of sampling and the IgG oligosaccharide profiles analysed. The quantitative profile of oligosaccharides released from the polyclonal IgG1,2,4 (Figure 4 and Table 2) are similar for each sample, however, under-galactosylation is evident, relative to normal polyclonal IgG . The oligosaccharide profile for the IgG3 paraprotein at presentation (FL-1) is similar to that of the polyclonal IgG preparation at first plateau early phase (FL-2), however, it is interesting to note that peak G is greater than peak F, the reverse of the ratio for the polyclonal IgG (Figure.5 and Table 2). This phenomenon is maintained for all samples analysed. There is evidence of increased galactosylation for the IgG3 paraprotein sample FL-2 with components D and H being increased. For the sample at first plateau mid phase (FL-3) hypergalactosylation is evident with components D and H accounting for 46% of total oligosaccharide. Galactosylation is decreased in the sample corresponding to disease progression (FL-4) relative to FL-3 but then increased again at second plateau (FL-5). Changes in the values of A+B+C+D reflect reduced fucosylation. It should be noted that the levels of paraprotein vary widely for the different samples.

DISCUSSION

The present data, together with those published previously[17], establish that in the disease multiple myeloma the polyclonal IgG is hypogalactosylated. The detail of the oligosaccharide analyses performed allow us to show that whilst there is evidence for a decreased galactosyltransferase activity other transferases maintain normal activity. Thus, the level of fucosylated and bisecting GlcNAc bearing oligosaccharides remains within the

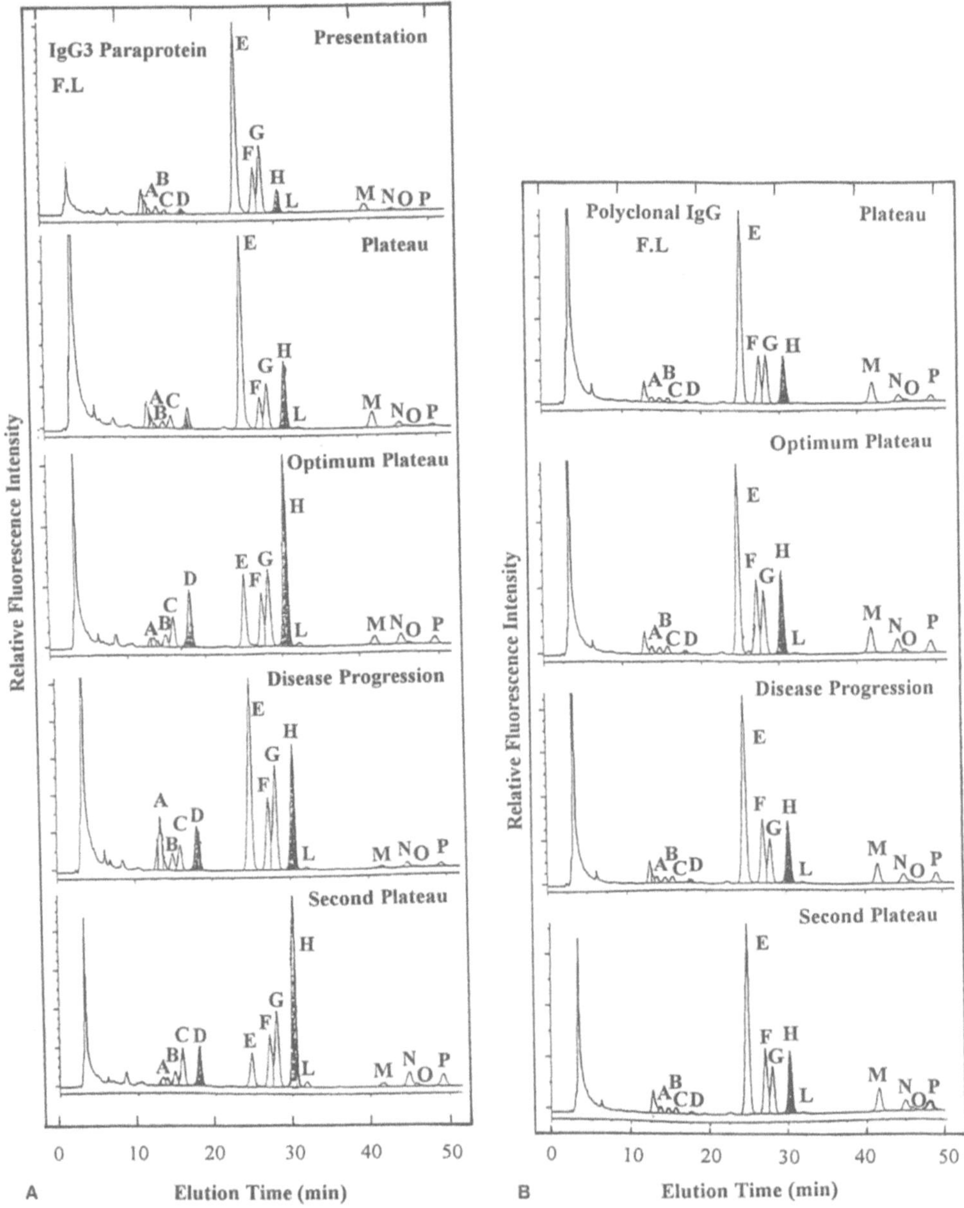

Figure 5. Comparison of oligosaccharide profiles of IgG3 paraprotein and residual polyclonal IgG1, 2 and 4 from patient F.L, no polyclonal sample was available on presentation. HPLC profiles were obtained from pyridylamino oligosaccharide derivatives obtained from these antibodies. Peaks A-P correspond to the structures shown in Figure 2.

100

Table 2. Percentage composition of pyridylamino derived neutral sugars structures at differing stages of the disease. The IgG3 paraprotein values are represented in bold type, whilst the residual polyclonal IgG1, 2 and 4 are represented in normal type. Dates of the serum samples and the corresponding paraprotein levels are noted above and below the table respectively.

Sample Date   6. 8. 1984   8. 1. 1986   19. 11 1988   8. 4. 1992   24. 9 1993

Oligosaccharide structure	Presentation (progressive-disease)		First Plateau Early Phase		First Plateau Mid Phase		Disease Progression		2nd Plateau	
A	**2.2**	-	**2.5**	4.2	**1.3**	3.6	**6.4**	1.3	**1.2**	4.5
B	**2.0**	-	**1.7**	1.0	**2.2**	1.2	**2.5**	1.2	**2.6**	1.2
C	**1.1**	-	**2.8**	1.0	**5.2**	1.5	**3.9**	1.4	**6.8**	1.3
D	**1.5**	-	**4.9**	0.5	**10.3**	0.8	**7.2**	1.0	**8.0**	0.8
E	**51.5**	-	**45.6**	46.6	**13.5**	36.1	**29.3**	42.6	**6.9**	40.8
F	**12.0**	-	**7.3**	11.5	**9.7**	14.4	**11.1**	14.6	**10.0**	14.0
G	**18.8**	-	**11.0**	12.0	**14.5**	12.4	**16.5**	10.3	**14.6**	10.2
H	**6.8**	-	**16.4**	12.0	**35.8**	16.5	**20.2**	14.9	**40.7**	13.9
L	**0.4**	-	**0.4**	0.1	**0.7**	0.3	**0.4**	0.5	**1.2**	0.3
M	**2.4**	-	**4.7**	6.2	**2.0**	6.0	**0.5**	5.8	**0.9**	6.2
N	**0.9**	-	**1.6**	2.3	**2.5**	3.5	**1.1**	2.8	**3.5**	3.0
O	**0.2**	-	**0.3**	0.6	**0.4**	0.6	**0.1**	0.7	**0.5**	0.7
P	**0.2**	-	**0.9**	1.9	**1.8**	3.1	**0.8**	3.0	**3.0**	3.1
TOTAL	**100**	-	**100**	100	**100**	100	**100**	100	**100**	100

IgG3 Paraprotein concentration.   53.6 mg/ml   11.6 mg/ml   6.2 mg/ml   23.7 mg/ml   14.2 mg/ml

normal range, suggesting fucosyltransferase and glucosaminytransferase III maintain normal activity. The oligosaccharide profile of the polyclonal IgG remained essentially un-perturbed throughout a period of nine years during which disease activity waxed and waned and regimens of cytotoxic therapy were administered and withdrawn.

Before considering the oligosaccharide profiles for the FL paraprotein samples it should be noted that the paraprotein concentrations differ widely between samples, Table 2, depending on clinical status. However, it is held that paraprotein levels do not accurately reflect tumour burden and that the paraprotein levels are, generally, modest compared to the number of neoplastic plasma cells present in bone marrow. Paraprotein samples FL-1 (53.6mg/ml) and FL-2 (11.6mg/ml) have G0 oligosaccharides as the dominant species, with a minimal change between the presentation and early plateau samples. 22 months later in the middle of the first plateau phase (FL-3), the glycosylation profile has changed dramatically with the dominant species being fully galactosylated oligosaccharides.

The patient was in remission for a period of 5.5yrs before disease progression was apparent and the paraprotein level had increased to 23.7mg/ml. The oligosaccharide profile of the IgG paraprotein was a relatively normal profile, which represents a reversion towards the profile of samples FL-1 and FL-2. Chemotherapy was instituted and 1.5yrs later the patient had achieved a second plateau and a paraprotein level of 14.2mg/ml. It is tempting to speculate that at presentation disease symptoms resulted from the presence of an aggressive (sub)clone of neoplastic plasma cells characterised by a deficit in its glycosylation machinery, as evidenced by the oligosaccharide profile of its IgG3 paraprotein. Chemotherapy was very successful in suppressing this subclone with the achievement of a plateau phase in which the paraprotein is the product of a subclone of plasma cells secreting IgG that is hypergalactosylated relative to the original paraprotein and normal IgG. At progression the original aggressive subclone is manifest but was successfully controlled with the institution of a second regimen of chemotherapy. The paraprotein at the second plateau (14.2mg/ml) may be the product of the same subclone as for the first plateau, by the criterion of the oligosaccharide profile and progression.

Alternatively, the glycoform profile may not be a clonal signature, but may fluctuate for all clones[18,19], whether associated with the paraprotein or not. Thus, the constant polyclonal profiles would then be the sum of the fluctuating monoclonal profiles.

The ratios for galactosylation on the man $\alpha(1\rightarrow6)$ arm versus the man $\alpha(1\rightarrow3)$ arm are given by the ratio of peaks B : C, F : G and N : O. It can be seen from the Figures 3, 4 and Table 1 that whilst for normal IgG a man $\alpha(1\rightarrow6)$ arm preference is apparent for the paraproteins reverse ratios are observed for B : C and F : G. This paraprotein under investigation is of the IgG3.G3m(b) allotype and it appears that the structure of this heavy chain influences the arm specificity of the GlcNAc $\beta(1\rightarrow4)$ galactosyltranferase. Interestingly, it has been reported that the arm preference for this transferase is man $\alpha(1\rightarrow 3)$ when the substrate is free oligosaccharide in solution[20,21]. It appears, therefore, that the Fc region of a majority of IgG molecules influences its own galactosylation so that the arm preference is reversed. Since IgG1 comprises ~65% of total serum IgG it is likely it is its galactosylation profile that is reflected in normal IgG preparations. We previously reported a man $\alpha(1\rightarrow3)$ preference for galactosylation of 4/4 IgG2 paraproteins. It is of further interest to note that a reduction in the man $\alpha(1\rightarrow3)$ preference is apparent for oligosaccharides having a bisecting N-acetylglucosamine residue[20,21]. Thus, this study demonstrates that the glycosylation profile of paraproteins may be useful indicators of disease activity and/or of clonal fluctuations, i.e. glycosyltransferase activity, in multiple myeloma and may be revealing underlying causes of progression. In addition, it may allow dissection of subtle structural characteristics that can influence glycosyltransferase activities.

REFERENCES

1. A. F. Williams, A. N. Barclay, The immunoglobulin super family- domains for cell surface recognition, Ann. Rev. Imm. 6: 381-405 (1988).
2. S. L. Morrison, In vitro antibodies- strategies for production and application, Ann Rev Immunol. 10: 239-265 (1992).
3. R. Jefferis and J. Lund, Molecular characterisation of IgG antibody Fc effector sites, In protein engineering of antibody molecules for prophylactic and therapeutic applications in man, M. Clarke, Ed. Academic press NY p115 (1995)
4. J. Lund, N. Takahashi, S. Hindley, R. Tyler, M. Goodall, and R. Jefferis, Glycosylation of human IgG subclass and mouse IgG2b heavy chains secreted by mouse J558L transfectoma cell lines as chimeric antibodies, Human Antibodies and Hybridomas. 4: 20-25 (1993).
5.J. Lund, N. Takahashi, H. Nakagawa, T. Bentley, S. Hindley, R. Tyler, M. Goodall and R. Jefferis, Control of IgG/Fc glycosylation: A comparison of oligosaccharides from chimeric human/mouse and mouse immunoglobulin G's, Mol Immunol. 30: 741-748 (1993).

6. R. Jefferis, J. Lund, H. Mizutani, H. Nakagawa, Y. Kawazo, Y. Arata, and N. Takahashi, A comparative study of N-linked oligosaccharide structures of human IgG subclass proteins, Biochem J. 268: 529-537 (1990).
7. T. Mizuochi, T. Taniguchi, A. Shimizu, and A. Kobata, Structural and numerical variations of the carbohydrate moiety of IgG, J. Immunol. 129: 2016-2020 (1982).
8. T. W. Rademacher, and R. A. Dwek, Prog Immun. 5: 95-112 (1983).
9. R. Malhotra, M. R. Wormald, P. M. Rudd, P. B. Fischer, R. A. Dwek and R. B. Sim, Glycosylation changes of IgG associated rheumatoid arthritis can activate complement via the MBP, Nature Medicine. vol 1 No 6: 599
10. R. B. Parekh, R. A. Dwek, B. J. Sutton, D. L. Fernandes, A. Leung, D. Stanworth, T. W. Rademacher, T. Mizuochi, T. Taniguchi, K. Matsuta, F. Takeuchi, Y. Nagano, T. Miyamoto and A. Kobata, Association of rheumatoid arthritis and primary osteoarthritis with changes in the glycosylation pattern of total serum IgG, Nature. 316: 452-457 (1985).
11. R. B. Parekh, D. A. Isenburg, B. M. Ansell, I. M. Roitt, R. A. Dwek, and T. W. Rademacher, Galactosylation of IgG associated oligosaccharides- reduction in patients with adult andjuvenile onset rheumatoid arthritis and relation to disease activity, Lancet. i: 966-969 (1988).
12. R. B. Parekh, D. A. Isenburg, G. Rook, I. M. Roitt, R. A. Dwek, and T. W. Rademacher, A comparative analysis of disease associated changes in the galactosylation of serum IgG, J. Autoimmunity. 2: 101-114 (1989).
13. G. A. W. Rook, J. Steele, R. Brearley, A. Whyte, D. A. Isenburg, N. Sumar, J. L. Nelson, K. B. Bodman, A. Young, I. M. Roitt, P. Williams, I. Scragg, C. J. Edge, P. D. Arkwright, D. Ashford, M. Wormald, P. Rudd, C. W. G. Redman, R. A. Dwek, and T. W. Rademacher, Changes in the IgG glycoform levels are associated with remission of arthritis during pregnancy, J. Autoimmunity. 779-794 (1991).
14. N. Takahashi, H. Nakagawa, K. Fujikawa, Y. Kawamura, and N. Tomiya, Three dimensional elution mapping of pyridylaminated N-linked neutral and sialyl oligosaccharides, Anal. Biochem. 226: 139 (1995).
15. S. Yamamoto, S. Hase, S. Fukuda, O. Sano, and T Ikenaka, Structures of the sugar chains of interferon-γ produced by human myelomonocyte cell line HBL-38, J. Biochem. 105: 547 (1989).
16. N. Tomiya, J. Awaya, M. Kurono, S. Endo, Y. Arata, and N. Takahashi, Analysis of N-linked oligosaccharides using a two dimensional mapping technique, Anal. Biochem. 171: 73 (1988)
17. M. Farooq, N. Takahashi, H. Arrol, M. Drayson, and R. Jefferis, Glycosylation of IgG antibody molecules in multiple myeloma, Glycoconjugate J. In press (1996).
18. D. R. Anderson, P. H. Atkinson and W. J. Grimes, Major carbohydrate structures at five glycosylation sites on murine IgM determined by high resolution ^{1}H-NMR spectroscopy, Arch. Biochem. Biophys. 13: 605-618 (1985).
19. R. J. Rothman, L. Warren, J. F. G. Vliegenhart, and K. J. Hard, Clonal analysis of the glycosylation of immunoglobulin G secreted by murine hybridomas, Biochemistry. 28: 1377-1384 (1989).
20. S. Fujii, T. Nishiura, A. Nishikawa, R. Miura and N. Taniguchi,Structural heterogeneity of sugar chains in immunoglobulin G, J. Biol Chem. 265: 6009-6018 (1990).
21. S. Narasimhan, J. C. Freed, and H. Schachter, Control of glycoprotein synthesis. Bovine milk UDP-galactose: N-acetylglucosamine β-4-Galactosyltransferase catalyzes the preferential transfer of galactose to the GlcNAcβ1,2Manα1,3- branch of both bisected and non-bisected complex biantennary asparagine-linked oligosaccharides, Biochemistry. 24: 1694-1700 (1985).

THE ROLE OF THE LECTIN CALNEXIN IN CONFORMATION INDEPENDENT BINDING TO N-LINKED GLYCOPROTEINS AND QUALITY CONTROL

John J.M. Bergeron[1], A. Zapun[1,3], W.-J. Ou[3], R. Hemming[1,3],
F. Parlati[2,3], P.H. Cameron[1], and D.Y. Thomas[1,2,3]

[1]Department of Anatomy and Cell Biology, McGill University, Montreal
Quebec, Canada, H3A 2B2
[2]Department of Biology, McGill University, Montreal
[3]Biotechnology Research Institute, NRC of Canada, Montreal

INTRODUCTION

It has often been speculated that one of the selective forces for the evolution of N-linked glycosylation is that of productive protein folding. Strong experimental support has now been put forward as a consequence of studies with the endoplasmic reticulum resident membrane protein calnexin and its luminal homologue calreticulin. These resident proteins, at least one of which (calnexin) is universal to all eukaryotes, act principally as lectins which recognize monoglucosylated intermediates of high mannose containing N-linked glycoproteins. A molecular chaperone apparatus consisting of the coupled actions of uridine diphosphate glucose: glycoprotein glucosyl transferase, calnexin/calreticulin and glucosidase II has been reconstituted using fully defined constituents in vitro. That this apparatus also functions as one of the regulatory mechanisms in the identification and triaging of misfolded glycoproteins (quality control) has also shown recent experimental support. From the perspective of calnexin N-linked glycosylation and more specifically monoglucosylation allows the temporal coupling of glycoprotein folding within the spatial confines of the endoplasmic reticulum and provides a mechanism to direct misfolded glycoproteins away from productive folding intermediates.

Calnexin was first uncovered as a consequence of attempting to identify integral membrane phosphoproteins which may be related to the structure and function of each organellar compartment. The basis of this undertaking was an accidental consequence of a totally unrelated project dealing with the physiological significance of ligand-mediated receptor internalization (reviewed in Baass et al., 1995). Since we wished to study the biochemical properties of the organelle (i.e. endosomes) harboring internalized receptor tyrosine kinases, a subcellular fractionation approach utilizing in vitro phosphorylation studies was a necessary first step. As expected isolated organelles harbored in vitro kinase activities which were easily distinguished from the more abundant in vitro tyrosine phosphorylation activity of internalized receptors (Khan et al. 1986; Kay et al. 1986; Wada et al., 1992). Nonetheless, in vitro kinase activity was observed in isolated organelles even in the absence of receptor internalization. What was noteworthy was that preparative procedures used to isolate different organelles revealed different phosphorylation patterns indicating either specificity in the kinases, substrates or both (Rindress et al. 1993). What was remarkable is that the repertoire of substrates which were phosphorylated were limited and when studying those phosphoproteins which were integral to the membrane of each organelle these phosphoproteins were few and distinct to each organelle. This now made the problem of their significance within biochemical reach.

Because of its relative ease of purification and abundance the rough Endoplasmic Reticulum (ER) was first selected. For biochemical ease in kinase identification both [γ-^{32}P]ATP and [γ-^{32}P]GTP were used. Since the simplest pattern was effected by [γ-^{32}P]GTP phosphorylation then purification of these integral membrane phosphoproteins was attempted (Fig. 1).

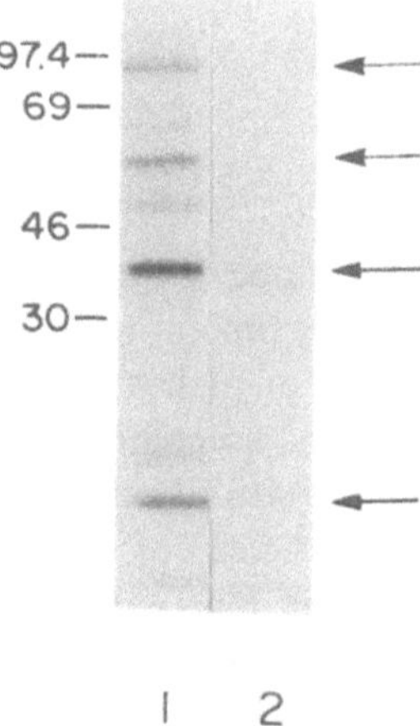

Figure 1. In vitro phosphorylation of stripped rough microsomes from dog pancreas. After incubation with [γ-^{32}P]GTP, 4 major proteins of 90, 56, 35 and 15 kDa are observed. Protease treatment of the intact membranes removes the signal; hence, the phosphorylation sites are cytosolically oriented (Wada et al., 1991).

Attention was particularly focused on a 35 kDa phosphoprotein since a protein of such molecular mass was implicated in a known ER function; i.e. recognition of the signal sequences of newly synthesized secretory proteins (Prehn et al., 1990; Görlich et al., 1990). By a combination of detergent partitioning protocols, ion exchange and gel filtration chromatography, a complex of 4 integral membrane proteins were coisolated of which 2 (90 and 35 kDa) represented the major integral membrane phosphoproteins of the ER of which one (35 kDa) was N-glycosylated. The two other integral membrane proteins (at ca. 25 kDa) were N-glycosylated but not phosphorylated. In hindsight, the now recognized properties of these integral membrane proteins have been informative.

ENDOPLASMIC RETICULUM TRANSLOCATION

Following their partial protein sequencing all 4 membrane proteins were cDNA cloned and their primary protein sequences deduced (Fig. 2).

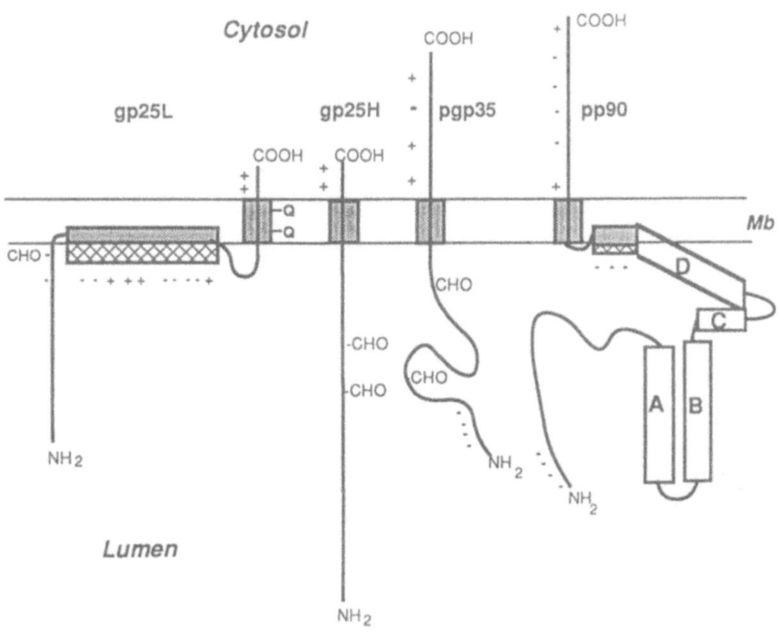

Figure 2. Topology of pgp35, gp25H, gp25L and pp90 (calnexin) in the ER membrane (Wada et al., 1991).

Immediate attention was focused on two of these, since coincidentally two laboratories had implicated the 35 kDa and one of the 25 kDa membrane glycoproteins as constituents of a membrane complex which interacted with signal sequences and were postulated to be constituents of the then unknown translocon. These initial proposals did not however withstand detailed experimental testing. Hence, using antibodies (in part provided by our group) to the 35 kDa membrane phosphoglycoprotein (misnamed by this time SSRα; i.e. the α-subunit of the signal sequence receptor) immunodepletion experiments with reconstituted proteoliposomes revealed no role for SSRα in signal sequence dependent protein translocation across the ER membrane (Migliaccio et al., 1992). Taken together with the absence of sequence related proteins to SSRα in the genome of the budding yeast (*Saccharomyces cerevisiae*) then a causal role in ER translocation of newly synthesized secretory proteins now appears unwarranted.

p88, CALNEXIN, IP90

A major leap forward in our understanding of the functional significance of the complex of Fig. 2 was realized by the fortuitous identification of the functional significance of the 90 kDa protein of Fig. 2 in the initial stages of glycoprotein folding in the ER. On the basis of its primary sequence similarity to calreticulin, the major luminal Ca^{2+} binding resident

resident protein of the ER, as well as its demonstrated high affinity binding to Ca^{2+}, we originally named the 90 kDa membrane phosphoprotein calnexin (from nexus implying a function of gathering or binding).

Clearly the precise biochemical function of calnexin was to be related to that of calreticulin (Figure 3). The two shared sequence similarity, the same location and both bound Ca^{2+} with high affinity. Although calreticulin had been originally identified in muscle and cDNA cloned and sequenced since 1989, little was known of its physiological function (Michalak et al., 1992). Indeed, initial studies proposed a cytosolic/ nuclear location and function (Michalak et al., 1994; Dedhar, 1994) even though calreticulin possesses a N-terminal signal sequence and a carboxyl terminal KDEL sequence common to all luminal ER resident proteins. Furthermore, calnexin possessed at its extreme carboxyl terminus the motif RKPRRE with the double arginine motif conceptually similar to the dilysine motif known to be common for the retention/retrieval of many ER proteins (Jackson et al., 1990). The function of calnexin and calreticulin had to be linked to their location in the ER the earliest compartment in secretory protein biogenesis.

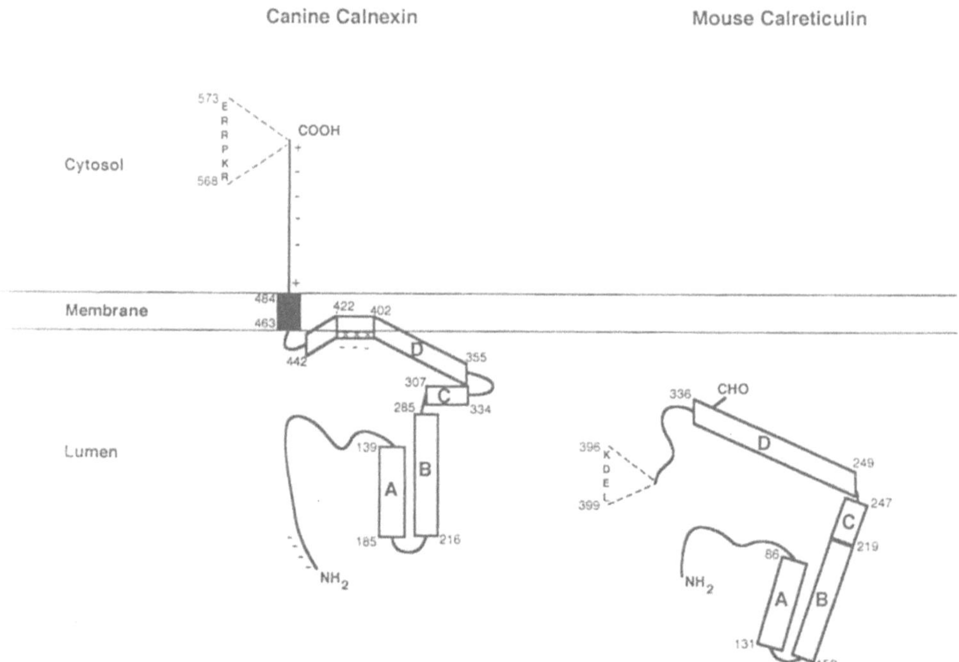

Figure 3. Schematic representation of the dog calnexin and mouse calreticulin. The designations A,B,C,D refer to regions of high sequence similarity.

An advance came from the work of Williams and colleagues. These investigators were interested in the biogenesis of MHC Class I molecules and were attempting to provide a molecular explanation for observation that Class I molecules synthesized in the absence of β2-microglobulin were retained in the ER. By the use of chemical crosslinking experiments they were able to uncover a polypeptide of ca. 88 kDa which associated with MHC Class I molecules almost immediately after ER translocation. This molecule termed p88 dissociated from MHC Class I heavy chain coincident with the association of β2-microglobulin (Degen and Williams, 1991). The protein was long-lived and a non N-glycosylated integral membrane protein, exactly the characteristics of calnexin!

Fortunately, a collaborative study was initiated and by peptide mapping led to the conclusive identification of p88 as calnexin (Ahluwalia et al., 1992). N-terminal sequencing coincidentally identified a protein, IP90, found independently by Brenner and colleagues to be in association with MHC Class I heavy chain, membrane Ig, and the α and β subunits of the T cell receptor, as identical to calnexin (Hochstenbach et al., 1992).

THE N-GLYCOSYLATION LINK

While initial studies may have pointed to a role for calnexin as a molecular chaperone specific to oligomeric membrane proteins (Hochstenbach et al., 1992), this point of view changed when several soluble monomeric proteins were identified to be associated with calnexin rapidly after their biosynthesis in the ER (Ou et al., 1993). However, specificity was uncovered (Table 1). Whereas both disulfide and nondisulfide bonded proteins were associated with calnexin, albumin the major nonglycosylated protein secreted by liver parenchymal cells did not associate with calnexin but all secretory proteins which were N-glycosylated did. The obvious experiment of blocking N-linked glycosylation with tunicamycin confirmed that in the absence of glycosylation the association of newly synthesized secretory glycoproteins with calnexin was abrogated (Ou et al., 1993).

Table 1

Protein	S-S Bonds	N-Glycosylation Sites	Calnexin Association
Albumin	17	0	-
α-fetoprotein	15	3	+
α1-antitrypsin	0	3	+
Transferrin	19	2	+

N-LINKED GLYCOSYLATION AND SECRETION

These observations now provided a rationale for several puzzling observations which had accumulated on the relationship between N-glycosylation and secretion. Firstly, it had been known from the studies of Williams et al. (1985) and Lodish and Kong (1984) that different proteins exited the ER at vastly different rates (Lodish, 1988). A clue to the mechanistic explanation was that this differential time of exit disappeared when N-glycosylation was inhibited with tunicamycin with secretion of all previously N-glycosylated proteins now corresponding to that of the slowest protein secreted (Lodish and Kong (1984). Furthermore, these observations could even be extended to specific modifications of N-linked glycans in the ER since inhibitors of glucose removal from N-linked glycans showed the same phenotype as tunicamycin treatment of cells in preventing the temporal differences in the egress of glycoproteins from the ER (Lodish and Kong, 1984). Indeed, this even led two groups to propose a specific receptor in the ER membrane which would recognize glucosylated N-linked glycans (Lodish and Kong, 1984; Yeo et al., 1989). Calnexin was clearly the candidate!

THE MONOGLUCOSYLATION CYCLE

A conceptual leap forward occurred when Helenius and colleagues tackled the problem. Armed with the experience which followed detailed studies of the phenomenon of quality-control' in the ER (Hurtley and Helenius, 1989) and the role of ER resident proteins in secretory protein folding they proposed a detailed model accounting for most of the properties of calnexin (Figure 4).

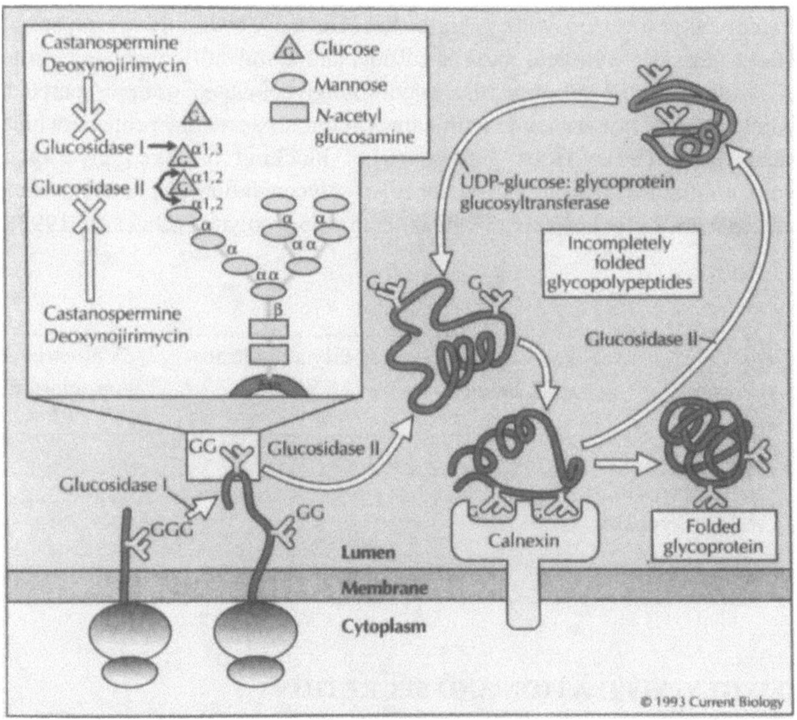

Figure 4. Detailed model of calnexin as a lectin specific for binding to monoglucosylated glycoproteins during glycoprotein maturation in the ER (Hammond and Helenius, 1993).

As summarized in their seminal paper (Hammond et al., 1994), calnexin most likely was a lectin specific for monoglycosylated N-linked glycoproteins with UDP glucose glycoprotein glucosyl transferase, an enzyme discovered by Parodi and colleagues (Parodi et al., 1984; Zapun et al., 1996) as the enzyme which sensed the correct folding pathway of a glycoprotein and which via monoglucosylation would re-`present' the newly synthesized glycoprotein to calnexin. Hence, calnexin would be expected to bind a newly synthesized glycoprotein after the sequential action of glucosidase I and II most likely in molecular proximity to the translocon. Indeed, the studies of Chen et al. (1995) have revealed that for VSV"G" glycoprotein that monoglucosylation dependent binding to calnexin occurred prior to completion of the polypeptide chain. More recently, it has been observed (Li et al., 1996) that the signal sequence of HIV gp120 when expressed in Sf9 cells is cleaved after gp120 glycosylation and binding to calnexin. Since signal peptidase is a *bona fide* constituent of the translocon this further implicates calnexin as initially interacting with newly synthesized glycoproteins in close proximity to the translocon. After initial interaction between calnexin and monoglucosylated N-linked glycoproteins near the translocon, a second opportunity to interact with calnexin would occur consequent to the action of the ER luminal resident protein uridine diphosphate glucose glycoprotein glucosyltransferase. This remarkable enzyme has the property of monoglucosylating denatured but not fully folded glycoproteins (Parodi et al., 1984).

Several studies have supported this model including the lack of binding of substrates to calnexin in glucosidase deficient cell lines (Kearse et al., 1994; Ora et al., 1995) and a cell free system using microsomes in which binding to calnexin was dependent on monoglucosylation and whereby dissociation from calnexin required glucosidase II (Hebert et al., 1995).

This has now led to a fully defined reconstituted system which conclusively supports a model illustrated in Figure 6. Here, RNase B in its denatured state was monoglucosylated with purified UGGT and tested for binding to calnexin. Since denatured RNase B can also be renatured to a native state with protein disulfide isomerase monoglucosylated RNase B could be refolded. Both the denatured and fully folded RNase B bound calnexin provided they were monoglucosylated. Taken together with the studies of Ware et al. (1995) who demonstrated directly the selective binding of Glc1Man9GlcNAc2 oligosaccharides by calnexin and our own work showing the direct binding to calnexin of monoglucosylated high mannose oligo-saccharides released from RNase B by endoH then calnexin is clearly a lectin. Furthermore, endoH treatment alone is sufficient for complete dissociation of monoglucosylated RNase B from calnexin (Zapun et al., 1996). This differs from the conclusion of Ware et al. (1995; Figure 5) that a stable association of calnexin with glycoprotein folding intermediates is via protein-protein interactions.

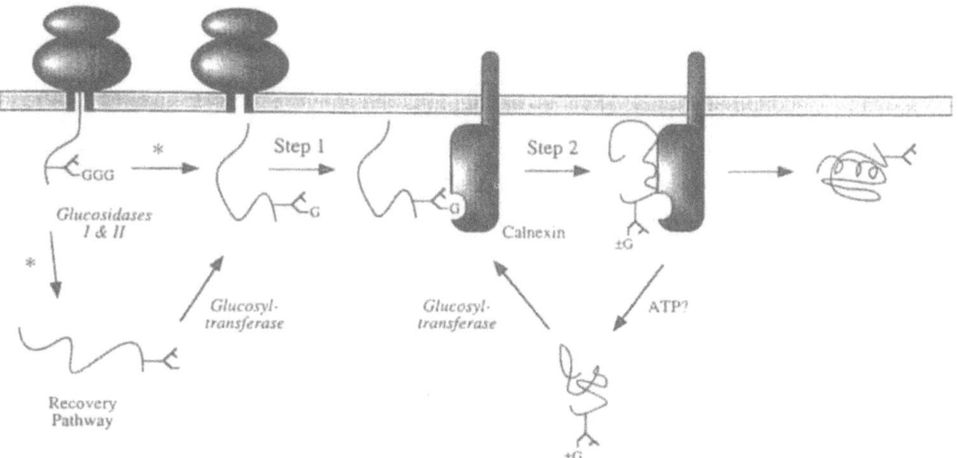

Figure 5. Model of Ware et al. (1995) depicting a two step model for presentation of N-linked glycoproteins to calnexin via monoglucosylation but of stable association via polypeptide motifs of substrate glycoproteins.

We have argued that such interactions may not exist in vivo and are non-specific (Zapun et al., 1996).

Remarkably, the enzyme glucosidase II (and even protein disulphide isomerase) has restricted access to monoglucosylated RNase B when bound to calnexin. However, when the ratio of calnexin to monoglucosylated RNase B is reduced (but still with calnexin in excess) then glucosidase II effectively removes the Glc α1-3 Man linkage. These and other studies showing competition of binding of monoglucosylated RNase B with Glc1 containing oligosccharides or small Glc1 containing glycopeptides (Zapun et al., 1996) have indicated that the interaction of calnexin with monoglucosylated substrates is very dynamic. Hence, the model which most accurately, in our opinion, indicates the mechanism of retention of calnexin with glycoprotein folding intermediates is represented in Figure 6.

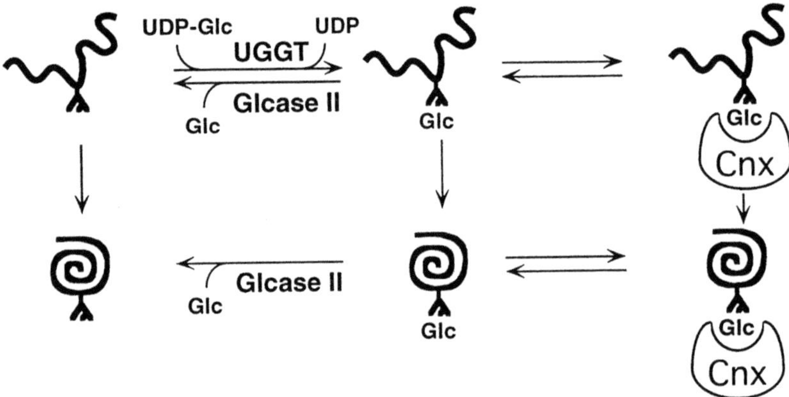

Figure 6. Model of Zapun et al. (1996) in which the dynamic interaction of monoglucosylated substrates is indicated. In this model the action of uridine diphosphate glucose: glycoprotein glucosyl transferase (UGGT) is indicated as well as the action of glucosidase II on monoglucosylated glycoproteins dissociated from calnexin.

CALNEXIN IN QUALITY CONTROL

Calnexin is expressed in yeast. Although in the budding yeast, *Saccharomyces cerevisiae*, it is expressed with a transmembrane domain but without a cytosolic tail. Targeted gene disruption revealed quite surprisingly in this budding yeast that the gene was non-essential. However, quality control was compromised, in that misfolded proteins were not able to exit the ER (Parlati et al., 1995a). Indeed, in this same budding yeast, it has also been demonstrated that calnexin is involved in the machinery for presentation of misfolded proteins to the ER associated degradation apparatus (McCracken and Brodsky, 1996). In mammalian cells, misfolded glycoproteins are retained for long periods via the remonoglucosylation cycle (Figure 6) in association with calnexin (Ou et al., 1993; Le et al., 1994; Pind et al., 1994). Furthermore, for two of these, i.e. mutated α1 antitrypsin and the ΔF508 mutant of CFTR this coincides with the polyubiquitinylation and presentation of these mutated molecules to the cytosolic proteosome for degradation (Ward et al., 1995; Jensen et al., 1995; Qu et al., 1996). Hence, the membrane association of calnexin and its topological location may be key to this quality control phenomenon. Remarkably, in the fission yeast *S. pombe* calnexin is essential for viability perhaps indicating a more stringent link to quality control (Parlati et al., 1995b).

PHOSPHORYLATION OF CALNEXIN

With the exception of calnexin in *Saccharomyces cerevisiae* most other eukaryotic species reveal a cytosolic tail for this molecule. In all mammalian cells, this cytosolically exposed tail is phosphorylated with casein kinase II a candidate for at least one of the kinases responsible for its phosphorylation (Ou et al., 1992), although it is too early to speculate the physiological significance of this phosphorylation (see Capps and Zuniga, 1994).

THE GP25L FAMILY

Finally, the 4th protein originally identified in association with calnexin, gp25L has now been identified as part of a larger family of membrane proteins implicated in ER to Golgi transport (Schimmoller et al., 1995; Fiedler et al., 1996; Rothman et al., 1996). The physiological significance of the coisolation of calnexin with gp25L points to a role downstream of calnexin in defining the microdomain in the ER selected for egress to the Golgi apparatus perhaps with gp25L family members serving as receptors for secretory cargo (Schimoller et al., 1995). However, more detailed experimentation awaits the testing of these speculations.

References

Ahluwalia, N., Bergeron, J.J.M., Wada, I., Degen, E., and Williams, D.B. 1992. The p88 molecular chaperone is identical to the endoplasmic reticulum membrane protein, calnexin. *J. Biol. Chem.* 267:10914-10918.

Baass, P.C., DiGuglielmo, G.M., Authier, F., Posner, B.I., and Bergeron, J.J.M. 1995. Compartmentalized signal transduction by receptor tyrosine kinases. Trends Cell Biol. 5:465-470.

Burns, K., Atkinson, E.A., Bleackley, R.C., and Michalak, M. 1994. Calreticulin: from Ca^{2+} binding to control of gene expression. Trends Cell Biol. 4:152-154.

Capps, G.G., and Zuniga, M.C. 1994. Class I histocompatibility molecule association with phosphorylated calnexin. J. Biol. Chem. 269:11634-11639.

Chen, W., Helenius, J., Braakman, I., Helenius, A. 1995. Cotranslational folding and calnexin binding during glycoprotein synthesis. Proc. Natl. Acad. Sci. USA. 92:6229-6233.

Dedhar, S. 1994. Novel functions for calreticulin: interaction with integrins and modulation of gene expression? Trends Biochem. Sci. 19:269-271.

Degen, E., and Williams, D.B. 1991. Participation of a novel 88-kD protein in the biogenesis of murine class I histocompatibility molecules. J. Cell Biol. 112:1099-1115.

Fiedler, K., Veit, M., Stamnes, M.A., and Rothman, J.E. 1996. Bimodal interaction of coatomer with the p24 family of putative cargo receptors. Science. 273:1396-1399.

Görlich, D., Prehn, S., Hartmann, E., Herz, J., Otto, A., Kraft, R., Wiedmann, M., Knespel, S., Dobberstein, B. and Rapoport, T.A. 1990. The signal sequence receptor has a second subunit and is part of a translocation complex in the endoplasmic reticulum as probed by bifunctional reagents. J. Cell Biol. 111:2283-2294.

Hammond, C., Braakman, I., and Helenius, A. 1994. Role of N-linked oligosaccharide recognition, glucose trimming, and calnexin in glycoprotein folding and quality control. Proc. Natl. Acad. Sci. USA. 91:913-917.

Hammond, C, and Helenius, A. 1993. A chaperone with a sweet tooth. Current Biology. 3:884-886.

Hebert, D.N., Foellmer, B., and Helenius, A. 1995. Glucose trimming and reglucosylation determine glycoprotein association with calnexin in the endoplasmic reticulum. Cell. 81:425-433.

Hochstenbach, F., David, V., Watkins, S., Brenner, M.B. 1992. Endoplasmic reticulum resident protein of 90 kilodaltons associates with the T- and B-cell antigen receptors and major histocompatibility complex antigens during their assembly. Proc. Natl. Acad. Sci. USA. 89:4734-4738.

Hurtley, S.M., and Helenius, A. 1989. Protein oligomerizaton in the endoplasmic reticulum. Annu. Rev. Cell Biol. 5:277-307.

Jackson, M.R., Nilsson, T., and Peterson, P.A. 1990. Identification of a consensus motif for retention of transmembrane proteins in the endoplasmic reticulum. EMBO J. 9:3153-3162.

Jensen, T.J., Loo, M.A., Pind, S., Williams, D.B., Goldberg, A.L., and Riordan, J.R. 1995. Multiple proteolytic systems including the proteasome, contribute to CFTR processing. Cell. 83:129-135.

Kay, D.G., Lai, W.H., Uchihashi, M., Khan, M.N., Posner, B.I., and Bergeron, J.J.M. 1986. Epidermal growth factor receptor kinase translocation and activation in vivo. *J. Biol. Chem.* 261:8473-8480.

Kearse, K.P., Williams, D.B., and Singer, A. 1994. Persistence of glucose residues on core oligosaccharides prevents association of TCR α and TCRβ proteins with calnexin and results specifically in accelerated degradation of nascent TCR α proteins within the endoplasmic reticulum. EMBO J. 13:3678-3686.

Khan, M.N., Savoie, S., Bergeron, J.J.M. and Posner, B.I. 1986. Characterization of rat liver endosomal fractions: In vivo activation of insulin-stimulable receptor kinase in these structures. *J. Biol. Chem.* 261:8462-8472.

Le, A., Steiner, J.L., Ferrell, G.A., Shaker, J.C. and Sifers, R.N. 1994. Association between calnexin and a secretion-incompetent variant of human α1-antitrypsin. J. Biol. Chem. 269:7514-7519.

Li, Y., Bergeron, J.J.M., Luo, L., Ou, W-J., Thomas, D.Y., and Kang, C.Y. 1996. Effects of inefficient cleavage of the signal sequence of HIV-1 gp120 on its association with calnexin, folding, and intracellular transport. Proc. Natl. Acad. Sci. USA. 93:9606-9611.

Lodish, H.F. 1988. Transport of secretory and membrane glycoproteins from the rough endoplasmic reticulum to the Golgi: A rate-limiting step in protein maturation and secretion. J. Biol. Chem. 263:2107-2110.

Lodish, H.F., and Kong, N. 1984. Glucose removal from N-linked oligosaccharides is required for efficient maturation of certain secretory glycoproteins from the rough endoplasmic reticulum to the Golgi complex. J. Cell Biol. 98:1720-1729.

McCracken, A.A., and Brodsky, J.L. 1996. Assembly of ER-associated protein degradation in vitro: dependence on cytosol, calnexin, and ATP. J. Cell Biol. 132:291-298.

Michalak, M., Milner, R.E., Burns, K., Opas, M. 1992. Calreticulin. Biochemical J. 285:681-692.

Migliaccio, G., Nicchitta, C.V., and Blobel, G. 1992. The signal sequence receptor, unlike the signal recognition particle receptor, is not essential for protein translocation. J. Cell Biol. 117:15-25.

Ora, A., and Helenius, A. 1995. Calnexin fails to associate with substrate proteins in glucosidase-deficient cell lines. J. Biol. Chem. 270:26060-26062.

Ou, W-J., Thomas, D.Y., Bell, A.W., and Bergeron, J.J.M. 1992. Casein kinase II phosphorylation of signal sequence receptor α and the associated membrane chaperone calnexin. J. Biol. Chem. 267:23789-23796.

Ou, W-J., Cameron, P.H., Thomas, D.Y. and Bergeron, J.J.M. 1993. Association of folding intermediates of glycoproteins with calnexin during protein maturation. Nature. 364:771-776.

Parlati, F., Dominguez, M., Bergeron, J.J.M., and Thomas, D.Y. 1995a. *Saccharomyces cerevisiae* CNE1 encodes an endoplasmic reticulum (ER) membrane protein with sequence similarity to calnexin and calreticulin and functions as a constituent of the ER quality control apparatus. J. Biol. Chem. 270:244-253.

Parlati, F., Dignard, D., Bergeron, J.J.M., and Thomas, D.Y. 1995b. The calnexin homologue cnx1[+] in Schizosaccharomyces pombe, is an essential gene which can be complemented by its soluble ER domain. EMBO J. 14:3064-3072.

Parodi, A.J., Mendelzon, D.H., Lederkremer, G.Z., and Martin-Barrientos, J. 1984. Evidence that transient glucosylation of protein-linked Man9GlcNAc2, Man8GlcNAc2 and Man7GlcNAc2 occurs in rat liver and phaseolus vulgaris cells. J. Biol. Chem. 259:6351-6357.

Pind, S., Riordan, J.R., and Williams, D.B. 1994. Participation of the endoplasmic reticulum chaperone calnexin (p88, IP90) in the biogenesis of the cystic fibrosis transmembrane conductance regulator. J. Biol. Chem. 269:12784-12788.

Prehn, S., Herz, J., Hartmann, E., Kurzchalia, T.V., Frank, R., Roemisch, K., Dobberstein, B., and Rapoport, T.A. 1990. Structure and biosynthesis of the signal-sequence receptor. Eur. J. Biochem. 188:439-445.

Qu, D., Teckman, J.H., Omura, S., and Perlmutters, D.H. 1996. Degradation of a mutant secretory protein, α1-antitrypsin Z, in the endoplasmic reticulum requires proteasome activity. J. Biol. Chem. 271:22791-22795.

Rothman, J.E., and Wieland, F.T. 1996. Protein sorting by transport vesicles. Science. 272:227-234.

Rindress, D., Lei, X., Ahluwalia, J.P.S., Cameron, P.H., Fazel, A., Posner, B.I., and Bergeron, J.J.M. 1993. Organelle specific phosphorylation: Identification of unique membrane phosphoproteins of the endoplasmic reticulum and endosomal apparatus. *J. Biol. Chem.* 268: 5139-5147.

Schimmoller, F., Singer-Kruger, B., Schroder, S., Kruger, U., Barlowe, C., and Riezman, H. 1995. The absence of Emp24p, a component of ER-derived COPII-coated vesicles, causes a defect in transport of selected proteins to the Golgi. EMBO J. 14:1329-1339.

Yeo, K.T., Yeo, T.K., and Olden, K. 1989. Bromoconduritol treatment delays intracellular transport of secretory glycoproteins in human hepatoma cell cultures. Biochem. Biophys. Res. Comm. 161:1013-1019.

Wada, I., Rindress, D., Cameron, P.H., Ou, W-J., Doherty, II, J-J. Louvard, D., Bell, A.W., Dignard, D., Thomas, D.Y., and Bergeron, J.J.M. 1991. SSR α and associated calnexin are major calcium binding proteins of the endoplasmic reticulum membrane. *J. Biol. Chem.* 266:19599-19610.

Wada, I., Lai, W.H., Posner, B.I., and Bergeron, J.J.M. 1992. Asociation of the tyrosine phosphorylated epidermal growth factor receptor with a 55-kD tyrosine phosphorylated protein at the cell surface and in endosomes. J. Cell Biol. 116:321-330.

Ward, C.L., Omura, S., and Kopito, R.R. 1995. Degradation of CFTR by the ubiquitin-proteasome pathway. Cell 83:121-127.

Ware, F.E., Vassilakos, A., Peterson, P.A., Jackson, M.R., Lehrman, M.A., and Williams, D.B. 1995. The molecular chaperone calnexin binds Glc1Man9GlcNAc2 oligosaccharide as an initial step in recognizing unfolded glycoproteins. J. Biol. Chem. 270:4697-4704.

Williams, D.B., Swiedler, S.J., and Hart, G.W. 1985. Intracellular transport of membrane glycoproteins: Two closely related histocompatibility antigens differ in their rates of transit to the cell surface. J. Cell Biol. 101:725-734.

Zapun, A., Petrescu, S.M., Rudd, P.M., Dwek, R.A., Thomas, D.Y., and Bergeron, J.J.M. 1996. Conformation independent binding of monoglucosylated ribonuclease B to calnexin. Cell (in press).

GLYCOSYLATION AND INFLAMMATION

IMMUNODETECTION OF GLYCOSYLTRANSFERASES: PROSPECTS AND PITFALLS

Eric G. Berger, Peter Burger, Lubor Borsig, Martine Malissard, Kristina Mrkoci Felner, Steffen Zeng and André Dinter

Institute of Physiology
University of Zurich
CH-8057 Zurich, Switzerland

INTRODUCTION

The key element to understand biosynthesis and function of the countless oligosaccharide structures expressed on cell surfaces and soluble glycoconjugates are the enzymes involved in their construction: the glycosyltransferases. Their investigation relies on three basic pillars: i) the biochemical which comprises methods for their purification and assays for enzymic activity; ii) the cell biological which is usually based on antibodies providing means to study their subcellular localization, trafficking and tissue distribution and iii) the approach based on recombinant DNA technology which provides insights in regulation of gene expression, structural data and approaches to test structure/function relationships of heterologously expressed enzymes.

The purpose of this overview is to give an account on the use of immunochemical techniques to investigate cell biological aspects of glycosyltransferases. Such studies have been initiated by specific antibodies to β1,4galactosyltransferase (galT), the hitherto most thoroughly studied glycosyltransferase. These antibodies have become available by the end of the seventies and proved useful for the immunocytochemical localization of galT to the Golgi apparatus[1] (see Fig. 1) where it exerts its function as anticipated on the basis of activity measurements in Golgi fractions[2]. Thus an easy and straightforward procedure to track Golgi structural elements was made available and used by a variety of investigators to approach fundamental problems of the cell biology of the Golgi apparatus (see below).

GLYCOSYLTRANSFERASES AS ANTIGENS

Domain Structure of Mammalian Glycosyltransferases

Mammalian glycosyltransferases are notoriously weak antigens. Molecular cloning of homologous enzymes among different species not only led to the establishment of the domain structure of glycosyltransferases[3] which is remarkably similar throughout this class of enzymes but also pointed to their interspecies homology. Fig. 2 shows a scheme of the general domain

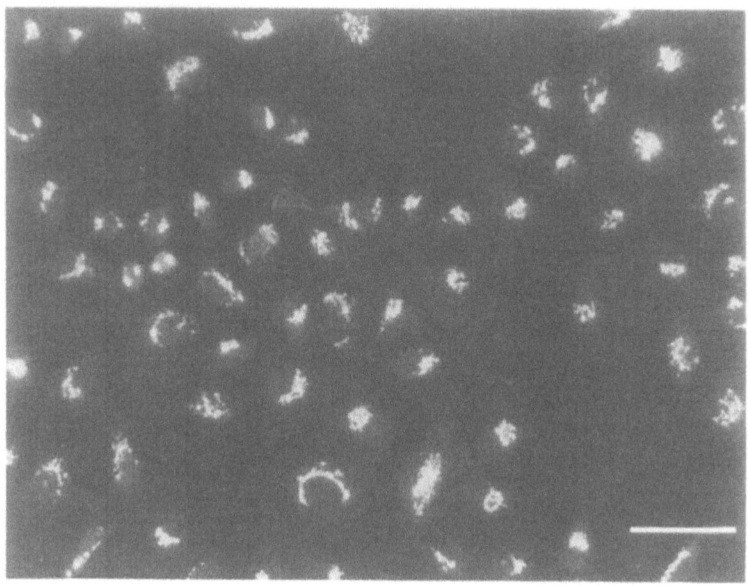

Figure 1. Immunofluorescent staining of the Golgi apparatus in HeLa cells
Polyclonal antibodies raised against human galT[1] were used to stain the Golgi apparatus in HeLa cells. This organelle appears as a concise or half-moon shaped juxtanuclear structure. Bar 50 μm.

structure of glycosyltransferases: These enzymes are monomers spanning the membrane and bearing a short cytoplasmic and a bulky luminal domain where the catalytic site is located. Adjacent to the luminal membrane side a poorly defined stem region carries the catalytic domain thought to be more peripherally located. Since no X-ray diffraction analysis of any glycosyltransferase has been reported yet, the catalytic domain can not be delineated with certainty but for some glycosyltransferases, site-directed mutagenesis has provided some clues to locate amino acids involved in catalysis. Selected examples are cited for galT[4-6], α2,6sialyl-T[7] and α1,3fucosyltransferase III[8](fucT-III). Comparison among species, however, also demonstrates that most of the substitutions are confined to the putative stem region and that little variability occurs in the distal third of the peptide. This observation allows two predictions, e.g. that the most conserved part belongs to the catalytic site which is usually least immunogenic[9] and that immunogenic epitopes are concentrated in the stem region. This assumption has been substantiated by injecting into rabbits two different truncated forms of galT, a long form (containing the 339 C-terminal amino acids) comprising the stem and the catalytic domain and a short form (containing 197 C-terminal amino acids) from which the putative stem region was removed. Only the long form was able to elicit an antiserum while practically no immunogenic response was observed with the short form[10]. Similar observations were made with α2,6sialyltransferase[11]. The species specificity of the stem region sequences also explains the high species specificity of the antisera. In our experience cross-reactivities among species are more the exception than the rule. In the case of the mouse monoclonal antibodies raised against human galT[12] only little crossreactivity with simian galT has been observed and none with any other commonly investigated species, such as mouse, rat or cow.

Native Glycosyltransferases as Antigens

In order to obtain monospecific antisera, a high degree of biochemical purification of the antigen is needed which often turns out to become a very tedious procedure. This explains why relatively few immunochemical data on glycosyltransferases have been obtained. Table 1 gives an overview on polyclonal and monoclonal antibodies that have been described to a va-

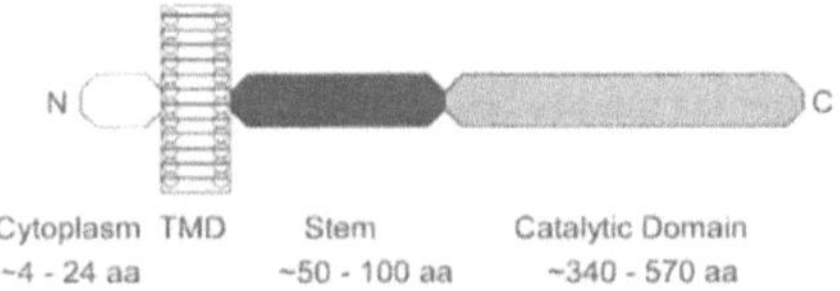

Figure 2. Common domain structure of glycosyltransferases
Glycosyltransferases are, with few exceptions, monomeric type II membrane proteins. The transition of the stem to the catalytic domain is arbitrary and is based on the more conserved structure on the C-terminal side. Dimerization may occur and limited proteolysis close to the transmembrane domain (TMD) may solubilize the enzymes *in vivo*.

riety of mammalian glycosyltransferases. It is always possible that minor contaminants present in the antigen preparation may be more immunogenic and thereby disturb the immunochemical tests. Such an observation has been made in the case of galT purified from human breast milk: this enzyme co-purifies with IgG on the two sequential affinity chromatographies on N-acetylglucosaminyl- and α-lactalbumin-Sepharose[13]; only a final purification step on a protein A column removes traces of contaminating IgG. Binding of galT to IgG has also to be considered when the nowadays popular proteinA-glycosyltransferase constructs are expressed in mammalian cells, concentrated on IgG-beads and assayed for activity[14]. These IgG beads may contain galT activity adsorbed from the medium.

Equally worrysome are antibodies directed against carbohydrate epitopes which are expressed on native glycosyltransferases. Again this aspect has been thoroughly investigated in the case of galT. When purified from human breast milk, this enzyme is extremely heterogeneous by charge[15], a fact mostly attributable to posttranslational modifications including glycosylation. These glycans carry blood-group epitopes of the corresponding donor and induce blood-group-specific antibodies which then recognize their cognate epitopes on tissue sections[16]. Conversely, protein epitopes may be shielded by glycans and render protein-specific antisera non-reactive. This was observed in the case of soluble galT expressed in *S. cerevisiae* where the enzyme was polymannosylated preventing antibodies from binding[17].

Recombinant Glycosyltransferases as Antigens

Heterologous expression of glycosyltransferases offers a versatile tool to overcome most of the drawbacks encountered with native enzymes. Purification has become an easy task by expressing glycosyltransferases tagged with his$_6$ in *E. coli*[18] or even higher eukaryotes[19] followed by purification on a Nickel-column. Other epitopes could also be used for purification although the poly-his-tag appears to be especially suitable as it seems not to be antigenic *per se*. In fact, other constructs such as fusion proteins have been used. As mentioned above galT has been fused with β-galactosidase to form the β-galactosidase-galT fusion protein which can easily be expressed in *E .coli* and purified from preparative gels by virtue of its size[10]. This strategy can be followed in two ways: Either antisera raised against the native enzyme can be purified on a fusion-protein column to obtain protein-epitope specific antibodies[20] or the fusion-protein itself can be used as an antigen. While the first approach may, in theory, lead to preparations containing some antibody species reacting with combined carbohydrate-protein epitopes, the second also contains protein-specific antibodies which are normally inaccessible for steric hindrance by glycans and/or folding. In addition, depending on the nature of the fusion protein, antibodies to the irrelevant part of the fusion protein are also present. Recent data showed that some of these antibodies appeared to react with the hinge region of the fusion protein: This evidence was based on the surprising result that antibodies to the fusion protein β-galactosidase-galT purified on a β-galactosidase column still recognized the Golgi apparatus in human cells by immunofluorescence; this signal could then be abolished by native human galT[21]. A similar phenomenon has been observed with a newly raised antiserum to a fusion protein consisting of β-galactosidase-α2,3(N)sialyltransferase[21].

Peptide Antigens Derived from Specific Portions of Glycosyltransferases

This approach has been used by Shur and associates to track the cytoplasmic tail of galT by raising an antibody to the N-terminal 13 amino acids of the long form of murine galT[22]. More recently, this approach was exploited to raise antibodies to fucTs; these enzymes comprise 5 different genetic forms (Fig. 3) of which those designated as fucT-III, fucT-V and fucT-VI are homologous (for review see ref. 23). Antibodies to the entire protein are expected to crossreact. Indeed, polyclonal antisera raised against native soluble recombinant fucT-VI reacted with fucT-V and -III (in preparation) while a polyclonal antiserum raised against a peptide occurring only in fucT-V was monospecific for fucT-V[24]. Thus, peptide specific antibodies were raised against those portions of fucT-V and -VI which had the lowest similarity, both located in the putative stem region. While a peptide antiserum against fucT-V was monospecific to fucT-V, a peptide antiserum to fucT-VI recognized recombinant FucT-VI and crossreacted strongly with an unknown native protein located in secretory granules of human endothelial cells[25], a finding which is currently further investigated. None of the peptide antibodies crossreacted with another fucT thus providing valuable tools to investigate tissue expression of these enzymes.

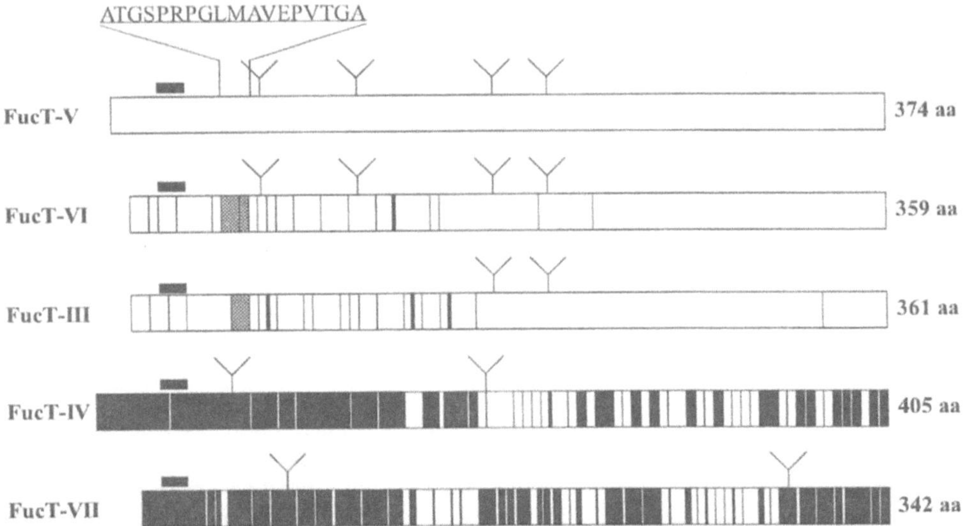

Figure 3. The α1,3fucosyltransferase family
Alignment of the 5 α1,3fucosyltransferases shows regions of homology (white), deletions (grey) and sequence differences (black) with respect to fucT-V. The transmembrane domain is indicated by a horizontal black bar. Putative N-glycosylation sites are indicated by antennaes.
This scheme exemplifies the general domain structure: On the C-terminal side (right), the sequences are most conserved and thus believed to represent the catalytic domain. Antibodies are usually generated to epitopes located on the N-terminal half. An antibody to a specific peptide of FucT-V (shown on the top) was shown to be monospecific to fucT-V[24].

APPLICATION OF ANTIBODIES FOR CELL BIOLOGICAL STUDIES ON GLYCOSYLTRANSFERASES

Immunocytochemical localization

Antibodies proved useful for two main applications, e.g. localization and trafficking of glycosyltransferases. These enzymes were commonly located to the Golgi apparatus which can be, at the light microscopical level, unequivocally defined in all cultured cells. Considering the long lasting controversy on the polymorphic appearances and even the existence of this organelle before the advent of electron microscopy when morphologists had to rely on a wealth of different staining recipes to visualize the Golgi apparatus[26], the advance due to immunocytochemical staining should not be underestimated. Novel techniques for immunoelectron microscopy were introduced at the beginning of the eighties and involved ultra thin cryosectioning or embedding of tissue in Lowicryl K40[27] which in many cases proved suitable to preserve antigenicity. Labelling of antigenic sites was then successfully done with colloidal gold complexed with a second antibody or protein A[28]. However, immunoperoxidase methods also proved suitable to localize galT[29] and N-acetylglucosaminyltransferase I[30]. The initially observed confinement of galT to the trans side of the Golgi apparatus[31] has been confirmed by various groups[29,32] and extended to other glycosyltransferases[33]. An overall scheme of the present concepts on how glycosyltransferases may be arranged across the Golgi stack in HeLa cells has recently been reviewed[34]; a summary is given on fig. 4.

The difficulty of raising antibodies to native glycosyltransferases prompted the widespread use of tags to which specific antibodies are available, such as the myc epitope. These tagged glycosyltransferases have been stably expressed in various cell lines at a level thought to approximate that of endogenously expressed transferases[33]. These cells were used to study topogenesis and mechanisms of retention of various glycosyltransferases (for review see ref. 35). A word of caution, however, is appropriate: Not only are levels of expression difficult to control in heterologous systems, but also species specificity of the transfected enzyme and the host cells were often not matched. The use of tags to identify the heterologous enzymes may also generate variable, often unpredictable results as recently shown by Storrie and associates[36].

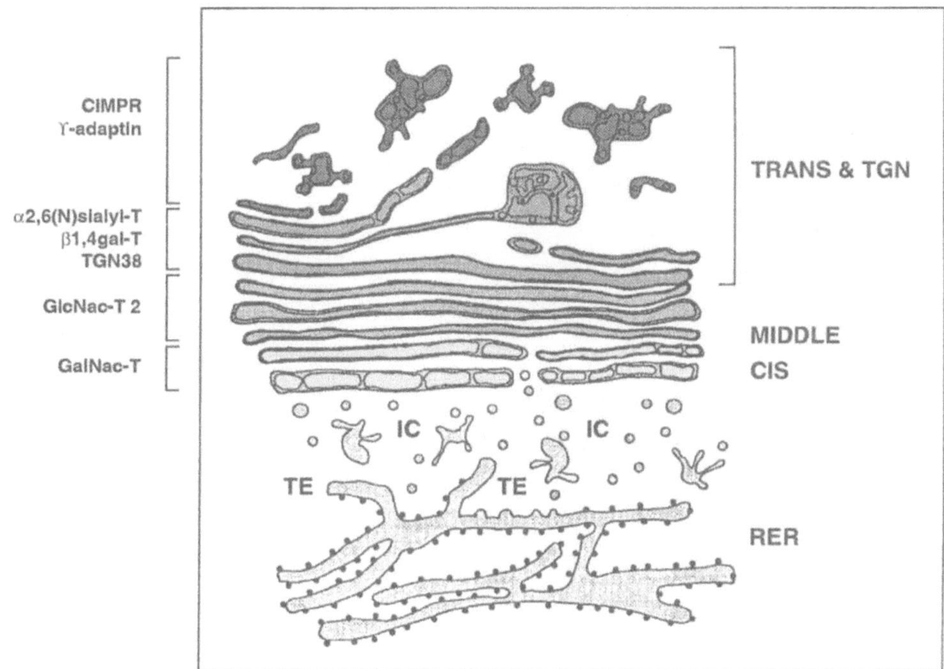

Figure 4. Current view on the compartmentalization of endogenous Golgi markers
The scheme (adapted from ref 34) relates to HeLa cells as a paradigm. There is evidence of cell-type specific differences. CIMPR (Cation-independent mannose 6 phosphate receptor), TGN (trans Golgi network), IC (intermediate compartment), TE (transitional elements), (RER) rough endoplasmic reticulum.

Flow Cytometric Analysis of Glycosyltransferases

Recently, a method was developed to assess relative amounts of glycosyltransferases in fixed and permeabilized suspension cells by flow cytometry[37]. It was found that the flow cytometric histogram originating from indirect fluorescent labeling of glycosyltransferases corresponds to a monomodal peak to which a mean fluorescence intensity value is easily assigned. Two glycosyltransferases have been investigated which co-localize by confocal laser scanning microscopy (CLSM) to the Golgi apparatus: galT and α2,6sialyltransferase. The former exhibited in a B lymphoblastoid JY cell line some surface fluorescence whereas no surface staining was revealed for the latter. This method was applied to relative quantification of galT in PMA stimulated T cells[38] which resulted in doubling of its expression. As published previously, Golgi perturbing agents such as monensin, dramatically dissociates galT from the Golgi apparatus[39] in reversible manner. We conducted similar experiments in Jurkat cells and assessed the level of galT both by flow cytometry and CLSM (Fig. 5). Golgi staining by using antibodies to galT disappeared completely whereas the flow cytometric signal remained virtually unchanged; also Golgi staining using antibodies to α2,6sialyltransferase was not affected. This approach may prove useful in investigation of B lymphocytes of patients affected by rheumatoid arthritis in which N-galactosylation of IgG appears to be impeded[40]. Since little or no galT deficiency was observed in these cells, the possibility remains that galactosylation is impeded by agents that interfere with compartmental organization of the Golgi apparatus.

Trafficking of Glycosyltransferases

These studies usually rely on metabolic labeling followed by immunoprecipitation of the labeled enzyme and its identification by SDS-polyacryamide gel electrophoresis/fluorography. The first study on intracellular transport and maturation of a glycosyltransferase concerned galT[41]. These studies described the different molecular forms of galT found during its topoge-

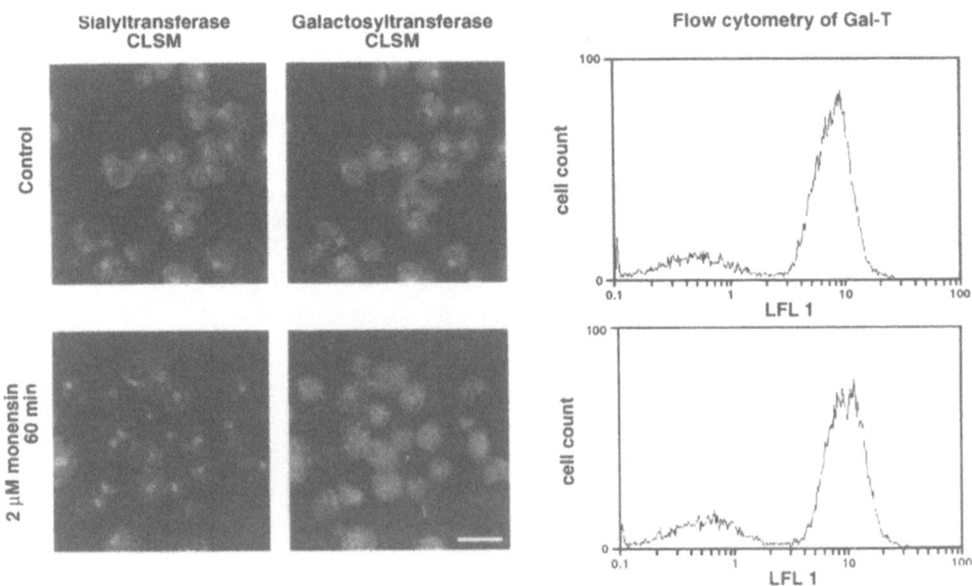

Figure 5. Immunofluorescent and flow cytometric staining of galT in Jurkat cells.
This experiment shows the previously described differential response of galT and sialyl-T to monensin treatment. While galT as visualized by confocal laser scanning microscopy (CLSM) changes from its compact juxtanuclear position to a diffuse distribution, sialyl-T remains virtually unchanged[39]. As expected, the number of antigenic sites as determined by flow cytometry remains constant.

nesis and release from the cells: Two forms were identified (45 and 47 kDa, resp.) which both matured to 54 kDa forms and which were converted to 52 kDa forms upon secretion. These studies inferred on the kinetics of intracellular transport of resident Golgi proteins: Since transport of galT from the ER to the Golgi apparatus is rapid (20 min), its presence upstream of its site of accumulation remains below the detection limit by immunofluorescence. The same holds for its release downstream of the *trans Golgi cisternae*. Thus, only galT retained at its site of steady-state accumulation can be visualized by immunocytochemical techniques. Interestingly, galT has also been shown to be phosphorylated at serine residues. These were no longer detectable on the soluble form released to the supernatant and are, thus, believed to be located at the cytoplasmic domain of the enzyme[42]. Although this finding has not been explored further, it suggests different behaviours of the two forms since the shorter form lacks serine residues on the cytoplasmic face. Indeed, treatment of HepG2 cells with uncoupling agents that lower cytoplasmic ATP levels affects part of galT as shown by immunofluorescent staining (our unpublished data).

Besides galT α2,6sialyltransferase has been investigated in a similar way in rat hepatoma cells. The main difference to galT was the fact that only one form could be identified and only 12% of the enzyme was secreted to the supernatant while the remaining enzyme was degraded intracellularly. An important difference is stability of Golgi association of this enzyme, since little dissociation is seen after treatment with monensin or uncoupling agents. Recent data of Ma and Colley revealed another interesting difference of sialyl-T with respect to galT: As suggested before, sialyl-T forms dimers at a late stage of maturation. These dimers were found to be disulfide bonded and inactive and were suggested to exert a lectin-like activity[43].

Tissue Distribution of Glycosyltransferases

Provided that mono- and protein-specific antibodies to glycosyltransferases are available, their tissue-specific expression can be investigated thus procuring complementary information to Northern blots. Such studies are available in a very limited number only; some them are referred to in Table 1. It is apparent that the presence of the products of glycosyltransferases as investigated by lectins does not always reflect the presence of the corresponding glycosyltransferase. The reasons for this discrepancy are not clear but one must bear in mind that in tissues glycan structures may occur which have not been analyzed and binding to which may reflect unknown specificities.

Tracing Golgi Structural Elements by Using Antibodies to GalT

Availability of specific reagents to galT prompted studies on basic cell biological problems concerning the Golgi apparatus. The key element was the finding that the steady-state distribution of galT in cultured cells was confined to a compact juxtanuclear location, the Golgi apparatus[1]. Its absence at other sites along the secretory pathway as discussed above (including the plasma membrane in almost all cells investigated) affords the conclusion that galT immunoreactive sites represent (at least) part of the Golgi apparatus and nothing else. The main findings using antibodies to galT may thus be summarized as follows: First, immunostaining of galT was used to track Golgi elements during mitosis: the Golgi apparatus fragments in numerous clusters of vesicles which are partitioned to the daughter cells in an equal fashion; after telophase these Golgi vesicles reform and build up cisternal stacks again[44]. Considering the differences of galT and sialyl-T in their response to Golgi disturbing agents the question arises to which extent studies on mitosis conducted with galT specific reagents must be re-evaluated using reagents to other Golgi proteins. Clearly, further studies are warranted to answer this question. Second, since the early studies on the Golgi apparatus[26] its juxtanuclear location close to the centrosome (or microtubular-organizing center, MTOC) has been observed to be a constant feature. This finding could unequivocally be confirmed by showing the close association of the MTOC with Golgi elements as visualized using antibodies to galT[45]; treatment with microtubule depolymerizing agents such as Colcemid or nocodazole[46] led to scattering of small, galT containing vesicles throughout the cytoplasm. This process was entirely reversible as Golgi stacks reassembled after removal of these agents along microtubules by dynein motors[47]. The close association of the Golgi apparatus with the MTOC was also shown in a model of cell differentiation, i.e. the formation of myotubes from myoblasts[29]. Third, another drug which provided further clues to Golgi biogenesis and vesicular traffic is brefeldin A[48]. Elucidation of its function was also in part based on visualization of the microtubule-guided back-flow of Golgi structural elements stained by antibodies to galT[49].

PROSPECTS

Antibodies to glycosyltransferases will exert a useful complementary function to Northern blots since they will delineate subcellular location of the respective enzyme at the level of single cells and cell-type specific expression by immunohistochemical localizations. In order to meet the high requirements of the antibodies in terms of affinity and specificity, they need to be thoroughly characterized. Antibodies should be protein-specific to epitopes which do not crossreact with related enzymes. Whether these requirements are best fulfilled by monoclonal or polyclonal antibodies to selected peptides must be decided on a case by case basis. In addition, these antibodies may also serve a useful function in diagnosis and investigation of carbohydrate-deficient glycoprotein syndromes. These clinical entities rely on defined glycosylation deficiencies, some of them attributable to specific glycosyltransferases as exemplified by the deficiency of N-acetylglucosaminyltransferase II[50].

ACKNOWLEDGMENTS

This work was supported in part by grants of the Swiss National Science Foundation to EGB.

REFERENCES

1. E.G. Berger, T. Mandel, and U. Schilt, Immunohistochemical localization of galactosyltransferase in human fibroblasts and HeLa cells, *J. Histochem. Cytochem.* 29:364 (1981).
2. R. Bretz, H. Bretz, and G.E. Palade, Distribution of terminal glycosyltransferases in hepatic Golgi fractions, *J. Cell Biol.* 84:87 (1980).
3. J.C. Paulson, and K.J. Colley, Glycosyltransferases. Structure, localization, and control of cell type-specific glycosylation, *J. Biol.Chem.* 264:17615 (1989).
4. Y. Wang, S.S. Wong, M.N. Fukuda, H.Y. Zu, Z.D. Liu, Q.S. Tang, and H.E. Appert, Identification of functional cysteine residues in human galactosyltransferase, *Biochem. Biophys. Res. Commun.* 204:701 (1994).
5. H.Y Zu,. M.N. Fukuda, S.S. Wong, Y. Wang, Z.D. Liu, Q.S. Tang, and H.E. Appert, Use of site-directed mutagenesis to identify the galactosyltransferase binding sites for UDP-galactose, Biochem. Biophys. Res. Commun. 206:362 (1995).
6. E.E. Boeggeman, P.V. Balaji, and P.K. Qasba, Functional domains of bovine beta- galactosyltransferase, *Glycoconjugate J.* 12:865 (1995).
7. A.K. Datta, and J.C. Paulson, The sialyltransferase "sialylmotif" participates in binding the donor substrate CMP-neuac, *J. Biol. Chem.* 270, 1497 (1995).
8. E.H. Holmes, Z.H Xu, A.L. Sherwood, and B.A. Macher, Structure-function analysis of human alpha 1->3fucosyltransferases - a GDP-fucose-protected, n-ethylmaleimide-sensitive site in fuct-III and fuct-V corresponds to ser(178) in fuct-IV, *J. Biol. Chem.* 270:8145 (1995).

9. B. Cinader, Enzyme-antibody interactions , *in* Methods in Immunology Immunochemistry IV, C.A. Williams, M.W. Chase, eds, Academic Press , New York, 313 (1977).

10. G. Watzele, R. Bachofner, and E.G. Berger, Immunocytochemical localization of the Golgi apparatus using protein-specific antibodies to galactosyltransferase, *Eur. J. Cell Biol.* 56:451 (1991).

11. K. Grimm, *Thesis,* University of Zurich, (1995).

12. E.G. Berger, E. Aegerter, T. Mandel , and H.-P. Hauri, Monoclonal antibodies to soluble, human milk galactosyltransferase (lactose synthase A protein), *Carbohyd. Res.* 149:23 (1986).

13. J.R. Wilson, M.M. Weiser, B. Albini, J.R. Schenck, H.G. Rittenhouse, A.A. Hirata, and E.G. Berger, Co-purification of soluble human galactosyltransferase and immunoglobulins, *Biochim. biophys. Res. Commun.* 105:737 (1982).

14. R.D. Larsen, V.P. Rajan, M.M. Ruff, J. Kukowska-Latallo, R.D. Cummings, and J.B. Lowe, Isolation of a cDNA encoding a murine UDPgalactose: beta-D-galactosyl--N-acetyl-D-glucosaminide alpha-1,3-galactosyltransferase expression cloning by gene transfer, *Proc. Natl Acad. Sci. USA* 86:8227 (1989).

15. A.C. Gerber, I. Kozdrowski, S.R. Wyss, and E.G. Berger, The charge heterogeneity of soluble human galactosyltransferases isolated from milk, amniotic fluid and malignant ascites, *Eur. J. Biochem.* 93:453 (1979).

16. R.A. Childs, E.G. Berger, S.J. Thorpe, E. Aegerter, and T. Feizi, Blood-group-related carbohydrate antigens are expressed on human milk galactosyltransferase and are immunogenic in rabbits, *Biochem. J.* 23:605 (1986).

17. M. Malissard, L. Borsig, S. DiMarco, M.G. Grütter, U. Kragl, C. Wandrey, and E.G. Berger, Recombinant soluble beta-1, 4-galactosyltransferases expressed in *Saccharomyces cerevisiae* - Purification, characterization and comparison with human enzyme, *Eur. J. Biochem.* 239:340 (1996).

18. T.T.M. Nguyen, D.A. Hinton, and B.D. Shur, Expressing murine beta 1, 4-galactosyltransferase in HeLa cells produces a cell surface galactosyltransferase-dependent phenotype, *J. Biol. Chem.* 269: 28000 (1994).

19. M. Malissard, L. Borsig, E.G. Berger, in preparation.

20. D.J. Taatjes, J. Roth, J. Weinstein, and J.C. Paulson, Post-Golgi apparatus localization and regional expression of rat intestinal sialyltransferase detected by immunoelectron microscopy with polypeptide epitope-purified antibody, *J. Biol. Chem.* 263:6302 (1988).

21. P.C. Burger, M. Lötscher, R. Kleene, B. Kaissling, and E.G. Berger, in preparation.

22. A. Youakim, D.H. Dubois, and B.D. Shur, Localization of the long form of beta-1, 4-galactosyltransferase to the plasma membrane and Golgi complex of 3T3 and F9 cells by immunofluorescence confocal microscopy, *Proc. Natl Acad. Sci. USA* 91:10913 (1994).

23. R. Mollicone, A. Cailleau, and R. Oriol, Molecular genetics of H, Se, Lewis and other fucosyltransferase genes, *Transfus. Clin. Biol.* 2:235 (1995).

24. L. Borsig, R. Kleene, A. Dinter, and E.G. Berger, Immunodetection of alpha 1-3 fucosyltransferase (FucT-V), *Eur. J. Cell Biol.* 70:42 (1996).

25. L. Borsig, E.G. Berger, and R. Moser, Immunodetection of FucT-VI in transfected COS cells and detection of a crossreactive antigen in Weibel-Palade bodies in human umbilical vein endothelial cells (HUVECs), USGEB, Experientia 52, Abstract (1996).

26. G.C Hirsch, Form- und Stoffwechsel der Golgi-Körper. Protoplasma-Monographien Vol. 18. Gebrüder Borntraeger, Berlin. (1939).

27. E. Carlemalm, W. Villiger, J.A. Hobot, J.D. Acetarin, and E. Kellenberger, Low temperature embedding with Lowicryl resins: two new formulations and some applications, *J. Microsc.* 140:55 (1985).

28. J. Roth, Light and electron microscopic localization of antigenic sites in tissue sections by the protein A-gold technique. *Acta Histochem. Suppl.* 29:9 (1984).

29. A. M. Tassin, M. Paintrand, E.G. Berger, and M. Bornens, The Golgi apparatus remains associated with microtubule organizing centers during myogenesis, *J. Cell Biol.* 101:630 (1985).

30. W.G. Dunphy, R. Brands, and J.E. Rothman, Attachment of terminal N-acetylglucosamine to asparagine-linked oligosaccharides occurs in central cisternae of the Golgi stack, *Cell,* 40:463 (1985).

31. J. Roth, and E.G. Berger, Immunocytochemical localization of galactosyltransferase in HeLa cells: Co-distribution with thiamine pyrophosphatase in trans-Golgi cisternae, *J. Cell Biol.* 92:223 (1982).

32. J.W. Slot, and H.J. Geuze, Immunoelectron microscopic exploration of the Golgi complex, *J. Histochem. Cytochem.* 31:1049 (1983).

33. C. Rabouille, N. Hui, F. Hunte, R. Kieckbusch, E.G. Berger, G. Warren, and T. Nilsson, Mapping the distribution of Golgi enzymes involved in the construction of complex oligosaccharides, *J. Cell Sci.* 108:1617 (1995).

34. J. Roth, Compartmentation of glycoprotein biosynthesis, *in* Glycoproteins, J. Montreuil, H. Schachter, and J.F.G. Vliegenthart, eds, Amsterdam, Elsevier, 287 (1995).

35. R. Kleene, E.G. Berger, The molecular and cell biology of glycosyltransferases, *Biochim. Biophys. Acta* 1154:283 (1993).

36. W. Yang, R. Pepperkok, P. Bender, T.E. Kreis, and B. Storrie, Modification of the cytoplasmic domain affects the subcellular localization of Golgi glycosyltransferases, *Eur. J. Cell Biol.* 71:53 (1996).

37. K. Mrkoci-Felner, E. Niederer, and E.G. Berger, Flow cytometric detection of the Golgi apparatus using antibodies to glycosyltransferases, submitted (1996).

38. S. Lemaire, C. Derappe, V. Pasqualetto, K.Mrkoci, E.G. Berger, M. Aubéry, and D. Néel, Human T lymphocyte activation is associated with an increase of the βgalactosyltransferase, submitted (1996).

39. E.G. Berger, K. Grimm, T. Bächi, H. Bosshart, R. Kleene, and M. Watzele, Double immunofluorescent staining of alpha 2, 6 sialyltransferase and beta 1, 4 galactosyltransferase in monensin-treated cells: evidence for different Golgi compartments?, *J. Cell Biochem.* 52:275 (1993).

40. A. Alavi, and J. Axford, Evaluation of beta -galactosyltransferase in rheumatoid arthritis and its role in the glycosylation network associated with this disease. *Glycoconjugate J.* 12:206 (1995).

41. G.J.A.M. Strous, and E.G. Berger, Biosynthesis, intracellular transport, and release of the Golgi enzyme galactosyltransferase (lactose synthetase A protein) in HeLa cells, *J. Biol. Chem.* 257:7623 (1982).

42. G.J. Strous, P. van Kerkhof, J. Fallon, and A.L. Schwartz, Golgi galactosyltransferase contains serine-linked phosphate, *Eur. J. Biochem.* 69:07 (1987).

43. J.Y. Ma, and K.J. Colley, A disulfide-bonded dimer of the Golgi beta-galactoside alpha 2, 6-sialyltransferase is catalytically inactive yet still retains the ability to bind galactose, *J. Biol. Chem.* 271:7758 (1996).

44. G. Warren, Membrane partitioning during cell division, *Annu. Rev. Biochem.* 62:323 (1993).

45. J. Wehland, M. Henkart, R. Klausner, and I.V. Sandoval, Role of microtubules in the distribution of the Golgi apparatus: effect of taxol and microinjected anti-alpha-tubulin antibodies, *Proc. Natl Acad. Sci. U S A* 80:4286 (1983).

46 W.C. Ho, V.J. Allan, G. Vanmeer, E.G. Berger, and T.E. Kreis, Reclustering of scattered Golgi elements occurs along microtubules, *Eur. J. Cell Biol.* 48:250 (1989).

47 T. Kreis, Role of microtubules in the organisation of the Golgi apparatus, *Cell Motility Cytoskel.* 15:67 (1990).

48. J. Lippincott-Schwartz, J.G. Donaldson, A. Schweizer, E.G. Berger, H.P. Hauri, L.C. Yuan, and R.D. Klausner, Microtubule-dependent retrograde transport of proteins into the ER in the presence of Brefeldin A suggests an ER recycling pathway, *Cell* 60:821 (1990).

49. G.J. Strous, E.G. Berger, P. van Kerkhof, H. Bosshart, B. Berger, and H.J. Geuze, Brefeldin A induces a microtubule-dependent fusion of galactosyltransferase-containing vesicles with the rough endoplasmic reticulum, *Biol. Cell* 71:25 (1991).
50. J. Jaeken, H. Schachter, H. Carchon, P. Decock, B. Coddeville, and G. Spik, Carbohydrate deficient: glycoprotein syndrome type II: a deficiency in Golgi localised n-acetylglucosaminyltransferase II, *Arch. Dis. Child.* 71:123 (1994).
51. J. Weinstein, U. de Souza-e-Silva, and J.C. Paulson, Purification of a Galbeta1-4GlcNAc alpha2-6 sialyltransferase and a Galbeta1-3(4)GlcNAc alpha2-3 sialyltransferase to homogeneity from rat liver, *J. Biol. Chem.* 257:13835 (1982).
52. H. Bosshart, and E.G. Berger, Biosynthesis and intracellular transport of alpha-2,6-sialyltransferase in rat hepatoma cells, *Eur. J. Biochem.* 208:341 (1992).
53. L.J. Melkerson-Watson, and C.C., Sweeley, Purification to apparent homogeneity by immunoaffinity chromatography and partial characterization of the GM3 ganglioside-forming enzyme, CMP-sialic acid-lactosylceramide alpha2.3-sialyltransferase (SAT-1), from rat liver Golgi, *J. Biol. Chem.* 266:4448 (1991).
54. N.L. Shaper, P.L. Mann, and J.H. Shaper, Cell surface galactosyltransferase: immunochemical localization, *J. Cell. Biochem.* 28:229 (1985).
55. J.T. Ulrich, J.R. Schenck, H.G. Rittenhouse, N.L. Shaper, and J.L. Shaper, Monoclonal antibodies to bovine UDP-galactosyltransferase. Characterization, cross-reactivity, and utilization as structural probes, *J. Biol. Chem.* 261:7975 (1986).
56. T. Suganuma, Subcellular localization of N-acetylglucosaminide beta1-4 galactosyltransferase revealed by immunoelectron microscopy - Authors' Response *J. Histochem. Cytochem.* 39:1440, (1991).
57. M. Uemura, T. Sakaguchi, T. Uejima, S. Nozawa, and H. Narimatsu, Mouse monoclonal antibodies which recognize a human (beta1-4)galactosyl-transferase associated with tumor in body fluids, *Cancer Res.* 52:6153 (1992).
58. J.I. Kawano, S. Ide, T. Oinuma, and T.T. Suganuma, A protein-specific monoclonal antibody to rat liver beta 1->4 galactosyltransferase and its application to immunohistochemistry, *J. Histochem. Cytochem.* 42:363 (1994).
59. H.J. Hathaway, R.B. Runyan, S. Khounlo, and B.D. Shur, Purification and characterization of avian betagalactosyltransferase: comparison with the mammalian enzyme, *Glycobiol.* 1:211 (1991).
60. S. Chatterjee, N. Ghosh, and S. Khurana, Purification of uridine diphosphate-galactose-glucosyl ceramide, beta1-4 galactosyltransferase from human kidney, *J. Biol. Chem.* 267:7148 (1992).
61. P.L. Smith, K.M. Gersten, B. Petryniak, R.J. Kelly, C. Rogers, Y. Natsuka, J.A. Alford, E.P. Scheidegger, S. Natsuka, and J.B. Lowe, Expression of the alpha(1,3)fucosyltransferase FucT-VII in lymphoid aggregate high endothelial venules correlates with expression of L-selectin ligands, *J. Biol. Chem.* 271:8250 (1996).
62. H. Kimura, T. Kudo, S. Nishihara, H. Iwasaki, N. Shinya, R. Watanabe, H. Honda, F. Takemura, and H.H Narimatsu, Murine monoclonal antibody recognizing human alpha(1,3/)fucosyltransferase, *Glycoconjugate J.* 126:802 (1995).
63. S. Yazawa, and K. Furukawa, Immunochemical properties of human plasma alpha1-2 fucosyltransferase specified by blood group H-gene, *J. Immunogenet.* 10:349 (1983).
64. T. White, U. Mandel, T.F. Orntoft, E. Dabelsteen, J. Karkov, M. Kubeja, S-I. Hakomori, and H. Clausen, Murine monoclonal antibodies directed to the human histo-blood group A transferase (UDP-GalNAc:Fuc alpha 1-2Gal alpha 1-3 N-acetylgalactosaminyltransferase) and the presence therein of N-linked histo-blood group A determinants, *Biochemistry* 29:2740 (1990).
65. J. Roth, P. Greenwell, and W.M. Watkins, Immunolocalization of blood group A gene specified alpha1,3N-acetylgalactosaminyltransferase and blood group A substance in the trans-tubular network of the Golgi apparatus and mucus of intestinal goblet cells, *Eur. J. Cell Biol.* 46:105 (1988).

66. J. Balsamo, R.S. Pratt, M.R. Emmerling, G.B. Grunwald, and J. Lilien, Identification of the chick neural retina cell surface N-acetylgalactosaminyltransferase using monoclonal antibodies, *J. Cell. Biochem.* 32:125 (1986).
67. F.L. Homa, T. Hollander, D.J. Lehman, D.R. Thomsen, and A.P. Elhammer, Isolation and expression of a cDNA clone encoding a bovine UDP-galNAc-polypeptide n-acetylgalactosaminyltransferase, *J. Biol. Chem.* 268:12609 (1993).
68. Y. Wang, J.L. Abernethy, A.E. Eckhardt, and R.L. Hill, Purification and characterization of a UDP-GalNAc-polypeptide N-acetylgalactosaminyltransferase specific for glycosylation of threonine residues, *J. Biol. Chem.,* 267:12709-12716, (1992).
69. J. Burke, J.M. Pettitt, D. Humphris, and P.A. Gleeson, Medial-Golgi retention of n-acetylglucosaminyltransferase I - contribution from all domains of the enzyme, *J. Biol. Chem.* 269:12049 (1994).

Antibodies to glycosyltransferases

Glycosyltransferase	E.C.	Species Tissue	Type of antibody	Ref
α2,6(N)sialyltransferase	2.4.99.1	rat liver	rabbit polyclonal	51
α2,6(N)sialyltransferase	2.4.99.1	rat liver	rabbit polyclonal	52
α2,6(N)sialyltransferase	2.4.99.1	human recomb. fusion protein with β-galactosidase	rabbit polyclonal	39
α2,3(N)sialyltransferase	2.4.99.6	human	rabbit polyclonal	in preparation
α2,3 sialyltransferase, (GM3 synthase)	2.4.99.9	Rat liver	mAB M12GC 7	53
β galactosyltransferase	2.4.1.22	human milk	rabbit polyclonal	1
β galactosyltransferase	2.4.1.22	human recombinant fusion protein with β-galactosidase	rabbit polyclonal	10
β galactosyltransferase	2.4.1.22	human milk	mAB GT2/36/118 (and others)	12
β galactosyltransferase	2.4.1.22	bovine milk	rabbit polyclonal	54
β galactosyltransferase	2.4.1.38		mAB DH162	55
β galactosyltransferase	2.4.1.38	mouse F9 embryonal carcinoma cells	mAB GTF4	56
β galactosyltransferase	2.4.1.38	human ovarian cancer effusion fluid	mAb's (several)	57
β galactosyltransferase	2.4.1.38	rat liver microsomes	mAB GTL2	58
β galactosyltransferase	2.4.1.38	murine-his$_6$-tagged fusion protein	rabbit polyclonal	18
β galactosyltransferase	2.4.1.38	N-terminal peptide of long form of murine galT	rabbit polyclonal	22
β galactosyltransferase	2.4.1.38	chicken serum	rabbit polyclonal	59
Lactosylceramide synthetase UDP Galactose: Glucosyl Ceramide β1-4 galactosyltransferase	E.C. 2.4.1.?	human kidney	rabbit polyclonal	60

α1,3 fucosyltransferase V	E.C. 2.4.1.152?	Peptide of stem region (human enzyme)	rabbit polyclonal	24
		human recombinant fusion protein with β-galactosidase		24
α1,3 fucosyltransferase VI	E.C. 2.4.1.152?	human recombinant enzyme expressed in CHO cells	rabbit polyclonal	in preparation
α1,3 fucosyltransferase VII	E.C. 2.4.1.152?	murine recombinant enzyme expressed in *E.coli* as fusion protein with β-galactosidase	rabbit polyclonal	61
α1,3 fucosyltransferase III	E.C. 2.4.1.65	human recombinant expressed in E. coli	mAB FTA 1-16	62
α1,2 fucosyltransferase Blood group H specific	E.C. 2.4.1.69	human plasma	rabbit polyclonal	63
N-acetylgalactosaminyltransferase Blood group A specific	E.C.2.4.1.40	human lung enzyme	mAB WKH-1	64
N-acetylgalactosaminyltransferase Blood group A specific	E.C.2.4.1.40	human plasma	rabbit polyclonal	65
N-acetylgalactosaminyltransferase, specificity not determined	E.C.	chick neural retinal cells	6H5	66
N-acetylgalactosaminyltransferase, polypeptide specific	E.C. 2.4.1.41	Bovine colostrum	chicken polyclonal	67
N-acetylgalactosaminyltransferase, polypeptide specific	E.C. 2.4.1.41	Porcine submaxillary glands	rabbit polyclonal	68
N-acetylglucosaminyltransferase I	E.C. 2.4.1.101	rabbit liver	mAB T1-1C4 and T1-2C1	30
N-acetylglucosaminyltransferase I	E.C. 2.4.1.101	rabbit, recombinant luminal domain expressed in *E. coli* as fusion protein with glutathione S-transferase	sheep polyclonal	69

CYTOKINE AND PROTEASE GLYCOSYLATION AS A REGULATORY MECHANISM IN INFLAMMATION AND AUTOIMMUNITY

Philippe Van den Steen[1], Pauline M. Rudd[2], Raymond A. Dwek[2], Jo Van Damme[1] and Ghislain Opdenakker[1,2,3]

[1]The Rega Institute, Laboratory of Molecular Immunology, University of Leuven, Minderbroedersstraat 10, B-3000 Leuven, Belgium
[2]The Glycobiology Institute, Department of Biochemistry, University of Oxford, South Parks Road, OX1 3QU, Oxford, U.K.

SUMMARY

Cytokines are locally produced hormones that alert the innate and specific immune systems. Many cytokines induce, enhance and govern the traffic of leukocytes. An important mechanism in cell trafficking and migration through endothelial basement membranes and connective tissues is the cytokine-regulated production of matrix degrading proteases. The latter include the serine proteinases of plasminogen activation and metalloproteinases such as collagenases, stromelysins and gelatinases. Many cytokines and all known matrix proteinases are glycoproteins and thus occur as sets of glycoforms. The relation between structures and functions of these glycoproteins has already been probed extensively at the protein level but not yet at the carbohydrate level. Attached oligosaccharides target the cytokines and proteinases to specific cellular receptors and matrix binding sites. In addition, a number of cytokines possess lectin-like functions and may thus interact with carbohydrates of the host or parasites. These intermolecular interactions influence for instance the compartmentalisation, the cell- and tissue-specific distribution and the pharmacokinetics of cytokines and proteinases. Attempts were done to deduce structure-function rules for the intramolecular effects of carbohydrates on cytokines and matrix proteinases. The relatively voluminous N-linked sugars downmodulate the specific activities of enzymes and cytokines. Because in host stress reactions (infection, inflammation, trauma) N-linked glycosylation is less efficient, glycosylation may constitute an important regulatory mechanism in the cytokine network and in multi-enzyme cascades.

1. INTRODUCTION

When the natural defence barriers of the host are destroyed by viruses, bacteria or other etiologic agents, e.g. trauma, cytokines are locally produced and alert the immune system. Some cytokines possess direct antimicrobial activities or induce other cytokines

[3]correspondence: Rega Institute, Leuven; FAX: 32(0)16 33 73 40

which assist in the elimination of, for instance, viruses and bacteria. Interleukin-1β (IL-1β) induces in connective tissue cells the local production of other cytokines including interferon-β (IFN-β) and chemotactic factors. When IL-1 enters the circulation, it results in endocrine effects such as fever, alterations in blood cell counts and acute phase reactants and the induction of colony stimulating factors (CSF) with hematopoietic effects. The interferons have direct antiviral effects; the chemokines for granulocytes (IL-8 and other CXC-chemokines) and for mononuclear cells (MCPs and other CC-chemokines) recruit phagocytes which destroy the infectious agents and which function as antigen presenting cells (APCs) for the specific arm of immunity (Figure 1). As a result, a ***network of cytokines*** with inductive and amplifying regulatory interactions is generated (Opdenakker et al., 1989). In addition to direct stressing effects on host cells (e.g. fever by IL-1, IL-6 and interferons), cytokines activate the neuro-endocrine system which results in stress neurotransmitters and stress hormones and altered cell metabolism.

Many naturally occurring cytokines are glycoproteins. Information on the primary protein structure of cytokines has mainly been obtained by cDNA-cloning and by tedious glycoprotein purification and sequencing experiments. The glycoform nature of natural cytokines has not yet been addressed in depth; most structural data on the carbohydrates

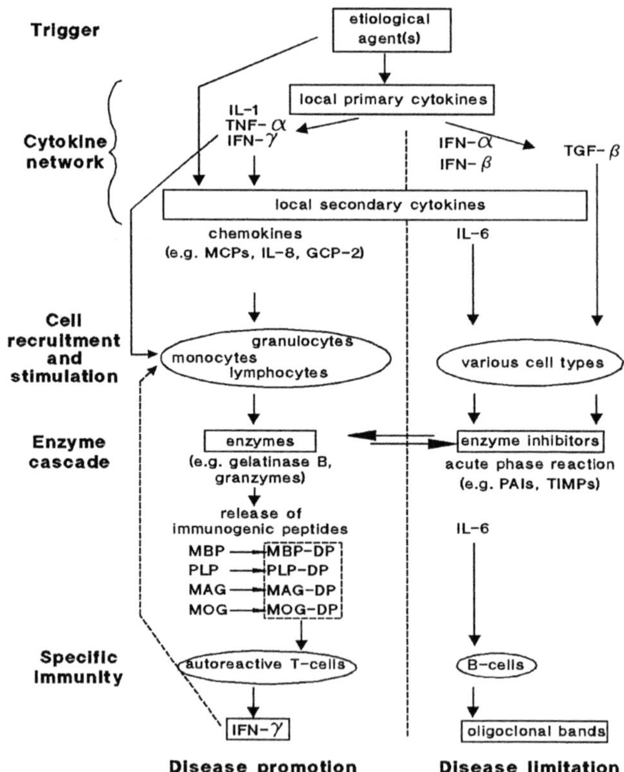

Figure 1. Cytokines and proteases in inflammation and autoimmunity
The etiological agent induces a network of interacting cytokines and a cascade of proteases which are regulated at the transcriptional level, by secretion, by activation and by inhibitors. The protease balance determines whether matrix components are proteolysed into degradation products (DP). These DP might directly or after processing by APCs lead to antigen-specific activation of T lymphocytes. IFN-γ, thus produced, sustains a positive feedback loop. Interference at several levels of this model is possible and applicable in the treatment of autoimmune diseases and pathologies with excess tissue damage. Molecules are indicated in boxes; cells are shown as ellipses. Modified from Opdenakker and Van Damme (1994).

of cytokines were obtained from recombinant glycoform populations (expressed in mammalian cells or other glycosylating heterologous systems). Because of the difficulties for the production and purification and hence the scarcity of (human) natural cytokines only limited data exist of natural glycoforms. No complete site-specific carbohydrate sequence analysis has been accomplished on any natural cytokine.

As any other glycoprotein, cytokines or proteases can have one or more glycosylation sites. The N-linked sites are anchored to the attachment sequon Asn-Xaa-Ser/Thr, the O-linked sugars are attached to serine or threonine residues. Any site can be occupied or not. However, when the intermediate residue (Xaa) is a proline in the sequon for asparagine-linked glycosylation, this site is not occupied. Beyond the multiplicity of sites and their possible occupancy, the extensive enzymatic glycosylation machinery of a cell (the Golgi apparatus and the endoplasmic reticulum constitute about one fourth of the cytoplasm) will usually deliver different sugars at each site. This leads to size-, and charge-heterogeneity of the glycoproteins. Cytokines and proteases thus always occur as populations of glycoforms (Rademacher et al., 1988). It is clear, however, that the site-specific glycoform population is not random. It seems that protein information, different from the sequon, contributes to the signalling information and determines which and how glycoforms can be modified by the glycosylating enzymes.

One of the focuses of our research is the regulation of extracellular proteolysis by cytokines in inflammation, cancer and autoimmune diseases (Masure and Opdenakker, 1989; Opdenakker and Van Damme, 1992; 1994). Inflammation-promoting cytokines (e.g. tumor necrosis factor-α, TNF-α; IL-1; cell recruiting chemokines, IL-12) induce the production of matrix degrading proteolytic enzymes. Disease-limiting cytokines (e.g. transforming growth factor-β, TGF-β; IL-6; IL-10) induce protease inhibitors. This results in a ***balance*** between enzymes and inhibitors which is defined as the ***protease load***. Furthermore, matrix constituents are synthesised by fibroblasts under the control of cytokines. Most of the matrix-degrading enzymes are produced by cells as inactive precursors (zymogens or pro-enzymes) and are activated by proteolysis by other matrix enzymes. Completion of matrix degradation depends on a series of proteases, a protease cascade. This situation is similar to the enzyme cascade of blood clotting. All enzymes of the extracellular matrix degradation cascade, identified to date, are glycoproteins. The protease load is thus dependent on at least four levels of regulation all of which are under the control of cytokines: transcriptional regulation, secretion, activation and inhibition (Figure 1)(Opdenakker and Van Damme, 1992; 1994). The fifth level of control, glycosylation will be discussed below.

One of the effects of the cytokine network and the matrix protease cascade is degradation of matrix components such as connective tissue collagen and myelin in the central nervous system. This degradation allows the chemoattracted leukocytes to reach the inflammatory focus and makes it for malignant cells possible to invade into tissues. In the case of a chronic inflammation or autoimmune process matrix components are constantly degraded into small peptides (with the same sequence but perhaps another conformation). The cytokine-regulated production of matrix proteases thus generates remnant epitopes which may be at the basis of autoimmunity (Opdenakker and Van Damme, 1994) (Figure 1). The "Remnant Epitope Generates Autoimmunity"- or REGA-model has helped to understand the early events in tissue-specific autoimmune diseases such as rheumatoid arthritis and multiple sclerosis and has been predictive for the development of autoimmune disease treatment strategies: the direct use of disease-limiting cytokines (e.g. IFN-β in multiple sclerosis), antagonists of disease-promoting cytokines (e.g. monoclonal antibodies against TNF-α in rheumatoid arthritis) and matrix protease inhibitors (e.g. hydroxamates, D-penicillamine, tetracyclines and novel matrix protease inhibitors are effective in various animal models for multiple sclerosis). Another aspect, which is not addressed here but which is relevant for the development of specific (auto)immunity, is the observation that many of the remnant epitopes, presented and processed by the APCs, are glycopeptides. For instance, it has been demonstrated that, in the immunodominant domain of type II collagen, the presence of the OH-lysine-coupled carbohydrate is essential for the T cell response (Michaëlsson et al., 1994).

2. CYTOKINE GLYCOSYLATION

Functional differences between recombinant or naturally glycosylated and enzymatically deglycosylated or bacterially expressed cytokines have recently been reviewed (Opdenakker et al., 1995). Two general conclusions can be drawn. First, in early studies and in most cases where only one cytokine function was studied no difference in biological activity between glycosylated and aglycosyl cytokine was found. In later studies and in those cases, where multiple functions were tested for one particular cytokine, usually altered specific activities were addressed to differently glycosylated forms of the (glyco)protein. Second, the rather voluminous N-linked carbohydrates have an attenuating effect on the biological functions leading to lower specific activities. The effect of O-linked sugars on cytokine activities may be enhancing or downregulating.

In most cases, the influence of glycosylation on specific activity is not spectacular: usually a difference of 2- or 3-fold between the glycosylated and aglycosyl forms. These obvious small differences become, however, significant when they are amplified in a cascade (see also the next section on enzymes). Indeed a two-fold increase in a cascade chain with n elements leads to a 2^n-fold amplification. It is also relevant to notice here that many cytokines possess lectin-like domains which might function in multimerisation and in the generation of high local concentrations by binding to glycosylated extracellular matrix molecules (George, 1994; Opdenakker et al., 1995). Typical examples of cytokines with lectin activity are tumour necrosis factor-α (Lucas et al., 1994) and interleukin-2 (Sherblom et al., 1989).

The ideal substrates to study the function(s) of attached oligosaccharides are individual glycoforms. These were, however, only recently obtained from the relatively simple natural glyco-enzyme, bovine pancreas ribonuclease (Rudd et al., 1994). Cytokines (and proteases) have only been studied as glycoform mixtures, in most cases they were expressed in eukaryotic cells and were quite heterogenous. The heterogeneity (by protein staining visible as doublets, multimers or even a (glyco)protein ladder) (Opdenakker et al., 1995) might be due to an altered peptide backbone or to changes in the glycosylation (Dijkmans et al., 1987). For an example of the carbohydrate structure of recombinant IFN-γ and a compositional analysis of the oligosaccharides of natural IL-6 the reader is referred to the studies by Mutsaers et al. (1986) and Parekh et al. (1992), respectively.

The workload of such studies is high and an additional complication is that the glycoform structure needs to be investigated in each particular case. Indeed, glycosylation patterns are cell-specific. This and the observation that cell culture conditions (e.g. the presence of cytokines) also change the spectrum of glycoforms produced (Pos et al., 1990), increase the complexity of structural analysis (Rademacher et al., 1988).

Finally, not all cytokines are N-glycosylated. It is remarkable that the cytokines without N-linked glycosylation, for instance the α-interferons and the CXC-chemokines, are derived from clustered multi-gene families. It is unusual to observe that the spectral molecular character of cytokines with N-linked sugars, is "compensated" in these latter cases by a spectrum of genetically determined protein variants (Opdenakker et al., 1995) and by some variation through (less heterogenous) O-linked sugars (Adolf et al., 1991).

2.1. N-linked glycosylation attenuates cytokine activity

Table 1 summarises the functional differences between glycoforms of cytokines with N-linked glycosylation. Differences in specific activities were observed for the cytokines IL-6 (Tanner et al., 1990), GM-CSF (Cebon et al., 1990), IL-4 (Thor an Brian., 1992) and IL-1ß (Livi et al., 1991; Fleer et al., 1991). In all these instances the presence of carbohydrates lowered the activities. One of the best examples is IL-1β (Livi et al., 1990, 1991). It indicates the extent to which carbohydrates can attenuate the function (at

Table 1. Effect of glycosylation on cytokine activities

Name	Form	Species	Glycosylation	Assay	Effect on activity	Reference
IL-1β	rec. (yeast)	human	N	Thymocyte proliferation and IL-2 production	↓ sixfold	Livi et al., 1991
GM-CSF	natural (lymfocytes)	human	N	Thymidine incorporation, superoxide anion production, receptor-affinity	↓ sixfold	Cebon et al., 1990
IL-3	natural (lymfocytes)	murine	N	Thymidine incorporation	No effect	Ziltener, 1993
IL-4	natural (D10) (splenocytes)	murine	N	Proliferation of NK-cells	↓ > twofold	Thor et al., 1992
IL-5	rec. (Sf9)	murine	N	Lymphoid cell Thymidine incorporation	Not essential for activity	Kunimoto et al., 1991
IL-5	rec. (Xenopus)	human	N	Receptor binding, B-cell growth and differentiation	Not essential for activity expression	Tominaga et al., 1990
IFN-γ	rec. (CHO)	human	N	Antiviral activity	No effect ?	Riske et al., 1991
IL-6	rec. (various)	human	N	Hybridoma growth assay and Ig-production enhancement	Differences in activity of glycoforms	Tanner et al., 1990
IL-5	rec. (CHO)	human	N	IgM secretion from BCL$_1$-cells	↓ 2.8-fold	Kodama et al., 1993
			O	IgM secretion from BCL$_1$-cells	↓ 10-fold	Kodama et al., 1993
			Sia	IgM secretion from BCL$_1$-cells	↓ 5-fold	Kodama et al., 1993
G-CSF	rec. (*E. coli*, CHO)	human	O	thermal and pH stability	Stability↑	Hasegawa et al., 1993; Oh-eda et al., 1990
	rec. (*E. coli*, CHO)	human	O	neutrophil colony formation	colony stimulating activity ↑	Nissen et al., 1994
IL-2	rec. (C127 and *E. coli*)	human	O	CTLL proliferation	? (↑)	Riske et al., 1991
MCP-1	rec. (baculovirus-Sf9, *E. coli*)	human	O	THP-1 chemotaxis	? (↑)	Ishii et al., 1995

rec., recombinant; the species indicates the origin of the gene, between brackets the cell type used for expression of glycosylated cytokines. N, Asn-linked N-glycosylation; O, Ser/Thr-linked O-glycosylation. ↓, down-modulation of biological activity by the attachment of sugars; ↑, up-regulation; ?, dicrepancy between abstract and shown data.

least 5-fold decrease in specific activity) and that the surrounding protein can determine the glycosylation site occupancy. Mature natural human IL-1ß glycoforms were first isolated as a 22 kDa molecule and were experimentally shown to have alanine as the NH$_2$-terminal residue and Asn$_7$ as a potential N-linked site (Van Damme et al., 1985). Yeast expression resulted in the production of unglycosylated (17 kDa) and glycosylated (22 kDa) IL-1ß. When the signal peptide cleavage site was mutated to form a longer distance between the NH$_2$-terminus and the N-linked site, the glycosylation became more efficient and the glycoform population as a whole less active. Processing of natural glycosylated IL-1ß from its 31 kDa precursor is accomplished by the interleukin-1 converting enzyme (ICE) (Cerretti et al., 1992, Thornberry et al., 1992). The clipping of preformed IL-1β leads to a glycosylated IL-1ß glycoform spectrum. The IL-1β, produced by *de novo* synthesis in the stressful event of a trauma, of an infection or other pathological condition, may carry less (or other) oligosaccharides and may not only possess an altered tissue and cell targeting through carbohydrate receptors (Drickamer, 1991), but also might have a significantly increased specific activity.

2.2. O-linked glycosylation influences the activity of signalling molecules

One general rule for the presence of O-linked glycosylation on cytokine activity does not yet exist. A possible explanation is the lack of our knowledge about the existence of different recognition patterns or "modules" for O-linked glycosylation. Indeed, with one known exception (O-fucosylation in epidermal growth factor domains of proteinases which is dependent on a Gly-Gly-Ser sequon), the protein motifs or sequons which are the recognition signals for the attachment of particular O-linked sugars need to be discovered. Some of the reasons for the slow progress in defining rules for the functions of O-linked sugars have been reviewed (Opdenakker et al., 1993; 1995): the need for manifold and sensitive assay systems and the basic understanding that protein glycosylation can not be probed by *in vitro* mutagenesis. Too many early studies appeared showing "no effect" of carbohydrates. Fortunately, by the development of novel reagents and technology and by establishment of sensitive assay systems more examples have unfolded which show the importance of glycosylation on the bioactivity of cytokines (see table 1).

O-linked glycosylation is important for the biological activity of interleukin-5 (IL-5). Although it was originally thought that IL-5 glycosylation was not important for its bioactivity (Kunimoto et al., 1991; Tominaga et al., 1990), a more recent report illustrated that both O- and N-linked sugars influence the activity of IL-5 (Kodama et al., 1993). Congruent with the above mentioned rule, removal of the N-linked sugar results in a 2.8-fold increase of the biological activity (and a decrease in thermostability). Desialylation increases the activity 5-fold and removal of the O-linked sugars yields a drastic (10-fold) increase in its activity. IL-5 is thus an example in which the O-linked glycosylation decreases the biological activity.

Granulocyte-colony stimulating factor (G-CSF) has one O-linked sugar which influences the molecular stability (Hasegawa et al., 1993) and the colony stimulating capacity of G-CSF (Nissen et al., 1994). Lenogastrim is a recombinant form of human G-CSF produced in Chinese hamster ovary (CHO) cells and hence is glycosylated (one O-link on Thr$_{133}$). This form possesses colony stimulating activity at a 16-fold lower dosis when compared with filgastrim, the aglycosyl form of human G-CSF produced in bacteria. At concentrations which are active *in vivo*, lenogastrim retains an at least 2-fold higher specific activity than filgastrim.

For two other cytokines, IL-2 and monocyte chemotactic protein-1 (MCP-1), it is less clear whether the O-linked glycosylation has an influence on the bioactivity. In contrast to the study by Marchese et al. (1990), Riske et al. (1991) mention that the activity of IL-2 with or without O-linked glycosylation is similar. However, from the data of the latter study a 2.3-fold difference in activity seems to be apparent. IL-2 produced in eukaryotic C127 cells (with O-linked sugars) seems to possess a higher specific activity than IL-2 produced in *Escherichia coli*. In a study of human MCP-1, Ishii et al. (1995) report that the O-linked sugars do not influence the activity, but the bacteria-derived MCP-1 without O-links seems to need a higher concentration for half-maximal activity than the MCP-1 produced in a baculovirus expression system. Without addressing the important issue of glycoprotein quantification (Opdenakker et al., 1993; 1995) in such analyses, a minimal conclusion of these studies is that structure function analysis of the role of carbohydrates in glycoproteins needs extensive and sensitive testing.

In conclusion, O-linked glycosylation has a clear influence on the activity and the stability of some cytokines: sometimes the O-linked sugars increase and in other cases they decrease the activity. The effect of O-linked glycosylation is not evident in all investigated molecules. This illustrates the difficulties encountered with such studies but also is indicative for need for more and sensitive assay systems.

3. MATRIX DEGRADING ENZYMES

It is clear that many enzyme cascades are involved in host defence mechanisms. In trauma, the clotting cascade is in action. In infarction and inflammation the fibrinolytic cascade, with plasmin as the key player, not only lyses the fibrin in the clot or in the capsule around the inflammatory focus (to enable migrating phagocytes to digest the tissue debris), it also activates a whole series of regulated matrix proteinases (Opdenakker and Van Damme., 1992). The complement cascade and the proteolytic enzymes which activate cytokines (e.g. ICE and the TNF-activating metalloproteases) constantly attract the attention of the immunologists and will not be discussed here. We here focus the attention on matrix degrading enzymes because they are crucial for leukocyte migration, because they yield the remnant epitopes generating autoimmunity and because they are used by cancer cells in invasion and metastasis. Last but not least, glycosylation constitutes the fifth regulation mechanism of these enzymes. The latter is exemplified here.

3.1. Effect of N-linked sugars on enzymes

The effects of N-linked glycosylation on enzyme activities have been well established with a number of examples. The first indications that removal of N-linked sugars from tissue plasminogen activator (t-PA) enhances its activity were confirmed in a number of other studies (Opdenakker et al, 1986; Parekh et al., 1989; Wittwer et al., 1989) and were among the first indications of the concept that N-linked sugars usually downmodulate the activity of enzymes. This rule was corroborated with several other examples some of which have already been mentioned. The study of bovine pancreas ribonuclease excels in this aspect because of the use of individual glycoforms to address the structure-function relation of attached oligosaccharides (Rudd et al., 1994; 1995).

In the cascade of matrix component degradation, the amplification effects of aglycosyl forms is evident for two enzymes: tissue plasminogen activator and plasmin(ogen)(Mori et al., 1995). Because plasmin catalyses the activation of collagenase and stromelysin from the respective pro-enzymes, a considerable effect may already be expected on this activation in stress situations of the host. Because the matrix metalloproteinases collagenase, stromelysin and the gelatinases A and B all carry N-linked sugars, it is a challenge to establish, for these enzymes too, sensitive assay systems to probe the role of the attached oligosaccharides on substrate conversion. In this respect, the substrate zymography technique to assay gelatinase activities has been useful to show the presence of N-linked sugars on natural human and mouse gelatinase B (Masure et al., 1993), and to show an effect of the attached O-linked sugars on the association (multimerisation) of gelatinase B. However this type of analysis has not yet enabled to document alterations in the specific activity between glycosylated and aglycosyl forms of the enzyme.

3.2. Effects of O-linked glycosylation on the activity of secreted enzymes

For only a small number of enzymes the influences of O-linked glycosylation on the catalytic activities have been described. Two examples are here discussed: thrombomodulin and lactase-phlorizin hydrolase (LPH). Although one example (LPH) is not directly linked to inflammation research or immunology, it indicates how the functional analysis of the carbohydrates in matrix proteinases might progress.

Human LPH is an intestinal brush border membrane enzyme which hydrolyses lactose. Two glycotypes LPH_N (with N-links) and $LPH_{N/O}$ (with N- and O-links) have been defined (Naim and Lentze, 1992). Both forms possess the same affinity for the substrate (same K_m), but $LPH_{N/O}$ has a higher catalytic velocity (V_{max}, four times higher than in LPH_N). The O-linked sugar induces a higher enzyme activity by fastening the catalysis. The difference in O-glycosylation may be related to the differentiation stage of the intestinal epithelial cells because the differentiation goes along with alterations in the expression levels of the Golgi-transferases; a possible function of the N-linked sugars consists in preventing aggregation because both forms aggregate after removal of the N-glycosylation.

Thrombomodulin (TM) induces a 70-fold acceleration of the inactivation of single chain urokinase plasminogen activator (scu-PA) by thrombin and thus interacts with the above mentioned matrix protease cascade. In addition, TM acts as a cofactor in the thrombin-dependent formation of activated protein C. TM possesses several N-linked oligosaccharides and a chondroitin sulphate glycosaminoglycan (CSGAG) which is O-linked to Ser_{492} (Parkinson et al., 1992). This serine is located in a consensus sequence for the attachment of CSGAG which is immediately carboxyterminal from the binding site of thrombin. The CSGAG consists of about 20 disaccharide units and is sulphated which results in a strong anionic character. This anionic site probably functions as a secondary binding site for thrombin which has a positively charged groove, the "anion-binding exosite". The presence of the CSGAG enhances the affinity for thrombin and it has also been documented that CSGAG (not bound to TM) enhances the activity of thrombin, albeit at a 1000-fold higher concentration (de Munk et al., 1993). Finally, the CSGAG enhances the activation of protein C by the thrombin-TM complex (Parkinson et al., 1992; de Munk et al., 1993). The O-linked CSGAG of TM is thus a determining factor in the activities of TM.

In general, we can conclude that the O-linked glycosylation of (matrix) enzymes can influence the enzymatic activity. This may be mediated by an influence on the affinity of the substrate or by increasing the catalytic velocity. From the mentioned examples it appears that O-linked glycosylation can upregulate the activity of enzymes. This is in sharp contrast to the observation that the voluminous N-linked sugars usually downmodulate the activities of enzymes and signal molecules.

4. CONCLUSIONS

The cytokine network initiates the host defence and alerts the specific immune system. Many cytokines are glycosylated and this glycosylation plays not only a role in the tissue partition and pharmacokinetics of cytokines, it also forms a control mechanism of the specific activity. Cytokine-regulated proteolysis is essential for leukocyte migration and for the generation of remnant epitopes in autoimmune diseases. Extracellular proteolysis of matrix components is achieved by a cascade of glyco-enzymes which are regulated at five levels: transcription, secretion, activation, inhibition and, finally, by glycosylation. N-linked sugars have always a downmodulation effect on enzymes (and cytokines); O-linked sugars may have an up- or downmodulating effect. In stress situations of the host, in which N-linked glycosylation is less efficient, the activities of cytokines and enzymes are thus considerably increased which results in an increased metabolism and substrate turnover. Glycosylation is thus of crucial importance in innate immunity and host defence.

ACKNOWLEDGEMENTS

The present study was supported by the Cancer Foundation of the General Savings and Retirement Fund (A.S.L.K.), the Belgian Cancer Association, The Charcot Foundation, The European Community (BMH 4-CT96-0893) and the "Fonds voor Wetenschappelijk Onderzoek Vlaanderen".

References

Adolf, G.R., Kalsner, I., Ahorn, H., Maurer-Fogy, I., and Cantell, K., 1991, Natural human interferon-α2 is O-glycosylated, *Biochem. J.* 276: 511-518.

Cebon, J., Nicola, N., Ward, M., Gardner, I., Dempsey, P., Layton, J., Durhrsen, U., Burgess, A.W., Nice, E., and Morstyn, G., 1990, Granulocyte-macrophage colony stimulating factor from human lymphocytes. The effect of glycosylation on receptor binding and biological activity, *J. Biol. Chem.* 265:4483-4491.

Cerretti, D.P., Kozlosky, C.J., Mosley, B., Nelson, N., Van Ness, K., Greenstreet, T.A., March, C.J., Kronheim, S.R., Druck, T., Cannizzaro, L.A., Huebner, K., and Black, R.A., 1992, Molecular cloning of the interleukin-1ß converting enzyme, *Science* 256:97-100.

de Munk, G.A., Parkinson, J.F., Groeneveld, E., Bang, N.U., and Rijken, D.C., 1993, Role of the glycosaminoglycan component of thrombomodulin in its acceleration of the inactivation of single-chain urokinase-type plasminogen activator by thrombin, *Biochem. J.* 290:655-659. K., 1991, Clearing up glycoprotein hormones, *Cell* 67:1029-1032.

Dijkmans, R., Heremans, H., and Billiau, A., 1987, Heterogeneity of chinese hamster ovary cell-produced recombinant interferon-g, *J. Biol. Chem.* 262:2528-2535.

Drickamer, K., 1991, Clearing up glycoprotein hormones, *Cell* 67:1029-1032.

Fleer, R., Chen, X.J., Amellal, N., Yeh, P., Fournier, A., Guinet, F., Gault, N., Faucher, D., Folliard, F., Fukuhara, H., and Mayaux, J.-F., 1991, High-level secretion of correctly processed recombinant human interleukin-1 beta in Kluyveromyces lactis, *Gene* 107:285-295.

George, A.J.T., 1994, Surface-bound cytokines -a possible effector mechanism in bacterial immunity?, *Immunol. Today* 15:88-89.

Hasegawa, M.,1993, A thermodynamic model for denaturation of granulocyte colony-stimulating factor: O-linked sugar chain suppresses not the triggering deprotonation but the succeeding denaturation, *Biochim. Biophys. Acta.* 1203:295-297.

Ishii, K., Yamagami, S., Tanaka, H., Motoki, M., Suwa, Y. and Endo, N., 1995, Full active baculovirus-expressed human monocyte chemoattractant protein 1 with the intact N-terminus, *Biochem. Biophys. Res. Comm.* 206:955-961.

Kodama, S., Tsujimoto, M., Tsuruoka, N., Sugo, T., Endo, T., and Kobata, A., 1993, Role of sugar chains in the in-vitro activity of recombinant human interleukin 5, *Eur. J. Biochem.* 211:903-908.

Kunimoto, D.Y., Allison, K.C., Watson, C., Fuerst, T., Armstrong, G.D., Paul, W., and Strober, W., 1991, High-level production of murine interleukin-5 (IL-5) utilising recombinant baculovirus expression.

Purification of the rIL-5 and its use in assessing the biologic role of IL-5 glycosylation, *Cytokine* 3:224-230.

Livi, G.P., Ferrara, A.A., Roskin, R., Simon, Pl., and Young, Pr., 1990, Secretion of N-glycosylated human recombinant interleukin-1 alpha in Saccharomyces cerevisiae. *Gene* 88:297-301.

Livi, G.P., Lillquist, J.S., Miles, L.M., Ferrara, A., Sathe, G.M., Simon, Pl., Meyers, C.A., Gorman, J.A., Young, Pr., 1991, Secretion of N-glycosylated interleukin-1 beta in Saccharomyces cerevisiae using a leader peptide from Candida albicans. Effect of N-glycosylation on biological activity, *J. Biol. Chem.* 266:15348-15355.

Lucas, R., Magez, S., De Leys, R., Fransen, L., Scheerlinck, J.P., Rampelberg, M., Sablon, E., and De Baetselier, P., 1994, Mapping the lectin-like activity of tumor necrosis factor, *Science* 263:814-817.

Marchese, E., Vita, N., Maureaud, T., Ferrara, P., 1990, Separation by cation-exchange high-performance liquid chromatography of three forms of Chinese hamster ovary cell-derived recombinant human interleukin-2, *J. Chromatogr.* 504:351-358.

Masure, S., Nys, G., Fiten, P., Van Damme, J., and Opdenakker, G., 1993, Mouse gelatinase B: cDNA cloning, regulation of expression and glycosylation in WEHI-3 macrophages and gene organisation, *Eur. J. Biochem.* 218:129-141.

Masure, S., and Opdenakker, G. 1989, Cytokine-mediated proteolysis in tissue remodelling, *Experientia* 45:542-549.

Michaëlsson, E., Malmstrom, V., Reis, S., Engstrom, A., Burkhardt, H., and Holmdahl, R., 1994, T cell recognition of carbohydrates on type II collagen, *J. Exp. Med.* 180:745-749.

Mori, K., Dwek, R.A., Downing, A.K., Opdenakker, G., and Rudd, P.M., 1995, The activation of type 1 and type 2 plasminogen by type 1 and type 2 tissue plasminogen activator, *J. Biol. Chem.* 270:3261-3267.

Mutsaers, J.H.G.M., Kamerling, J.P., Devos, R., Guisez, Y., Fiers, W., and Vliegenthart, J.F.G., 1986, Structural studies of the carbohydrate chains of γ-interferon, *Eur. J. Biochem.* 156:651-654.

Naim, H.Y., and Lentze, M.J., 1992, Impact of O-glycosylation on the function of human intestinal lactase-phlorizin hydrolase. Characterization of glycoforms varying in enzyme activity and localization of O-glycoside addition, *J. Biol. Chem.* 267:25494-25504.

Nissen, C., Dalle Carbonare, V., and Moser, Y., 1994, In vitro comparison of the biological potency of glycosylated versus nonglycosylated rG-CSF, *Drug invest.* 7:346-352.

Oh-eda, M., Hasegawa, M., Hattori, K., Kuboniwa, H., Kojima, T., Orita, T., Tomonou, K., Yamazaki, T., and Ochi, N., 1990, O-linked sugar chain of human granulocyte colony-stimulating factor protects it against polymerization and denaturation allowing it to retain its biological activity, *J. Biol. Chem.* 265:11432-11435.

Opdenakker, G., Cabeza-Arvelais, Y., and Van Damme, J., 1989, Interaction of interferon with other cytokines, *Experientia* 45:513-520.

Opdenakker, G., Rudd, P.M., Ponting, C.P., and Dwek, R.A., 1993, Concepts and principles of glycobiology, *FASEB J.* 7:1330-1337.

Opdenakker, G., Rudd, P.M., Wormald, M., Dwek, R.A., and Van Damme, J., 1995, Cells regulate the activities of cytokines by glycosylation, *FASEB J.* 9:453-457.

Opdenakker, G., and Van Damme, J., 1992, Cytokines and proteases in invasive processes: molecular similarities between inflammation and cancer, *Cytokine* 4:251-258.

Opdenakker, G., and Van Damme J., 1994, Cytokine-induced proteolysis in autoimmune diseases, *Immunol. Today* 15:104-107.

Opdenakker, G., Van Damme, J., Bosman, F., Billiau, A., and De Somer, P., 1986, Influence of carbohydrate side-chains on activity of tissue-type plasminogen activator, *Proc. Soc. Exp. Biol. Med.* 182:248-257.

Parekh, R.B., Dwek, R.A., Thomas, J.R., Opdenakker, G., Rademacher, T.W., Wittwer, A.J., Howard, S.C., Nelson, R., Siegel, N.R., Jennings, M.G., Harakas, N.K., and Feder, J., 1989, Cell-type-specific and site-specific N-glycosylation of type I and type II human tissue plasminogen activator, *Biochemistry* 28:7644-7662.

Parekh, R.B., Dwek, R.A., Rademacher, T.W., Opdenakker, G., Van Damme, J., 1992, Glycosylation of interleukin-6 from normal human blood mononuclear cells, *Eur. J. Biochem.* 203:135-141.

Parkinson, J.F., Vlahos, C.J., Yan, S.C., and Bang, N.U., 1992, Recombinant human thrombomodulin. Regulation of cofactor activity and anticoagulant function by a glycosaminoglycan side chain, *Biochem. J.* 283:151-157.

Pos, O., van der stelt, M.E., Wolbink, G.J., Nijsten, M.W., van der Tempel, G.L., and van Dijk, W., 1990, Changes in the serum concentration and the glycosylation of human alpha 1-acid glycoprotein and alpha 1-protease inhibitor in severely burned persons: relation to interleukin-6 levels, *Clin. Exp. Immunol.* 82:579-582.

Rademacher, T.W., Parekh, R.B., and Dwek, R.A., 1988, Glycobiology, *Ann. Rev. Biochem.* 57:785-838.

Riske, F.J., Cullen, B.R., Chizzonite, R., 1991, Characterisation of human interferon-gamma and human interleukin-2 from recombinant mammalian cell lines and peripheral blood lymphocytes, *Lymphokine and Cytokine Res.* 10:213-218.

Rudd, P.M., Joao, H.C., Coghill, E., Fiten, P., Saunders, M.R., Opdenakker, G., and Dwek, R.A., 1994, Glycoformsmodify the dynamic stability and functional activity af an enzyme, *Biochemistry* 33:17-22.

Rudd, P.M., Woods, R.J., Wormald, M.R., Opdenakker, G., Downing, A.K., Campbell, I.D., and Dwek, R.A., 1995, The effects of variable glycosylation on the functional activities of ribonuclease, plasminogen and tissue plasminogen activator, *Biochim. Biophys. Acta* 1248:1-10.

Sherblom, A.P., Sathyamoorthy, N., Decker, J.M., and Muchmore, A.V., 1989, IL-2, a lectin with specificity for high-mannose glycopeptides, *J. Immunol.* 143:939-944.

Tanner, J.E., Goldman, N.D., and Tosato, G., 1990, Biochemical and biological analysis of human interleukin 6 expressed in rodent and primate cells, *Cytokine* 2:363-374.

Thor, G., Brian, A.A., 1992, Glycosylation variants of murine interleukin-4: evidence for different functional properties, *Immunology* 75:143-149.

Thornberry, N.A., Bull, H.G., Calaycay, J.R., Chapman, K.T., Howard, A.D., Kostura, M.J., Miller, D.K., Molineaux, S.M., Weidner, J.R., Aunins, J., Elliston, K.O., Ayala, J.M., Casano, F.J., Chin, J., Ding, G. J.-F., Egger, L.A., Gaffney, E.P., Limjuco, G., Palyha, O.C., Raju, S.M., Rolando, A.M., Salley, J.P., Yamin, T.-T., Lee, T.D., Shively, J.E., MacCross, M., Mumford, R.A., Schmidt, J.A., and Tocci, M.J., 1992, A novel heterodimeric cysteine protease is required for interleukin-1ß processing in monocytes, *Nature* 356:768-774.

Tominaga, A., Takahashi, T., Kikuchi, Y., Mita, S., Naomi, S., Harada, N., Yamaguchi, N., and Takatsu, K., 1990, Role of carbohydrate moiety of IL-5. Effect of tunicamycin on the glycosylation of IL-5 and the biologic activity of deglycosylated IL-5, *J. Immunol.* 144:1345-1352.

Van Damme, J., De Ley, M., Opdenakker, G., Billiau, A., De Somer, P., and Van Beeumen, J., 1985, Homogeneous interferon-inducing 22K factor is related to endogenous pyrogen and interleukin-1, *Nature* 314:266-268.

Wittwer, A.J., Howard, S.C., Carr, L.S., Harakas, N.K., Feder, J., Parekh, R.B., Rudd, P.M., Dwek, R.A., Rademacher, T.W., 1989, Effects of N-glycosylation on in vitro activity of Bowes melanoma and human colon fibroblast derived tissue plasminogen activator, *Biochemistry* 28:7662-7669.

Ziltener, H.J., 1993, Glycosylation does not affect *in vitro* biological activity of interleukin-3, *Cytokine* 5:291-297.

OCCURRENCE AND POSSIBLE FUNCTION OF INFLAMMATION-INDUCED EXPRESSION OF SIALYL LEWIS-X ON ACUTE-PHASE PROTEINS

Willem Van Dijk, Els C.M. Brinkman-Van der Linden, and
Ellen C. Havenaar

Department of Medical Chemistry
Faculty of Medicine, Vrije Universiteit
Van der Boechorststraat 7
1081 BT Amsterdam, The Netherlands

INTRODUCTION

The complex-type glycan structures of acute-phase glycoproteins (APPs) are subject to marked changes during acute and chronic inflammation (see for review Van Dijk et al., 1994). *In vivo* studies with mice transgenic for human α_1-acid glycoprotein (AGP) and *in vitro* studies with human hepatoma cell lines (Mackiewicz et al., 1991) and isolated human and rat hepatocytes (Pos et al., 1988; Pos et al. 1989; Van Dijk et al., 1991a; Van Dijk et al. 1991b) have indicated that these changes are not simply determined by changes in serum concentration or in secretion rates of acute-phase glycoproteins. The biosynthesis of the glycans in the hepatocytes of the liver appeared to be affected as part of the hepatic acute-phase reaction. Inflammatory cytokines, like IL-6 and IL-1α, and glucocorticoids are involved in the regulation of these processes.

This review summarizes the results obtained by our group in the last few years with respect to the inflammation-induced changes in the α3-fucosylation of acute-phase proteins resulting in an increased expression of the sialyl Lewis^x (sLeX) determinant in acute and chronic inflammation.

INFLAMMATION-INDUCED CHANGES IN GLYCOSYLATION OF APPs

The highly glycosylated acute-phase protein AGP (Schmid et al., 1977), also known as orosomucoid, has been used by us as a reporter glycoprotein for studies towards inflammation-induced changes in glycosylation of APPs. At least 12 glycoforms of AGP can be detected in normal human serum, each containing 5 glycans, but differing in degree of branching (di-antennary *versus* tri- or tetra-antennary glycans) as well as in degree of α3-fucosylation (LeX-type) and sialylation of the glycans (Bierhuizen et al., 1988; Van der Linden et al., 1994). Acute and chronic inflammatory processes induce large changes in the proportions of specific glycoforms of AGP in serum, as well as in the serum level of total AGP. This is exemplified by a transient increase in the serum levels of glycoforms expressing two or more diantennary glycans during the early phase of an acute-phase reaction (Pos et al. 1990a). Furthermore, strong increases in highly fucosylated glycoforms of AGP are apparent in the late phase of the acute-phase reaction (De Graaf et al., 1993), but also in sepsis (Brinkman-Van der Linden et al., 1996a), hyper-IgD syndrome (Havenaar et al., 1995) and rheumatoid arthritis (De Graaf et al., 1994; Havenaar et al., 1997). The changes in fucosylation were detected by crossed affinoimmunoelectrophoresis using the fucose-specific *Aleuria aurantia* lectin (AAL) as affinity component in the first dimension gel. It could be demonstrated by immunoblotting with the monoclonal anti-sLeX antibody CSLEX-1 that the fucosylated glycoforms expressed high levels of the sLeX determinant (De Graaf et al., 1993; Brinkman-Van der Linden et al., 1996a; Havenaar et al., 1995,1997). The level of expression of sLeX correlated with the changes in extent of retardation by AAL in CAIE. Non-sialylated LeX groups were not detectable on AGP by anti-LeX antibodies in any of the conditions studied, indicating that the majority of the α3-linked fucose residues are present as sLeX (Brinkman-Van der Linden, 1997a). The results obtained are represented schematically in Figure 1.

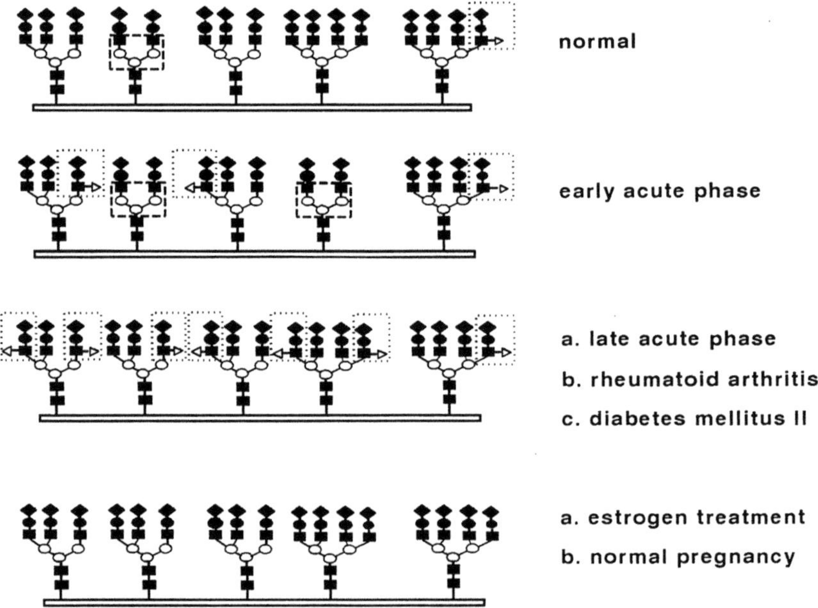

Figure 1. Schematic representations of the predominant glycoforms of AGP under various conditions. The composing sugars, Fuc, Gal, GlcNAc, Man and sialic acid, are respectively represented by triangles, filled circles, rectangles, open circles and diamonds. Broken and dotted lines are used to indicate diantennary glycans respectively, sLeX structures.

CHANGES IN FUCOSYLATION AND sLeX EXPRESSION ON OTHER APPs

Inflammation-induced changes in α3-fucosylation and sLeX expression are not restricted to AGP. Comparable changes were found for α₁-antichymotrypsin (ACT) and haptoglobin (HG) (Brinkman-Van der Linden, 1997a). Table 1 clearly shows that acute as well as chronic inflammation induces high increases in the plasma levels of fucosylated glycoforms of the APPs. However, AGP appears to be the most predomi-

Table 1. Serum levels of fucosylated glycoforms of AGP, ACT and HG under various conditions.

Condition	Fucosylated APP[1] (mg/ml)[2]		
	AGP	ACT	HG
Control	0.53±0.06	0.19±0.06	0.53±0.01
Acute trauma	1.90±0.14	1.08±0.08	1.62±0.07
Rheumatoid arthritis	1.78±0.17	0.53±0.02	2.05±0.10
Estrogen treatment	0.27±0.04	0.16±0.02	0.47±0.02

[1]Fraction of APP retarded by the fucose-specific lectin AAL in agar-gel electrophoresis (Brinkman-Van der Linden, 1997a). Means of three determinations are given using pooled sera from healthy individuals (n=75), from patients suffering from severe trauma in the second week of trauma (n=6), from patients suffering from rheumatoid arthritis (n=15), and from male-to-female transsexuals receiving oral estrogen treatment (n=8).

nant protein in plasma when only the fucosylated glycoforms with presumably two or more α3-linked fucose residues are taken into account (Table 2). This corresponds with the lower level of expression of sLeX on HG and especially ACT (not shown). The differences between the three proteins most likely reside in the degree of branching of their glycans, which is one of the parameters determining the specificities of at least the sialyltransferases involved in the synthesis of sLeX groups (Paulson et al, 1978; Joziasse et al. 1987). AGP is the only APP also containing tetraantennary glycans (Fournet et al. 1978; Bierhuizen et al., 1988; Laine et al., 1991, Dobryszycka, 1993; Hodges et al., 1979; Mega et al., 1980), whereas diantennary glycans predominate on ACT and especially HG (Brinkman-Van der Linden, 1997a). The dependency of (S)LEX expression on degree of branching is supported by the absence of antibody-detectable (S)LEX on another APP, α_1-protease inhibitor (PI) of which the diantennary glycan content is even higher than of HG (Brinkman-Van der Linden, 1997a).

Table 2. Serum levels of glycoforms of AGP, ACT and HG with 3 or more fucose residues under various conditions.

Condition	Fucosylated APP[1] (mg/ml)		
	AGP	ACT	HG
Control	0.40±0.06	0.03±0.00	0.04±0.00
Acute trauma	1.55±0.08	0.35±0.03	0.14±0.03
Rheumatoid arthritis	1.53±0.06	0.13±0.02	0.36±0.06
Estrogen treatment	0.20±0.03	0.02±0.01	0.02±0.01

[1]Fraction of APP retarded most strongly by the fucose-specific lectin AAL in agar-gel electrophoresis (Brinkman-Van der Linden, 1997a). See Table 1 for further details.

REDUCED sLeX EXPRESSION DURING ESTROGEN TREATMENT AND PREGNANCY

The branching of the glycans of AGP, HG, ACT and PI is increased and the α3-fucosylation and sLeX expression is decreased in women and male-to-female transsexuals, receiving oral estrogen treatment in comparison to non-treated or transdermally estrogen-treated individuals (Brinkman-Van der Linden, 1996c, 1997a). The difference between oral and transdermal treatment indicates that estrogen receptors on liver cells play a significant role in the observed effects. A physiological increase in the plasma levels of estrogen occurs during pregnancy. A pregnancy-induced increase in branching of the glycans of various APPs was observed in late pregnancy (see e.g. Raynes, 1982). In a longitudinal study during pregnancy we could confirm this effect and, in addition, detected that highly branched AGP and ACT glycoforms lacking diantennary glycans increased continuously during pregnancy (Havenaar et al., 1996). In contrast, the extent of α3-fucosylation of APPs decreased during pregnancy until week 25, and then started to rise again (Havenaar et al., manuscript submitted). Although the pregnancy-induced changes in fucosylation and branching are similar to those observed during oral treatment with estrogen, the fall and rise of the fucosylation suggest that in pregnancy other factors are also involved in the regulation of these phenomena besides estrogen, since its plasma levels are known to rise constantly during pregnancy.

HEPATIC FUCOSYLTRANSFERASE VI IS RESPONSIBLE FOR THE α3-FUCOSYLATION OF APPs

The regulation of changes in glycosylation of APPs is largely unknown. However, it has clearly been established that the changes have a biosynthetic origin residing in the epithelial cells of the liver (reviewed in Van Dijk et al., 1994, 1995). In a first approach to delineate the mechanisms involved in the pathophysiological variations in expression of sLeX on APPs, we have investigated which of the three possible fucosyltransferases (III, V and VI) are involved in the α3-fucosylation of these proteins in the human liver. To this aim sera of individuals with or without inactivated *FUT3* and/or *FUT6* gene(s), but with a functional *FUT5* gene, were analysed for the presence and extent of α3-fucosylated AGP, HG, and PI (Brinkman-Van der Linden et al., 1996b). No α3-fucosylated glycoforms of the APPs could be detected in any of the sera from individuals with the *FUT6* missense mutation Gly-739→Ala in double dose. An intrinsic resistance to fucosylation was excluded, since *in vitro* α3-fucosylation of AGP was possible with an α3/4-fucosyltransferase isolated from human milk. α3-Fucosylated glycoforms were found in all individuals with a functional *FUT6* gene. Thus, it could be concluded that in humans the product of the *FUT6* gene, fucosyltransferase VI, is responsible for the α3-fucosylation of the APPs in liver. Furthermore, it was concluded that no other fucosyltransferase genes take over when inactivating mutations in the *FUT6* gene are present.

POSSIBLE FUNCTIONS OF sLeX EXPRESSION ON AGP

Various immunomodulatory properties of APPs have been shown to be glycosylation or glycoform-dependent (reviewed in Van Dijk et al., 1995), indicating that changes in the plasma levels of specific APP glycoforms during the hepatic acute-phase reaction can be of physiological importance in the feed-back control of the inflammatory reactions. Potentially, changes in sLeX expression of AGP may affect lectin-type interactions between endothelial E-selectin and sLeX-bearing leukocytes in inflamed areas (De Graaf et al., 1993). In accordance with this hypothesis, we recently could demonstrate that highly fucosylated and sLeX-containing AGP glycoforms isolated from patient sera indeed can bind to E-selectin *in vitro* in a calcium-dependent way (Brinkman-Van der Linden 1997b), like the natural ligand for E-selectin ESL-1 (Steegmaler et al., 1995).

ACKNOWLEDGEMENTS

Part of this work was supported by the Dutch Organization for Scientific Research (NWO 900-512-164) and by the EC Biomed I concerted action EUROCARB.

REFERENCES

Bierhuizen, M.F.A., De Wit, M., Govers, C.A.R.L., Ferwerda, W., Koeleman, C., Pos, O., and Van Dijk, W., 1988, Glycosylation of three molecular forms of human α_1-acid glycoprotein having different interactions with Concanavalin A: Variations in the occurrence of bi-, tri-, and tetraantennary glycans and the degree of sialylation, *Eur. J. Biochem.* 175:387.

Brinkman-Van der Linden, E.C.M., Van Ommen, E.C.R., and Van Dijk, W., 1996a, Glycosylation of α_1-acid glycoprotein in septic shock: changes in degree of branching and in expression of sialyl Lewis[x] groups, *Glycoconj. J.* 13:1.

Brinkman-Van der Linden, E.C.M., Mollicone, R., Oriol, R., Larson, G., Van den Eijnden, D.H., and Van Dijk, W., 1996b, A missense mutation in the FUT6 gene results in the total absence of α3-fucosylation of human α_1-acid glycoprotein, *J. Biol. Chem.* 271:14492.

Brinkman-Van der Linden, E.C.M., Havenaar, E.C., Van Ommen, E.C.R., Van Kamp, G.J., Gooren, L.J.G., and Van Dijk W, 1996c, Oral estrogen treatment induces a decrease in expression of SLeX on AGP in females and male-to-female transsexuals, *Glycobiology* 6:407.

Brinkman-Van der Linden, E.C.M., 1997a, Inflammation-induced expression of sialyl Lewis[x] on α_1-acid glycoprotein: Occurrence, regulation and function, Chapter 5 in Ph.D. Thesis, Ponsen $ Looijen, Wageningen.

Brinkman-Van der Linden, E.C.M., 1997b, Inflammation-induced expression of sialyl Lewis[x] on α_1-acid glycoprotein: Occurrence, regulation and function, Chapter 7 in Ph.D. Thesis, Ponsen $ Looijen, Wageningen.

Dobryszycka, W., 1993, Haptoglobin: Retrospectives and perspectives, in: "Acute phase proteins: Molecular biology, biochemistry and clinical applications", A. Mackiewicz, I. Kushner, and H. Baumann, eds. Boca Raton, CRC press, 185.

De Graaf, T.W., Van der Stelt, M.E., Anbergen, M.G., and Van Dijk, W., 1993, Inflammation-induced expression of sialyl Lewis X-containing glycan structures on α_1-acid glycoprotein (orosomucoid) in human sera, *J. Exp. Med.* 177:657.

De Graaf, T.W., Van Ommen, E.C.R., Van der Stelt, M.E., Kerstens, P.J.S.M., Boerbooms, A.M.Th., and Van Dijk W., 1994, Effects of low-dose methotrexate therapy on the concentration and the glycosylation of α_1-acid glycoprotein in the serum of rheumatoid arthritis patients: A longitudinal study, *J. Rheumatol.* 21:2209.

Fournet, B., Montreuil, J., Strecker, G., Dorland, L., Haverkamp, J., Vliegenthart, J.F.G., Binette, J.P., and Schmid, K., 1978, Determination of the primary structures of 16 asialo-carbohydrate units derived from human plasma α_1-acid glycoprotein by 360-MHz ^{1}H NMR spectroscopy and permethylation analysis *Biochemistry* 17:5206.

Havenaar, E.C., Drenth, J.P.H., Van Ommen, E.C.R., Van der Meer, J.W.M., and Van Dijk, W., 1995, Elevated serum level and altered glycosylation of α_1-acid glycoprotein in hyperimmunoglobulinemia D and periodic fever syndrome: Evidence for persistent inflammation, *Clin. Immunol. Immunopathol.* 76:279.

Havenaar, E.C., Axford, J.F., Brinkman-Van der Linden, E.C.M., Alavi, A., Spector, T., and Van Dijk W., 1996, Effect of pregnancy on glycosylation of α_1-acid glycoprotein in rheumatoid arthritis, *Glycoconj. J.* 13:899.

Havenaar, E.C., Dolhain, R.J.E.M., Turner, G.A., Goodarzi, M.T., Van Ommen, E.C.R., Breedveld, F.C., and Van Dijk, W., 1997, Do synovial fluid acute-phase proteins from patients with rheumatoid arthritis originate from serum *Glycoconj. J.* 14:in press.

Hodges, L.C., Laine, R., and Kai Chan, S., 1979, Structure of the oligosaccharide chains in human α_1-protease inhibitor *J. Biol. Chem.* 254:8208.

Joziasse, D.H., Schiphorst, W.E.C.M., Van den Eijnden, D.H., Van Kuik, J.A., Van Halbeek, H., and Vliegenthart, J.F.G., 1987, Branch specificity of bovine colostrum CMP-sialic acid: Galβ1$\rightarrow$4-

GlcNAc-R $\alpha2{\rightarrow}6$-sialyltransferase. Sialylation of bi-, tri- and tetraantennary oligosaccharides and glycopeptides of the N-acetyllactosamine type *J. Biol. Chem.* 262:2025.

Laine, A., Hachulla, E., Strecker, G., Michalski, J.-C., and Wieruszeski, J.-M., 1991, Structure determination of the glycans of human-serum α_1-antichymotrypsin using ^{1}H-NMR spectroscopy and deglycosylation by N-glycanase *Eur. J. Biochem.* 197:209.

Mackiewicz, A., Pos, O., Van der Stelt, M.E., Yap, S.H., Kapcinska, M., Laciak, M., Dewey, M.J., Berger, F.G., Baumann, H., Kushner, I., and Van Dijk, W., 1991, Regulation of glycosylation of acute-phase proteins by cytokines in vitro, Chapter 9, in: "Affinity Electrophoresis: Principles and Application," J. Breborowicz, and A. Mackiewicz, A., eds., CRC Press, Boca Raton.

Mega, T., Lujan, E., and Yoshida, A., 1980, Studies on the oligosaccharide chains of human α_1-protease inhibitor. II Structure of oligosaccharides *J. Biol. Chem.* 255:4057.

Paulson, J.C., Prieels, J.P., Glasgow, L.R., and Hill, R.L., 1978, Sialyl- and fucosyltransferases in the biosynthesis of asparaginyl-linked oligosaccharides in glycoproteins. Mutual exclusive glycosylation by β-galactoside $\alpha2{\rightarrow}6$sialyltransferase and N-acetylglucosaminide $\alpha1{\rightarrow}3$-fucosyltransferase *J. Biol. Chem.* 253:5617.

Pos, O., Van Dijk, W., Ladiges, N., Linthorst, C., Sala, M., Van Tiel, D., and Boers, W., 1988, Glycosylation of four acute-phase glycoproteins secreted by rat liver cells *in vivo* and *in vitro*: Effects of inflammation and dexamethasone, *Eur. J. Cell Biol.* 46:121.

Pos, O., Boers, W., Moshage, H.J., Yap, S.H., Aarden, L.A., Van Gool, J., Brugman, A.M., and Van Dijk W., 1989, Effects of monocytic products, recombinant interleukin-1 and interleukin-6 on the glycosylation of α_1-acid glycoprotein: Studies with primary human hepatocyte and rats, *Inflammation* 13:415.

Pos, O., Van der Stelt, M.E., Wolbink, G.J., Nijsten, M.W.N., Van der Tempel, G.L., and Van Dijk, W., 1990, Changes in the serum concentration and the glycosylation of human α_1-acid glycoprotein and α_1-protease inhibitor in severely burned patients: relation to interleukin-6 levels, *Clin. Exp. Immunol.* 82:579.

Raynes, J., 1982, Variations in the relative proportions of microheterogeneous forms of plasma glycoproteins in pregnancy and disease, *Biomedicine* 36:77.

Schmid, K., Nimberg, R.B, Kimura, A., Yamaguchi, H., and Binette, J.P., 1977, The carbohydrate units of human plasma α_1-acid glycoprotein, *Biochim. Biophys. Acta* 492:291.

Steegmaler, M., Levinovitz, A. Isenmann, S., Borges, E., Lenter, M., Kocher, H.P., Kleuser, B., and Vestweber, D., 1995, The E-selectin ligand ESL-1 is a variant of the fibroblast growth receptor *Nature* 373:615.

Van Dijk, W., Van der Stelt, M.E, Salera, A., and Dente, L., 1991a, Effect of transgenic expression of human alpha1-acid glycoprotein (AGP) on the glycosylation of human and mouse AGP in various transgenic mouse sera, *Eur. J. Cell Biol.* 55:143.

Van Dijk, W., Pos, O., Van der Stelt, M.E., Moshage, H.J., Yap, S.H., Dente, L., Baumann, P., and Eap C.B., 1991b, Inflammation-induced changes in expression and glycosylation of genetic variants of $\alpha1$-acid glycoprotein (AGP); Studies with human sera, primary cultures of human hepatocytes and transgenic mice, *Biochem. J.* 276:343.

Van Dijk, W., Turner, G.A., and Mackiewicz, A., 1994, Changes in glycosylation of acute-phase proteins in health and disease: Occurrence, regulation and function, *Glycosyl. Disease* 1:5.

Van Dijk, W., Havenaar, E.C., and Brinkman-Van der Linden, E.C.M., 1995, α_1-Acid glycoprotein (orosomucoid): Pathophysiological changes in glycosylation in relation to its function, *Glycoconj. J.* 12:227.

Van der Linden, E.C.M., De Graaf, T.W., Anbergen, M.G., Dekker, R.M., Van Ommen, E.C.R., Van den Eijnden, D.H., and Van Dijk, W., 1994, Preparative affinity electrophoresis of different glycoforms of serum glycoproteins: Application for the study of inflammation induced expression of sialyl-Lewis[x] groups on alpha1-acid glycoprotein (orosomucoid), *Glycosyl. Disease* 1:45.

GLYCOSYLATION AND DISEASE

THE GLYCOSYLATION OF THE COMPLEMENT REGULATORY PROTEIN, HUMAN ERYTHROCYTE CD59

*#Pauline M. Rudd, "B. Paul Morgan, *Mark R. Wormald,
*David J. Harvey, "Carmen W. van den Berg, $Simon J. Davis,
~Michael A.J. Ferguson and *#Raymond A. Dwek

*Glycobiology Institute, Department of Biochemistry, University of Oxford,
South Parks Road, Oxford, OX1 3QU
"University of Wales College of Medicine, Department of Medical Biochemistry,
Heath Park, Cardiff, CF4 4XN
~Department of Biochemistry, University of Dundee, Dundee, DD1 4HN
$ Molecular Sciences Division, Nuffield Department of Clinical Medicine,
University of Oxford, John Radcliffe Hospital, Headington, Oxford, OX3 9DU, UK

INTRODUCTION

CD59 is a cell surface glycoprotein that binds to the complement proteins C8 and/or C9 in the nascent membrane attack complex, thereby protecting host cells from lysis (1). CD59 belongs to the Ly-6 superfamily (2) and is present on a wide variety of cell types, including leukocytes, platelets, epithelial and endothelial cells, placental cells and erythrocytes, where it is present at $2.5\text{-}5\times10^4$ copies/cell (3). In addition to its role in protecting host tissues from homologous complement, it has been proposed that CD59 mediates T-cell adhesive interactions by synergising with CD58 via direct interactions with CD2 (4,5) although this is controversial (6,7). It has also been proposed that CD59 participates in T cell activation pathways and platelet secretory responses (1).

CD59 is normally attached to the cell surfaces via a glycosylphosphatidylinositol (GPI) anchor (8) although a number of soluble forms have been found, for example in saliva, amniotic fluid, milk and urine (1). The complementary DNA sequence of the CD59 gene (8) has shown that the translated precursor contains 128 amino acids. The cleavage site for the anchor attachment is between Asn77 and Gly78 (9). The molecule contains 8 potential O-glycosylation sites at Thr10, 15, 29, 51, 52 and 60, and at Ser20 and 21. There is one fully occupied N-glycosylation site (Asn18CysSer) (8) which is completely conserved in all known CD59 sequences except rat CD59 which is glycosylated at the adjacent residue, Asn16 (10).

RESULTS

Analysis of GPI Anchor

Analysis of the glycans associated with the CD59 GPI anchor: The GPI neutral fraction was fractionated by Dionex high pH anion exchange chromatography into three species that eluted at 2.51 (91%), 3.07 (8%) and 3.72 (1%) Dionex Units (Du). Taken together with P4 chromatographic data and exoglycosidase digestions, these results indicate that the two major GPI neutral glycan species are Manα1-2Manα1-6Manα1-4AHM, (derived from Manα1-2Manα1-6Manα1-GlcN-R by deamination and reduction) and Manα1-2Manα1-6(GalNAcb1,4)Manα1-4AHM (derived from Manα1-2Manα1-6(GalNAcβ1,4)Manα1-GlcN-R) (Fig.1).

Characterisation of the PI moiety of the GPI anchor: The PI moiety released by nitrous acid deamination of CD59 was analysed by negative ion ES-MS. The data strongly suggest that the PI moiety of CD59 consists of approximately equal amounts of 1-O-(C18:1)alkyl-2-O-C(22:4)acylglycerol-3-HPO$_4$-1-(2-O-(C16:0)acyl)myo-inositol and 1-O-(C18:0)alkyl-2-O-(C22:4)acylglycerol-3-HPO$_4$-1-(2-O-(C16:0)acyl)myo-inositol. These so called acyl-PI structures are identical to the structures previously described for the GPI anchor of human erythrocyte acetylcholinesterase (11). The presence of the acyl (palmitoyl) group attached to the 2-position of the inositol ring explains the resistance of CD59 to the action of PIPLC enzymes which require a free 2-hydroxyl group on the inositol ring to participate in the cleavage reaction, i.e. nucleophilic attack of the phosphodiester phosphorus atom (12) (Fig. 1).

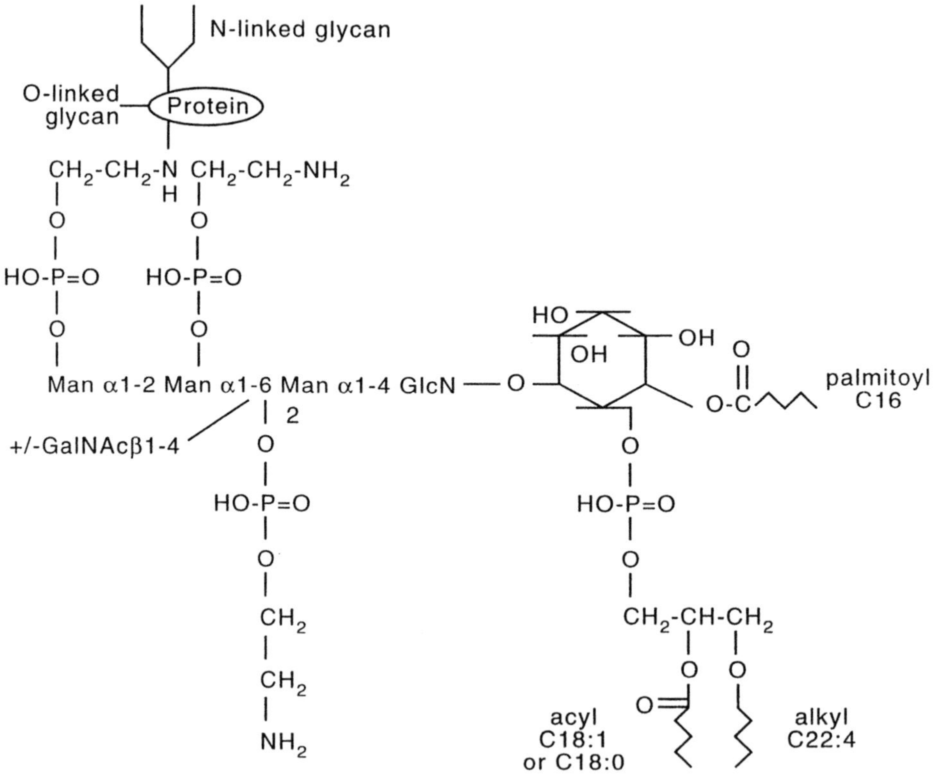

Figure 1.

Schematic drawing of the CD59 glycan anchor. The structures of the lipids and glycans are based on data derived in this paper. The inclusion of the ethanolamine groups in this model is solely by analogy with other mammalian GPI anchors.

GC-MS monosaccharide composition: The monosaccharide composition of the total CD59 oligosaccharides following methanolysis, re-N-acetylation and trimethylsilylation was determined by GC-MS. The detection of 6.1nmol of GalNAc/50mg of CD59 suggested the presence of O-linked glycans. The high proportion of N-acetyl glucosamine (34.8 nmol/50mg protein) and galactose (31.7 nmol/50mg protein) compared with mannose (5.2 nmol/50mg protein) suggested that CD59 was associated with a high percentage of multi-antennary and polylactosamine type oligosaccharides. The proportion of fucose (8.2 nmol/50mg protein) suggested that many structures contained outer arm as well as core fucose. Finally, the detection of 10 nmol/50mg protein of sialic acid residues indicated that many of the glycans were charged.

Comparison of the glycosylation profiles of human erythrocyte and human platelet CD59: The 2AB labelled sialylated N- and O-glycan pools released from human erythrocyte and human platelet CD59 were resolved by normal phase HPLC (Figure 2). The differences in these profiles, which compare the overall glycosylation of CD59 expressed in two different cell types, confirm previous findings that glycosylation is cell type specific. There are more than 40 peaks in each of these profiles indicating that platelet and erythrocyte CD59 are both expressed as a population of many different glycoforms which contain N- and O-linked glycans. The bar chart below the figure indicates the elution positions of standard 2AB labelled oligosaccharides.

Characterisation of the N-glycans attached to HuE CD59: The sialylated N-glycan pool from HuE CD59 was resolved into at least 36 peaks by normal phase HPLC (Fig. 2). Each peak was assigned an HPLC glucose unit (gu) value by comparison with the

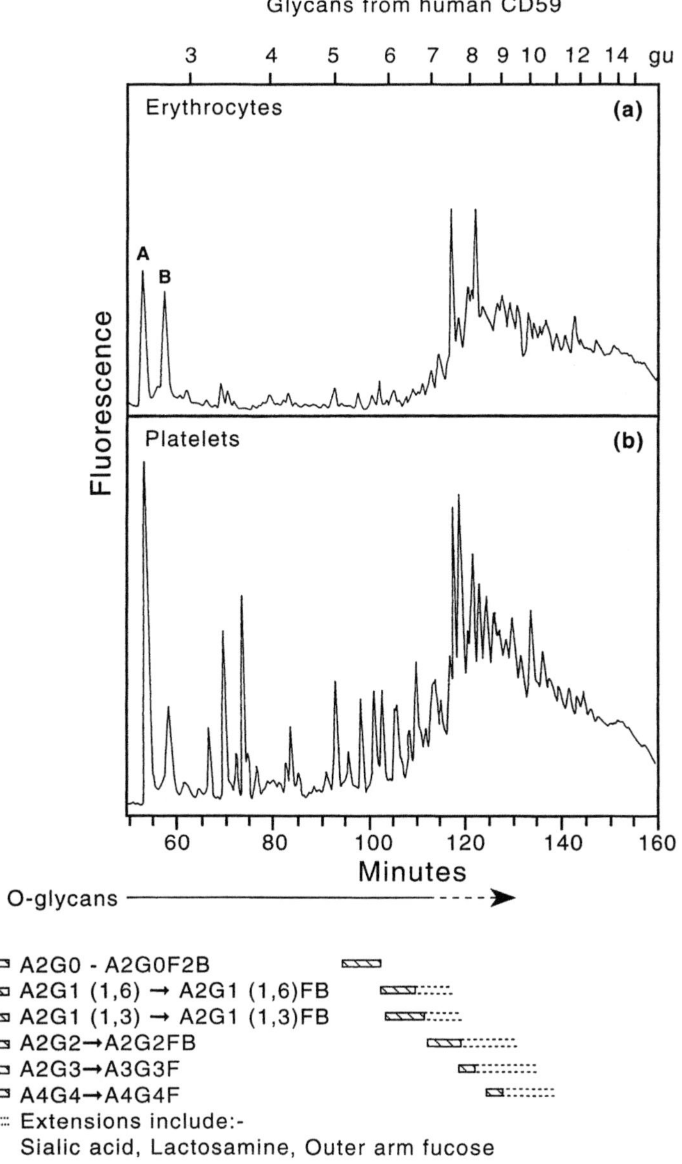

Figure 2.

Comparison of the normal phase HPLC profiles of 2AB labelled glycans released from CD59 isolated from (a) human erythrocytes (HuE) (b) human platelets. The elution positions of the dextran ladder and of the different classes of oligosaccharides are shown in the scale above and the bar chart below the figure.

elution positions of a standard 2AB-labelled dextran hydrolysate mixture (shown at the top of figure). Individual peaks were then assigned structures from their gu values using the elution positions of standard glycans and pre-determined incremental values for monosaccharide residues (13). The data indicated that human erythrocyte CD59 contained families of bi, tri, or tetra-antennary complex glycans, which were both neutral and sialylated; neither hybrid nor oligomannnose structures were detected.

MALDI MS of the desialylated N-glycan pool indicated that asialo CD59 contained at least 123 sugars associated with the one N-glycosylation site. Since MALDI MS determines only molecular masses this type of analysis does not give information with respect to isomers which arise from differences in linkage and arm specificity. These additional variations, as well as diversity arising from the addition of sialic acid residues to many of the neutral oligosaccharides, may be expected to increase the number of structures even further.

Simultaneous sequencing of aliquots of the total N-glycan pool with enzyme arrays: In view of the wide heterogeneity revealed by both the MALDI MS and HPLC analyses of the CD59 N-glycans no attempts were made to purify and analyse every individual sugar. Classically enzyme arrays and P4 GPC have been used to sequence individual sugars purified to >80% homogeneity from the glycan pool. In the case of CD59 this would have required an extremely lengthy and difficult separations procedure. Therefore a strategy was developed which involved the simultaneous sequencing of aliquots of the total glycan pool with enzyme arrays followed by HPLC analysis (Fig. 3). The glycan structures assigned by this novel strategy were confirmed as follows: (i) the sialylation status of the sugars was confirmed by WAX chromatography followed by normal phase HPLC (ii) further studies with exoglycosidase enzymes confirmed the structures of the bi-antennary non-extended complex glycans and the oligosaccharides containing core and/or outer arm fucose and (iii) glycans with lactosamine extensions were further characterised using endo b-galactosidase. The data indicated that CD59 contains (i) biantennary glycans with lactosamine extensions on both arms (ii) tetraantennary glycans with two and three lactosamine extensions and (iii) many larger polylactosamine structures present at low levels which were digested by EBG to A2G1(1,3)B and/or A3G1 (H4N5).

These latter structures are derived from tri-antennary glycans with extensions on two arms and/or biantennary glycans with bisecting GlcNAc and lactosamine extensions on one arm. The major core structures were A2G0 (30%), A2G0B/A3G0 (60%), A3G0B (5%).

Characterisation of the O-glycans attached to human erythrocyte CD59: The elution positions of A and B and the products of exoglycosidase digests of the O-glycan pool were consistent with the presence of two species of glycans, both containing the disaccharide, Galβ1,3GalNAc, but with sialic acid linked α2,3 to the Gal in one case (A) and α 2,6 to the GalNAc in the other (B) (Figure 2).

Location of possible O-glycosylation sites: Since O-glycans attached to CD59 have not been reported previously attempts were made to locate the glycosylation site/s in the protein by Edman degradation of intact CD59. The data, while not definitive, suggest that O-linked sugars may be variably attached to residues Thr51 and Thr52. Further studies are being undertaken to confirm the presence of the O-linked glycans, and to determine the linkage positions, since these structures have not been reported in earlier studies of CD59.

Molecular modelling studies of CD59: Molecular modelling studies using the X-ray crystallographic co-ordinates (14) and the enzyme binding site residues (15) indicate that PNGase F will only cleave N-glycosidic linkages when the sugar is in a conformation in which there are no interactions of the outer arm residues and the target protein. In this study, CD59 was incubated with PNGase F to probe for possible interactions of the outer arm residues with the protein. Analysis of CD59 by SDS PAGE showed that the protein migrated as a broad band with an apparent molecular weight of 20-25kD, consistent with the presence of a range of glycoforms. After incubation of native, undenatured CD59 with PNGase F, the protein migrated as a narrow band with an apparent molecular weight of approximately 15kD. This indicated that the glycoform population had been digested by the enzyme and that the Asn18 - sugar amide linkage is fully accessible to PNGase F.

Figure 4 shows a model of the CD59 structure, based on the solution structure (16), with a trisialylated tetraantennary complex N-glycan modelled at Asn 18 pointing away from the protein. The structure of the N-glycosidic linkage at Asn 18 was based on a study by Wormald et al., (17). Mono-sialylated GalGalNAc O- glycan has been modelled at Thr 51 and the glycan anchor attached to the C-terminal peptide residue. The amino acid residues (Trp40, Asp24, Arg53, Glu56) implicated in forming at least part of the active site of CD59 are highlighted, as are the potential O-glycosylation sites located at Thr 15, 51, 52 and 60 and at Ser 21.

Two structures are available for CD59, both derived from NMR solution studies (16, 18). Rotation and inspection of both structures indicates that Ser 20, which forms part of the N-glycosylation sequon, may not be readily accessible once the N-linked sugar is attached at Asn 18. However, the side chain of the adjacent residue, Ser 21, presents a possible site for O-glycosylation. Thr 15 is also close to Asn 18, but is accessible, as are the side chains of Thr 52 and Thr60. Thr 29 is inaccessible. In the structure determined by Fletcher et al. (16), the side chain of Thr51 is also accessible. Interestingly, in the structure determined by Kieffer et al., (18) a small difference in the conformation of the helix to which Thr51 is attached causes the side chain to be differently oriented so that, in this case, it is inaccessible. Thr 10 is partially hindered in both models. As a result of these observations it may be concluded that potential sites at Ser20, Thr10 and Thr29 are unlikely to be O-glycosylated.

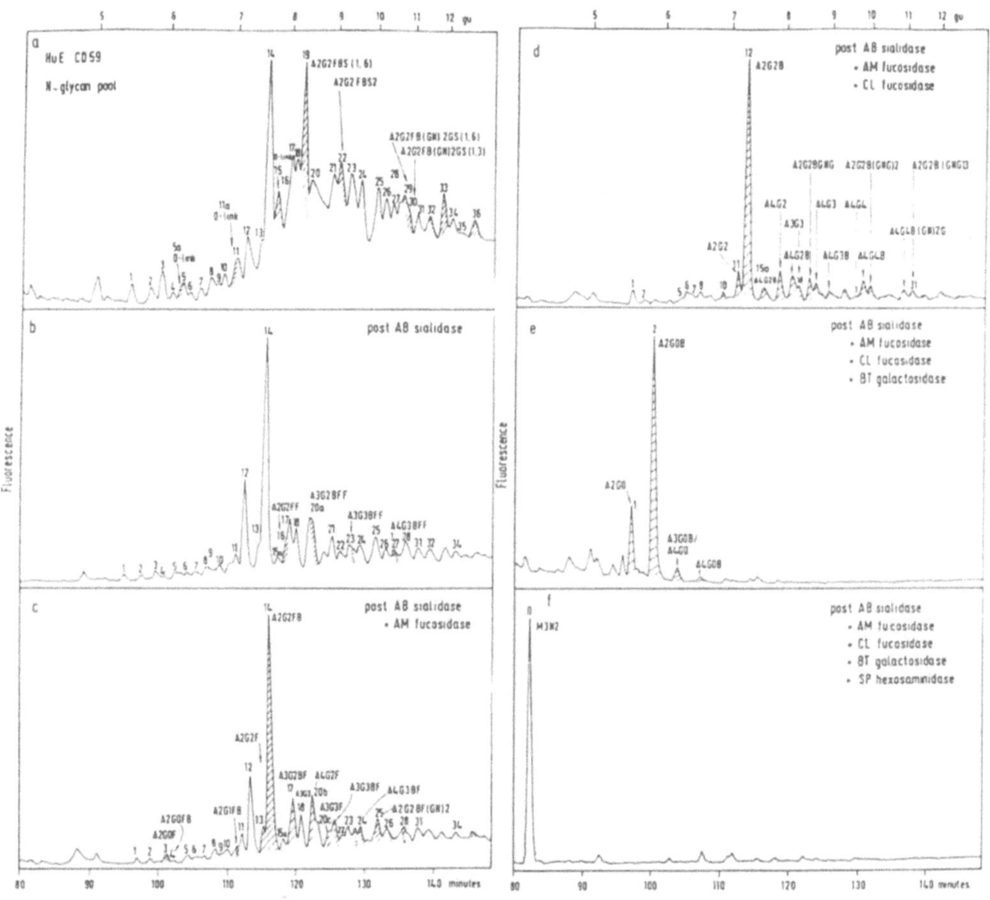

Figure 3a-f.

HPLC profiles of the N-glycan population of HuE CD59 simultaneously digested with a series of enzyme arrays. The HPLC analysis of the total glycan pool (a) and the products resulting from the digestion of five aliquots of the total CD59 glycan pool with a series of enzyme arrays (b-f). The shaded areas define the peaks which contain glycans which were subsequently digested by the additional enzyme present in the next array. The gu value of each peak was calculated by comparison with the dextran hydrolysate ladder shown at the top of the figure. Structures were assigned from the gu values, previously determined incremental values for monosaccharide residues (13) and the known specificity of the exoglycosidase enzymes.

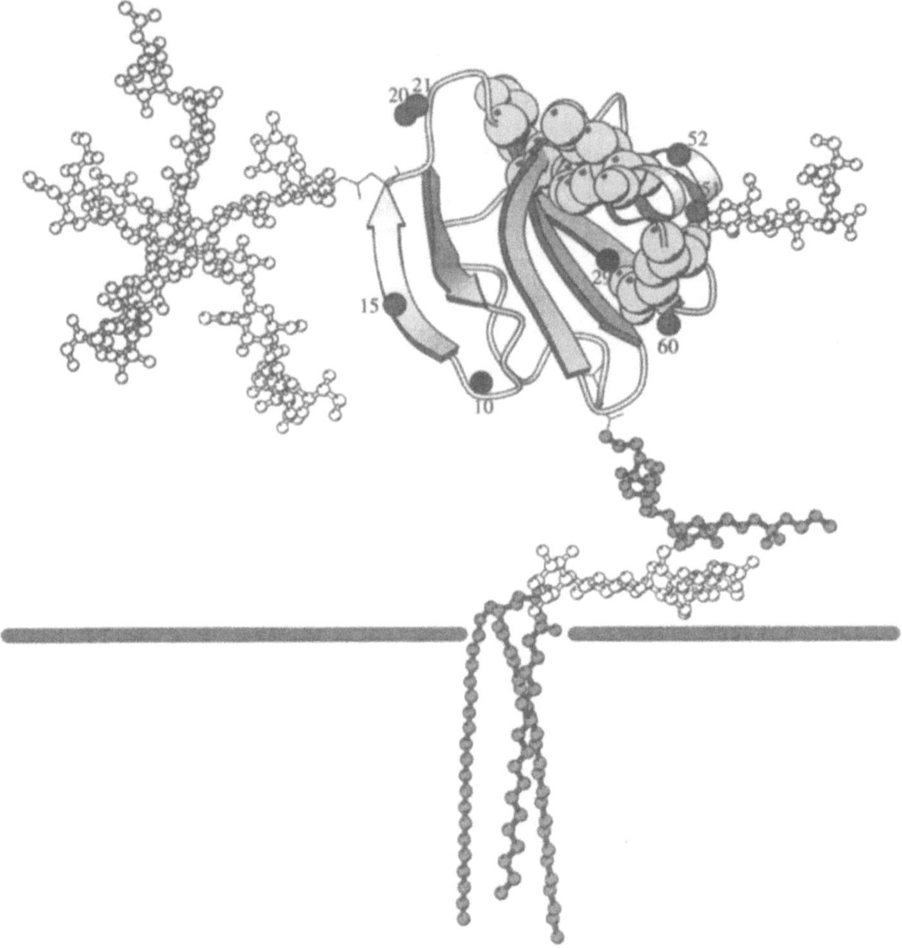

Figure 4.

A molecular model of CD59 based on the protein co-ordinates from the solution structure (14). The binding site residues (modelled with Van der Waals surfaces in grey cpk notation) are located at Trp40, Asp24, Arg53, Glu56 (23). The glycan anchor is modelled with a tri-mannosyl core, an ethanolamine bridge at Man3 and additional ethanolamine groups at Man1 and Man2. Two lipids are attached to inositol via phosphate and the third is attached directly to the inositol ring through an ester linkage. A trisialylated, tetra-antennary complex N-glycan is shown attached to Asn 18. The O-glycan, NeuNAc2,3Galb1-3GalNAc, is attached to Thr51 to indicate one of the possible linkage positions. Thr and Ser hydroxyl side chains in the protein are numbered. The sugar chains and the glycan contained within the GPI anchor are depicted in black 'ball and stick' notation.

DISCUSSION

The GPI anchored glycoprotein, CD59, consists of a heterogeneous mixture of more than 120 glycoforms. The full significance of glycosylation for the structure and function of CD59 can only be explored properly when its glycan population is viewed in the context of the complete structure of the molecule. With respect to its structure, CD59 is now the best defined cell surface molecule studied to date. The key findings of the analyses of the GPI anchor and the N- and O-linked glycans are discussed below, and an examination of a molecular model of CD59 suggests some possible general roles for the glycosylation of cell surface molecules.

A comparison of the GPI anchors attached to CD59 and acetylcholinesterase (AChE) suggests that structural features of GPI anchors may be cell type specific. This study has enabled the first comparison of GPI anchors expressed on two different glycoproteins in the same tissue, human erythrocytes. The majority (91 mol%) of the GPI anchors of CD59 were identical, in terms of PI and glycan structure, to that described for acetylcholinesterase (11,19,20).

The identical PI moieties in particular, which contain alkyl chains and C22:4 fatty acid components that are not common in mammalian PI pools (21), suggest that, in reticulocytes, the two proteins receive the same GPI precursor in exchange for their different COOH-terminal GPI signal peptides. The small proportion of CD59 anchors subsequently modified by the addition of β-GalNAc suggests that at least some human reticulocytes possess the β-GalNAc transferase required for this relatively common GPI modification. The difference in the level of GPI modification (9% for CD59 and 0% for AChE) expressed in the same cells presumably reflects factors imposed by the glycoprotein to which the anchor is attached. More comparative studies are needed to explore the measure of control which the individual protein exerts over the glycosylation of its own anchor.

The role of the palmitoyl group attached to the 2-position of the inositol ring in CD59 and AChE is unclear. Apart from making the anchors resistant to bacterial PIPLC enzymes (11), the presence of this third lipid chain may increase the stability of the interaction of these proteins with the lipid bilayer. For CD59, this may play a role in reducing the probability of CD59 leaving host erythrocytes and becoming incorporated into the membranes of complement-targets such as micro-organisms. A similar argument could be used for PIPLC resistant GPI anchored human erythrocyte decay accelerating factor (22). However, the GPI anchors of several erythrocyte proteins, including CD59, from some other mammals are PIPLC sensitive, suggesting that this is not a general mechanism for preventing cell-cell exchange of GPI anchored proteins.

Human erythrocyte CD59 consists of at least 120 different glycoforms - While the structure of the GPI anchor glycan was relatively homogeneous, MALDI MS, coupled with exoglycosidase digestions and HPLC, revealed that the N-glycan population was extremely heterogeneous. However, despite the apparent heterogeneity of the N-glycans, the major population of glycoforms consisted of a family of structures based on bisected, core fucosylated, biantennary glycans carrying varying numbers of lactosamine extensions. The relative abundance of the bisected biantennary sugar (H3N5F), increased after incubation with endob-galactosidase indicating that a high proportion of the larger sugars with the general structures H(n-1)N(n) and H(n-1)N(n)F were bisected bi-antennary sugars with polylactosamine extensions on one arm. Interestingly, most of the larger structures contained one or more fucose residues α1-3 linked to a GlcNAc residue.

The major single sugar (A2G2FB; 15% of the total sugars) contained both a bisecting GlcNAc residue and a core fucose. Various truncated structures lacking one or both of the galactose residues and the bisecting GlcNAc were also present (37% of the total sugars). In addition to the extended bi-antennary glycans, tri- and tetra-antennary sugars were also identified. The maximum size of the sugars could not be determined because their relative abundance decreased with increasing molecular weight and eventually fell below the detection limit of the MALDI experiment. The largest structures detected contained approximately 24 monosaccharide residues.

Human erythrocyte CD59 contains two potential O-glycosylation sites. A population of sialylated O-glycans was recovered from human erythrocyte CD59; the major species which were identified were two mono-sialylated forms of the disaccharide Galβ1,3GalNAc.

Molecular modelling suggests several roles for the glycans attached to CD59. The N-linked oligosaccharides (size range 3-6nm in length) attached to the disc-like extra-cellular region of CD59 (diameter approximately 3nm) project away from the protein domain in the plane of the active face and adjacent to the membrane surface (Fig. 4). The glycans do not appear to restrict access to proposed active site residues of human CD59 (Asp24, Trp 40, Arg 53 and Glu 56) located on the membrane distal surface of the extracellular domain (23).

However, the glycans would be expected to restrict the rotational freedom of the extracellular domain around axes parallel to the membrane which may, in turn, stabilise an exposed location for the active face. Removal of the conserved N-linked glycan might therefore reduce the affinity of CD59 for the membrane attack complex without eliminating it completely. The effects of removing the N-linked glycan might therefore be expected to depend on the density of expression of the glycoprotein at the cell surface and this may explain the observed variation in the activities of unglycosylated CD59 (23, 24). A discrepancy between the measured two dimensional affinity of the cell surface recognition molecules CD2 and LFA-3, and the two-dimensional affinity predicted on the basis of the measured three dimensional affinity (6), suggests that the mobility of the CD2 ligand binding site is restricted perpendicular to the cell surface (25).

It seems likely that N-glycans attached to the highly conserved, membrane proximal glycosylation site at the base of CD2 may contribute to this effect (26). This suggests that restricting the conformational space available to cell surface glycoproteins may be a key function of N-glycosylation.

Molecular modelling of the GPI-anchored glycoprotein Thy-1 (27) suggested that the glycoprotein sits directly on the membrane and that there are extensive interactions between the protein and the GPI glycan. In contrast to Thy-1, where the anchor is attached to a cysteine residue involved in an intra-molecular disulphide bond (28), CD59 contains a hydrophilic C-terminal sequence linking the protein to the GPI anchor. The protein component of CD59 would therefore have considerable dynamic freedom relative to the membrane and it seems likely that there would be very few non-covalent interactions between the protein and the anchor. This view is supported by NMR data from the analysis of the soluble form of CD59 which shows that the last few hydrophilic residues have a poorly defined structure. The bulky, hydrophilic glycans in the anchored form of CD59 would be expected to limit interactions with the lipid bilayer and in this way facilitate diffusion of the glycosylated protein in the membrane (Fig. 5).

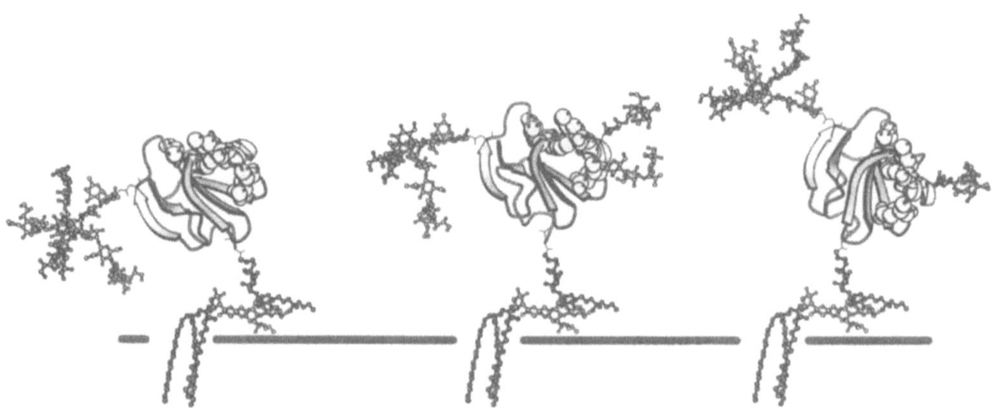

Figure 5.

A schematic figure showing the effect of CD59 glycosylation on the flexibility of the protein relative to the GPI anchor. Three different glycoforms are shown. Both the O- and N-linked oligosaccharides (size range 2-6nM) attached to CD59 (diameter approximately 3nM) restrict the conformational space available to the protein and limit its interaction with the lipid bilayer. The sugars may therefore orient the N-terminal active site of CD59 towards the C5b-9 complex, which is also inserted into the cell membrane. In addition, the heterogeneity of the sugars suggests that the glycans influence the geometry of the packing and it is unlikely that CD59 molecules will form a regular array on the cell surface.

The heterogeneity of the sugars suggests that the glycans influence the geometry of the packing and it is likely that they will also prevent the aggregation of CD59 molecules on the cell surface. By limiting such protein-protein interactions the glycans could influence the distribution of CD59 molecules at the cell surface where GPI anchored proteins may associate in micro-domains in dynamic equilibrium with isolated individual molecules. The large N-glycans may also be important in preventing proteolysis of the extracellular domain since N-glycosylation has been shown to increase the dynamic stability of a protein while different glycoforms variably increase its resistance to protease digestion (29).

CONCLUSION

In this study the structures of the N- and O-glycans of CD59 and the glycan and lipid components of its GPI anchor have been defined. The major findings are (i) HuE CD59 and acetyl cholinesterase (26) have essentially identical GPI anchors suggesting that erythrocytes synthesise only one type of PIPLC resistant GPI anchor; (ii) in contrast, the N-glycans are highly heterogeneous and their structure precludes erythrocyte CD59 as a precusor of the urinary form and (iii) HuE CD59 contains a limited range of O-glycans, but the O-glycosylation site/s have not yet been unambiguously identified.

Possible roles for the sugars were suggested from an examination of the molecular model which was constructed from the NMR solution structure (16) and the analyses reported in this study. Such roles may include stabilising an exposed location for the active face of CD59, protease protection, orientation of the molecule and spacing or packing of the CD59 molecule on the cell surface. The finding that HuE CD59 contains more than 120 complex type sugars associated with the N-glycosylation site is entirely consistent with these proposals.

Acknowledgements: The authors would like to thank Joshua Dwek for his valuable contribution to the molecular modelling studies. SJD is supported by the Welcome Trust. MAJF is supported by the Welcome Trust and is a Howard Hughes Medical Institute International Research Scholar.

REFERENCES

1. Walsh, L.A., Tone, M., Thiru, S., Waldman, H. (1992) Tissue Antigens 40, 213-220
2. Williams, A.F. (1991) Cell Biol Int Rep 15, 769-77
3. Brooimans, R.A., van der Ark, A.A.J., Tomita, M., van Es, L.A., and Daha, M.R. (1992) Eur. J. Immunol. 22, 791-796
4. Deckert, M., Kubar, J., Zoccola, D., Bernard-Pomier, G., Angelisoval, P., Horejsi, V. and Bernard, A. (1992) Eur. J. Immunol. 32, 2943-2947
5. Hahn W.B., Menu, E. Bothwell, A.L.M., Sims P.J., Bierer, B.E. (1992) Science 256, 1805-1807
6. van der Merwe, A., Barclay, N., Mason, D., Davies, E., Morgan, B.P., Tone, M., Krishnam, A.K.C., Lanelli, C., and Davis, S. (1994) Biochemistry 33, 10149-10160
7. Arulanandam, A.R.N., P. Moingeon, M.F. Concino, M.A. Recny, K. Kato, H. Yagita, S. Koyasu and E.L. Reinherz. (1993) J. Exp. Med. 177, 1439-1450
8. Davies, A., Simmons, D.L.., Hale, G., Harrison, R.A., Tighe, H., Lachmann, P.J. and Waldman, H. (1989) J. Exp. Med. 170, 637-654
9. Tomita, M., Tobe,T., Choi-Miura, N., Nakano, Y., Kusano, M. and Oda, E. (1991) Abstracts XIVth Int. Complement Workshop p.233
10. Rushmere, N.K., Harrison, R.A., van den Berg, C.W., Morgan, B.P. (1994) Biochem. J. 304:595- 601
11. Roberts, W.L., Santikarn, S., Reinhold, V.N. and Rosenberry, T.L. (1988a) J.Biol.Chem. 263, 18776- 18784
12. Volwerk J.J., Shashidhar, M.S., Kuppe, A., and Griffith, O.H. (1990) Biochemistry 29, 8056-8062
13. Guile, G.R., Rudd, P.M., Wing, D.R. and Dwek, R.A.(1996) Anal. Biochem 240 210-226
14. Norris, G.E., Stillman, T.J., Anderson, B.F. and Baker, E.N. (1994) Current Biology Structure 2, 1049-1059
15. van Rooey, P., Rao, V., Plummer, T.H. Jr, Tarentino, A.L. (1995) Biochemistry 33, 13989-13996
16. Fletcher, C.M., Harrison, R.A., Lachman, P.J. and Neuhaus, D. (1994) Current Biology Structure 2, 185-199.
17. Wormald, M.R., Wooten, E.W., Bazzo, R., Edge, C.J. Feinstein, A., Rademacher, T. W., Dwek, R.A. (1991) Eur. J. Biol. 198, 131-139
18. Kieffer, B., Driscoll, P.C., Campbell, I.D., Willis, A.C., van der Merwe, P.A. and Davis, S.J. (1994) Biochemistry 33, 4471-4482
19. Roberts, W.L. Myher, J.J., Kuksis, A., Low, M.G. and Rosenberry, T.L. (1988b) J.Biol.Chem. 263, 18784-18775
20. Deeg et al., (1992) J. Biol. Chem. 267, 18573-18580
21. Kerwin, J.L., Tuininga, A.R., Ericsson, L.H. (1994) J.Lipid Res. 35, 1102-1114
22. Walter, E.I., Roberts, W.L., Rosenberry, T.L., Ratnoff, W.D. and Medof, M.E. (1990) J. Immunol. 144, 1030-1036
23. Bodian, D.L., Davis, S.J., Rushmere, N.K. and Morgan, B.P. - submitted.
24. Ninomiya, H., Stewart, B.H., Rollins, S.A., Zhao, J., Bothwell, A.L.M. and Sims, P.J. (1992) J. Biol. Chem. 267, 8404-8410
25. Davies, A. and Morgan, B.P. (1993) Biochem. J. 295, 889-896
26. Dustin, M.L., Ferguson, L.M., Chan, P-Y., Springer, T.A., Golan, D.E. (1996) J.Cell Biol. 132, 3 456-474
27. Rademacher, T.W.R., Edge, C.J. and Dwek, R.A. (1991) Current Biology 1, 41-42

28. Homans, S.W., Ferguson, M.A.J., Dwek, R.A., Rademacher, T.W., Anand, R. and Williams, A.F. (1988) Nature 333, 269-272
29. Rudd, P.M., Joao, H.C., Coghill, E., Fiten, P., Saunders, M. R., Opdenakker, G. , Dwek, R.A. Biochemistry (1994) 33, 17-22

GLYCOSYLATION AND RHEUMATIC DISEASE

John S. Axford

Academic Rheumatology Unit
St. George's Hospital Medical School
Cranmer Terrace
London SW17 ORE
United Kingdom

INTRODUCTION

Oligosaccharide biochemistry has been an important field of research for many years and this area of basic science continues to flourish.

Glycobiology deals with the nature and role of carbohydrates in biological events (1,2,3). It is now clear the oligosaccharides may indeed be more than a decorative irrelevance when it comes to molecular mechanisms of diseases (4-11). The last decade has witnessed an exponential interest in the associations of oligosaccharides with disease mechanisms, and significant advances have been made in the design of carbohydrate based therapies and diagnostic techniques.

OLIGOSACCHARIDES SYNTHESIS

Recent work from many laboratories has highlighted the fact that oligosaccharides are mediators of biospecific information by virtue of their structural complexity. For example, the sialyl-Lewis[x] tEtrasaccharide, one of the crucial molecules involved in cell adhesion and trafficking (3,12,13). The mechanisms of their synthesis is therefore crucial. At the centre stage of the synthesis of glycoside bonds are the glycosyltransferases whose specificity for their substrates, the donor sugar nucleotides, and acceptor is precise (14). The catalytic function the glycosyltransferases exert is essentially irreversible and results in generation of nucleoside-diphosphates, which are further degraded to monophosphates, by phosphatases co-localised with glycosyltransferases. The nucleoside-monophosphate is then exchanged by an antiport mechanism across the Golgi membrane with newly made sugar-nucleotides synthesised in the cytoplasm.

The most important organelle involved in glycosylation is the Golgi apparatus (GA) originally described as "appareil reticulaire interne" by Camillo Golgi (1898) and then confirmed to be a morphological entity by electron microscopy (15). Its biochemical properties have been explored by fractionation using galactosyltransferase (GTase) as a marker enzyme. Immunoelectron microscopical localisation of this and other enzymes has clearly established the GA as the site of glycan chain elongation and termination (16,17). The enzyme function within the cisterne is well ordered and GTase has been found predominantly on the trans side of the GA. Recent work has indicated that there is a sequential arrangement of the glycosyltransferases which increases their biosynthetic efficiency (18). The sequential action of glycosyltransferases produces oligosaccharide and glycan structures that reflect the constitution of the multi-glycosyltransferase system (19) of a given cell type and implies that the targeting mechanism for glycosyltransferases along the secretory pathway is highly ordered (20).

The synthesis of a glycan specific for a given cell type thus depends on a variety of factors (21):
- The expression of functional glycosyltransferase at the proper location within the cell;
- Synthesis and translocation of sugar nucleotides to the biosynthetic micro-environmental of the Golgi lumen;
- The proper trafficking of intra-cellular acceptor substrates;
- Absence of toxic influences such as inhibitors of processing enzymes and glycosyltransferases.

GALACTOSYLTRANSFERASE

The glycoprotein UDPß1,4 galactosyltransferase (GTase) is an intracellular membrane-bound enzyme that can be localised to the Golgi apparatus (22), but may also be found on the cell surface (23) and in a soluble form in milk, amniotic fluid, cerebrospinal fluid, saliva, urine and serum (24).

The molecular role of GTase is to catalyse the transfer of galactose from UDP-galactose to an N-acetylglucosamine acceptor during oligosaccharide elongation, for example, in IgG glycosylation (25). The gene encoding human GTase is thought to be located in chromosome 9 (26), and it may specify more than one mRNA transcript (27).

GTase seems to play a multifunctional role in normal cell physiology and has been associated with sperm-egg binding (28), cell-cell recognition (29), embryonic maturation (30), and cell development (31).

IMMUNOGLOBULIN G

The glycoprotein IgG has a conserved N-linked glycosylation site in the Fc region, with variable glycosylation in the Fab depending on the presence or absence of the glycosylation motif in the variable region and whether it is conformationally possible. Investigation of the glycoforms of serum IgG has established significant differences in oligosaccharide structure between groups of patients with certain diseases compared with IgG from healthy controls. Different glycoforms arise due to the presence or absence of certain sugar residues on the oligosaccharide backbone (32). Figure 1 shows a composite structure of the N-linked oligosaccharide chain on human IgG and indicates how some structural variations can arise.

RHEUMATOID ARTHRITIS

Rheumatoid arthritis (RA) is a multisystem disorder in which immunological abnormalities characteristically result in symmetrical joint inflammation, articular erosions and extra-articular complications. It is the most common and disabling autoimmune arthritis, and genetic susceptibility is well defined (33).

AGALACTOSYL-IgG HAS A PATHOGENIC ROLE

We have used a series of IgG preparations differing in their content of oligosaccharide chains lacking galactose from 18 to 86% to determine whether changes in sugar content affect the binding of rheumatoid factor (34). Five of 16 monoclonal rheumatoid factors prepared from synovial tissue, from patients with juvenile or adult rheumatoid arthritis, bound better to IgG which was deficient in galactose (Fig.2). Six of the 16 rheumatoid factors from the same patients bound independently of the galactose content. Four of the 16 rheumatoid factors could not be absolutely grouped in this manner but seemed to demonstrate a preference for agalactosyl IgG. One rheumatoid factor bound better to fully galactosylated IgG. There was an association between enhanced binding to galactose-deficient IgG and monoreactivity and a very strong association between the functional affinity of the rheumatoid factors and the dependent binding.

164

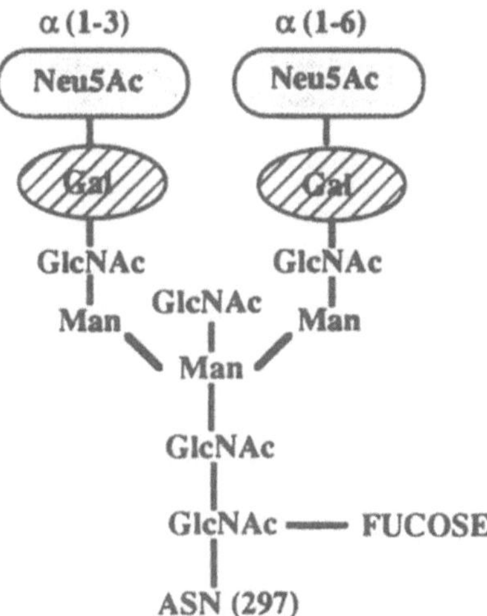

Figure 1.
The biantennary complex oligosaccharide structure found at the Fc region of human serum IgG.
There is a common pentasaccharide core containing two α-mannosyl residues attached to a β-mannosyl-di-N-acetylchitobiose unit. Terminal sialic acid (Neu5Ac) is usually found in the Fab region of IgG, whereas terminal galactose (Gal) and N-acetylglucosamine (GlcNAc) are more common in the Fc region. Fucose and the GlcNAc bisect are relatively uncommon in the Fc sugars.

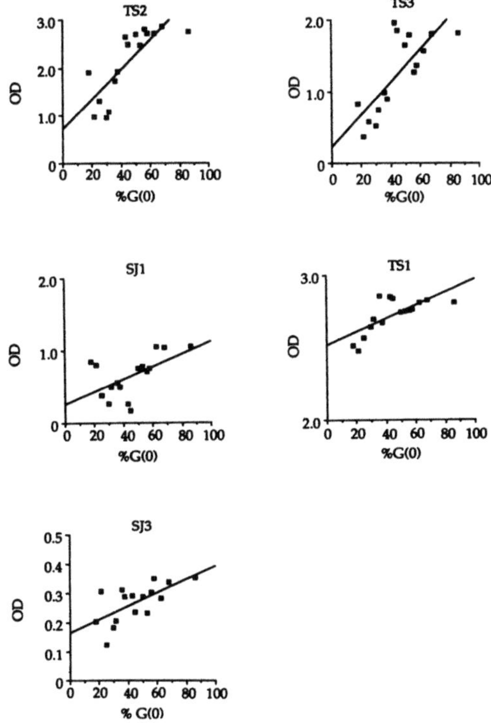

Figure 2.
Binding of five synovial tissue-derived IgM monoclonal RF's to IgG preparations varying in G(0). The horizontal axis represents G(0) and the vertical axis represents optical density at 450mm wavelength. A line of best fit is shown for each graph.

GLYCOSYLATION HOMEOSTASIS WITHIN RHEUMATOID ARTHRITIS LYMPHOCYTES IS THOUGHT TO BE ABNORMAL

To investigate potential mechanisms controlling protein glycosylation we have studied the interrelationship between lymphocytic GTase activity and serum agalactosylated immunoglobulin G levels (G(0) in healthy individuals and patients with rheumatoid arthritis and non-autoimmune arthritis (35) In RA there was reduced GTase activity (Fig.3) and increased G(0). A positive linear correlation between B and T cell GTase was found in all individuals (Fig.4). The relationship between GTase and G(0) was found to be positive and linear in the control population (Fig.5) and negative and linear in the RA population (Fig.6). These data describe a defect in RA lymphocytic GTase, with associated abnormal G(0) changes. A possible regulatory mechanism controlling galactosylation in normal cells is suggested. This is disrupted in RA, where the positive feedback between GTase and G(0) is lost. We suggest that these mechanisms are of relevance to the pathogenesis of RA, and that their manipulation may form part of a novel therapeutic approach.

IN WHAT WAY COULD ENZYMATIC CONTROL BE ABNORMAL?

We have investigated possible specific ß1,4 GTase isoenzyme changes in serum of patients with RA (36). Using solution phase isoelectric focusing, we have determined the isoenzyme profiles in GTase in healthy individuals (HI) (n=9) and patients with psoriatic arthritis (PsA) (n=9) and RA (n=8). GTase activity was determined using a previously reported assay, in which 3_H-galactose is transferred to ovalbumin.

Comparison of GTase isoenzyme activity profiles demonstrated that the RA group (Fig.7) was significantly different from both the PsA patients and HI (Fig.8) (p < 0.01). In 7 patients with RA, 8 HI and 6 PsA, two fractions showed distinct peaks of activity. The first peak of activity formed at pH 4.5 (range 4.3-4.7), pH 4.6 (range 4.4-5.0) and 4.65 (range 4.35-5.0), whilst the second peak of activity focused at pH 5.0 (range 5.0-5.1), pH 5.2 (range 5.0-5.4) and 5.20 (range 5.0-5.50) in RA, HI and PsA populations respectively. There was a significant shift in the PI position of the second peak when comparing the RA group to the healthy individuals (p = 0.006). A significant increase in GTase activity was noted in the RA second peak (p = 0.008) when compared to HI and PsA.

Two main peaks of serum GTase activity have therefore been identified and there is thus evidence that RA associated serum GTase isoenzymes occur. The RA GTase isoenzyme profile is significantly different from the HI and PsA as there is acidic skewing of the first peak, a prominent second peak concentrated over a narrow pH band (pH 4.97-5.11) and a loss of activity in the pH range 5.20-5.60. The significance of these findings now needs to be determined with reference to RA pathogenesis and the enzyme contained within these peaks fully characterised.

ARE THE RA ASSOCIATED GLYCOSYLATION ABNORMALITIES UNIQUE?

We have investigated the relationship between exposed galactose and N-acetylglucosamine on IgG in RA, juvenile chronic arthritis (JCA) and Sjögren's syndrome (SS) (37). This was achieved using IgG isolated from serum where the levels of galactose and N-acetylglucosamine (GlcNAc) were detected using biotinylated lectins. Galactose and GlcNAc on IgG from patients with RA and JCA are inversely related, but in contrast, in SS, galactose expression on IgG decreased while GlcNAc expression remained similar to normal levels (Fig.9). Alterations in IgG glycosylation are closely associated with the development of adult and juvenile chronic arthritis and SS, but the changes involved are different in RA compared with SS, suggesting that the precise pattern of exposed sugars is associated with different rheumatological diseases.

SUGAR PRINTING RHEUMATIC DISEASES IS POSSIBLE

To determine whether specific glycosylation profiles and hence pathogenic mechanisms are associated with rheumatic disease, detailed IgG oligosaccharide profiles were obtained from a spectrum of rheumatic conditions (38).

High performance liquid chromatography (HPLC) analysis was carried out to identify a total of 16 distinct IgG derived neutral oligosaccharide structures (Fig.10) from patients with RA (n=5), systemic lupus erythematosus(SLE) (n=10), primary Sjögren's syndrome (PSS) (n=6), AS (n=10), JCA (n=13), PsA (n=9) and healthy individuals (n=19).

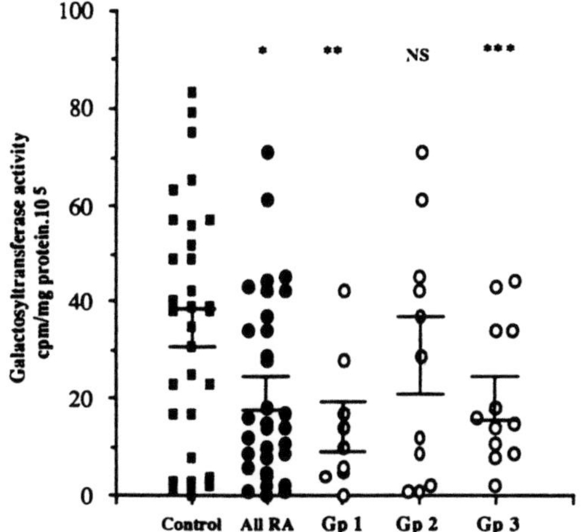

Figure 3.
B lymphocytic galactosyltransferase activity (mean cpm/mg) protein $10^5\pm$SEM) in the total RA population and divided according to drug therapy. Control: 35±4. All RA: 21±3. Gp1 (NSAID or no drugs): 14±5. Gp2 (SASP): 28±8. Gp.3 (other second/third line therapy): 21±4. Bars define the mean ±SEM. (Significance of difference from controls: *P<0.01; ** P<0.02; ***P<0.05).

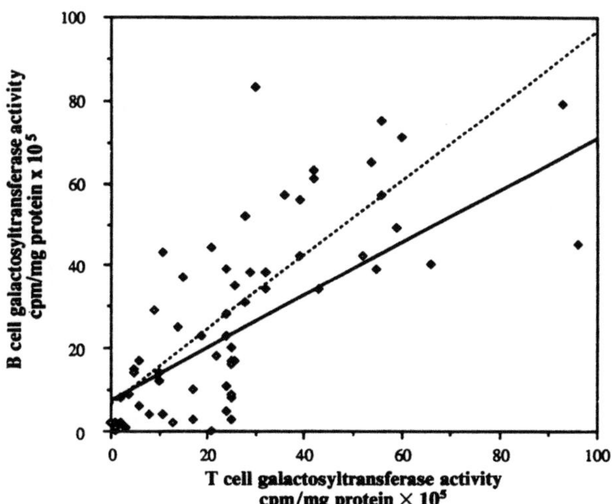

Figure 4.
Regression analysis of control (closed diamonds and broken line, r=0.770, P<0.001) and RA (open diamonds and unbroken line, e=0.691, P<0.001) paired B and T lymphocytic galactosyltransferase activities.

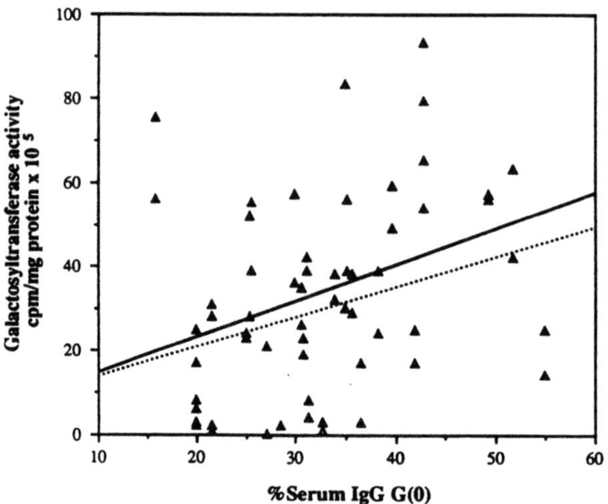

Figure 5.
Regression analysis of paired B (open triangles and unbroken line) and T (closed trianges and dotted line) lymphocytic galactosyltransferase activities and serum agalacto-immunoglobulin G (percent serum IgG G(0)) in the control population (r=0.263 and 0.211 for B and T cells, respectively, P<0.05).

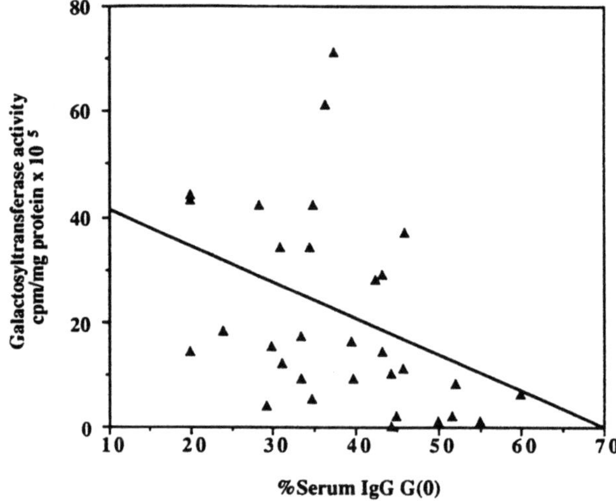

Figure 6.
Regression analysis of paired B lymphocytic galactosyltransferase activity and serum agalacto-immunoglobulin G values (percent serum IgG G(0) in the rheumatoid arthritis population (r+0.338, P<0.004). There is a significant difference (P<0.01) when this value is compared with the control regression value (r=0.263).

168

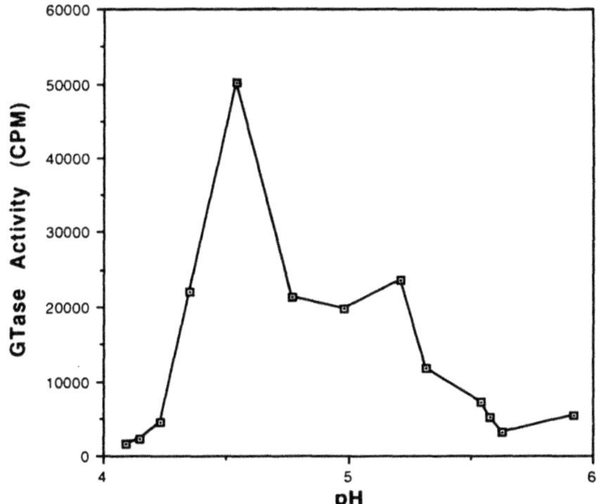

Figure 7.
A profile of GTase activity plotted against pH following solution phase isoelectric focusing of serum from a healthy individual. There is a large peak forming at pH 4.55 and much smaller peak at pH 5.20.

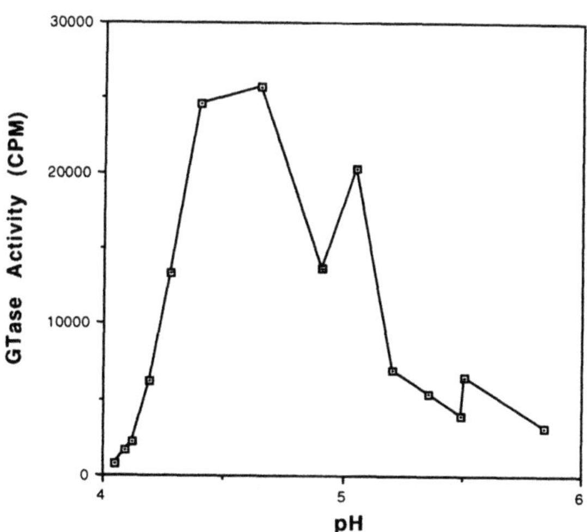

Figure 8.
This is a profile of GTase activity plotted against pH following solution phase isoelectric focusing of serum from a patient with RA. A large peak is seen at pH 4.65 and a smaller peak forms at pH 5.05. There is a left shift of the second peak, and it constitutes a significantly greater proportion of the total activity when compared to the healthy population.

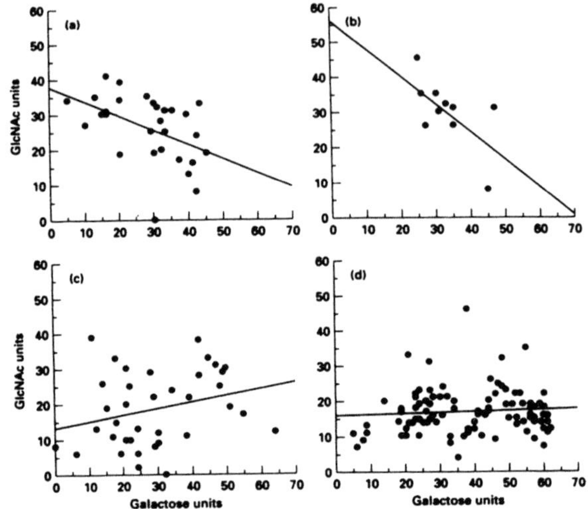

Figure 9.
(a) Correlation between galactose and GlcNAc levels on RA IgG (n=29)
(b) Correlation between galactose and GlcNAc levels on JCA IgG (n=11)
(c) Correlation between galactose and GlcNAc levels on SS IgG (n=37)
(d) Correlation between galactose and GlcNAc levels on normal IgG (n=100).

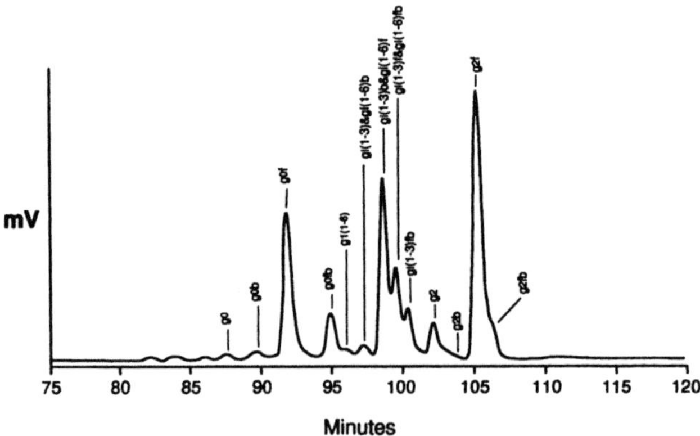

Figure 10.
An example of a normal phase HPLC profile from a healthy control, annotated to show the oligosaccharide structures represented by each peak. Normal phase resolves sixteen oligosaccharides into thirteen peaks. There are four agalactosyl and four digalactosyl oligosaccharide peaks and monogalactosyl oligosaccharides form five peaks with co-elution of different oligosaccharides according to their arm specificity.

The proportions of the different oligosaccharide structures in each disease group was significantly different from each other and the healthy controls (p<0.01). In particular, specific associations were found with each disease group with respect to the proportion of galactose (g0, g1, g2 depending on galactose residues), N-acetylglucosamine (b=bisecting), sialic acid and fucose (f). Oligosaccharide profiles comprising these oligosaccharide changes were disease specific, giving a unique sugar print for each disorder. For example, the RA sugar print showed an increase in g0, g0f, g0fb and reduction in g1(1,3)b, g1(1,6)f, g2f structures. In contrast, the SLE sugar print showed an increase in g1(1,6), g2, g2fb and reduction in g1(1,3)b, g1(1,6)f and g2f structures.

Disease specific IgG sugar prints have therefore been demonstrated. This suggests that disease specific glycosylation of IgG occurs during synthesis and may indicate that IgG is involved in different pathogenic mechanisms in each disease. It also raises the possibility that sugar prints may be useful in the diagnosis and differentiation of rheumatic diseases.

CONCLUSION

To date our work concerning the glycoimmunology of rheumatoid arthritis and other rheumatic diseases has demonstrated a significant defect in the galactosyltransferase enzyme that results in a profound change in the galactosylation of immunoglobulin G. This change has been demonstrated to be integrally associated with pathogenic mechanisms associated with inflammation in rheumatoid arthritis. It is hypothesised that there are major disruptions in glycosylation homeostasis associated with the pathogenesis of rheumatoid arthritis and that perhaps in part these involve the generation of unique galactosyltransferase isoenzymes that differ functionally from those present in healthy individuals. It is not thought that these changes are unique to rheumatoid arthritis but it is thought that there may be subtle changes in the disruption of glycosylation homestasis causing a unique sugar change to be associated with a number of other rheumatic diseases. This is referred to as **"sugar printing the rheumatic diseases"** and may be a concept both useful diagnostically and therapeutically.

REFERENCES

1. Dwek RA. Glycobiology: Towards undertanding the function of sugars. (Wellcome Trust Award Lecture 1994) Biochemical Society Transactions 23 1995

2. Feizi T. Demonstration by monoclonal antibodies that carbohydrate structures of glycoproteins and glycolipids are onco-developmental antigens. Nature 1985; 314: 53-57.

3. Crocker PR & Feizi T.Carbohydrate recognition systems: functional triads in cell-cell interactions. Curr Opin. Structur Biol 1996 (In Press).

4. Axford JS et al (1992) Second Jenner International Glycoimmunology Meeting. Ann Rheum Dis 1269-1285.

5. Axford JS et al (1994) Third Jenner International Glycoimmunology Meeting. Glycosylation & Dis 3: 197-227.

6. Axford JS (1993) Trends. Second Jenner International Glycoimmunology Meeting Immunology Today 14(3): 104-106.

7. Axford JS (1995) Trends. Third Jenner International Glycoimmunology Meeting Immunology Today 16(5): 213-215.

8. Axford JS (1991) Oligosaccharides: An optional extra or of relevance to disease mechanisms in rheumatology? J Rheumatol 18(8): 1124-1127.

9. Axford JS (1994) Glycosylation and rheumatic disease: more than icing on the cake. J Rheumatol 21(10): 1791-1795.

10. Alavi A & Axford JS (1996) Chapter: The glycosyltransferases from abnormalities of IgG glycosylation and immunological disorders. Edited by Isenberg DA & Rademacher TW. Published by John Wiley & Sons.

11. Alavi A & Axford JS. Glycoimmunology. Published by Plenum Publishing Co. Ltd 1995.

12. Varki A. Biological roles of oligosaccharides: all of the theories are correct. Glycobiology 1993;3(2): 97-130.

13. Feizi T. Oligosaccharides that mediate mammalian cell-cell adhesion. Curr Opin. Struct Biol. 1993;3: 701-710.

14. Hagopian A & Eylar EH. Glycoprotein biosynthesis: studies on the receptor specificity of the polypeptide: N-acetylgalactosaminyl-transferase from bovine submaxillary glands. Arch. Biochem Biophys 1968;128(2): 422-433.

15. Farquhar MG & Palade G. The Golgi apparatus (complex) - (1954-81) from artifact to center stage. J Cell Biol 1981;91(3): 77-103.

16. Roth J & Berger EG. Immunocytochemical localisation of galactosyltransferase in HeLa cells: co-distribution with thiamine pyrophosphatase in trans-Golgi cisternae. J Cell Biol 1982;93(1): 223-229.

17. Roth J. Taatjes DJ, Lucocq JM, Weinstein J, Paulson JC. Demonstration of an extensive trans-tubular network continuous with the Golgi apparatus stack that may function in glycosylation. Cell 1985;43(1): 287-295.

18. Rabouille C, Hui N, Hunte F, Kieckbusch R, Berger EG, Warren G, Nilsson T. Mapping the distribution of Golgi enzymes involved in the construction of complex oligosaccharides. J Cell Sci. 1995;108(4): 1617-1627.

19. Roseman S. The synthesis of complex carbohydrates by multi-glycosyltransferase systems and their potential function in intercellular adhesion. Chem Phys Lipids 1970; 5(1): 270-297.

20. Nilsson T & Warren G. Retention and retrieval in the endoplasmic reticulum and the Golgi apparatus. Curr Opin Cell Biol 1994;6(4): 517-521.

21. Berger EG & Thurnher M. Clues to the cell-specific synthesis of complex carbohydrates. News Physiol Sci. 1993;8: 57-60.

22. Berger EG, M Thurnher and U Muller. Galactosyltransferase and sialyltransferase are located in different subcellular compartments in HeLa cells. Exp Cel Res (EPB) 1987;173: 267-73.

23. Shur BD and S Roth. Cell surface glycosyltransferases. Biochim Biophys Acta 1975;415: 473-512.

24. Schachter H and L Rodén. The biosynthesis of animal glycoproteins. In: Metabolic Conjugation and Metabolic Hydrolysis; WH Fishman, Editor; Academic Press, New York. pp 1 - 149.

25. McGuire EJ, R Kerlin, JJ Cebra and S Roth. A human milk galactosyltyransferase is specific for secreted, but not plasma IgA. J Immunol 1989;143(9): 2933-8.

26. Shaper NL, JH Shaper, V Bertness, H Chang, IR Kirsch and Hollis GF. The human galactosyltransferase gene is on chromosome 9 at band p13. Somatic Cell Mol Genet 1986;12:

27. Russo RN, NL Shaper and JH Shaper. Bovine β 1,4 galactosyltransferase: two sets of mRNA transcripts encode two forms of the protein with different amino-terminal domains. In vitro translation experiments demonstrate that both the short and the long forms of the enzyme are type II membrane-bound glycoproteins. J Biol Chem 1990;265(6): 3324-31.

28. Shur, BD,and NG Hall. Sperm surface galactosyltransferase activities during in vitro compaction. J Cell Biol 1982;95: 567-573.
29. Bayna EM, RB Runyan, NF Scully, J Reichner, LC Lopez and BD Shur. Cell surface galactosyltransferase as a recognition molecule during development. Mol Cell Biochem 1986;72: 141-51.
30. Bayna ME, JH Shaper and BD Shur. Temporally specific involvement of cell surface β-1,4 galactosyltransferase during mouse embryo morula compaction. Cell 1988;53: 145-157.
31. Begovac PC and BD Shur. Cell surface galactosyltransferase mediates the initiation of neutite outgrowth from PC12 cells on laminin. J Biol Chem 1990;110(2): 461-70.
32. Parekh RB, Dwek RA, Sutton BJ etl al. Association of rheumatoid arthritis and primary osteoarthritis with changes in the glycosylation pattern of total serum IgG. Nature 1985;316:452-457.
33. Axford JS. Medicine Published by Blackwell Science 1996.
34. Soltys AJ, Hay FC, Bond A, Axford JS, Jones MG, Randen I, Thompson K & Natvig J. The binding of synovial tissue-derived human monoclonal immunoglobulin M rheumatoid factor to immunoglobulin G preparations of differing galactose content. Scan J Immunol 40(2): 135-143 (1994).
35. Axford JS, Sumar N, Alavi A, Isenberg DA, Young A, Bodman KB, Roitt IM. Changes in normal glycosylation mechanisms in autoimmune rheumatic disease. J Clin Invest 89(3): 1021-1031 (1992).
36. Pool AJ, Alavi A & Axford JS. β1,4-galactosyltransferase isoenzyme changes in serum of patients with rheumatoid arthritis. Br J Rheumatol 35(1) Suppl. 174 No.335 (1996).
37. Bond A, Alavi A, Axford JS, Youinou P & Hay FC. The relationship between exposed galactose and N-acetylglucosamine residues on IgG in RA, JCA and Sjögren's syndrome. Clin Exp. Immunol 105: 99-103 (1996).
38. Watson M, Rudd P, Dwek RA & Axford JS. Sugar printing rheumatic diseases: A potential method for diagnosis and differentiation using immunoglobulin G oligosaccharides. Arth Rheum 39(9) Suppl. S216 No. 1133 (1996).

IgA GLYCOSYLATION IN IgA NEPHROPATHY

Alice Allen and John Feehally

Department of Nephrology
Leicester General Hospital
Leicester LE5 4PW
United Kingdom

IgA NEPHROPATHY AND HENOCH-SCHÖNLEIN PURPURA

IgA nephropathy (IgAN) is the commonest form of glomerulonephritis in developed countries where renal biopsy is widely practiced[1]. IgAN is defined by the deposition of IgA in the glomerular mesangium. Most typically, patients present with episodes of visible haematuria, which often concide with respiratory or gastrointestinal infection. At this early stage renal function may be preserved but slowly progressive renal failure is common. Up to 30% of patients with IgAN will require renal replacement therapy by dialysis or transplantation within 20 years of presentation[2]. Since IgAN is characteristically diagnosed in children and young adults, many of these patients will develop renal failure in young or middle age. The formidable personal, social and financial implications of renal failure emphasise the importance of investigating fundamental pathogenic processes in IgAN with the long term aim of interrupting disease progression.

Henoch-Schönlein purpura (HSP) is a form of systemic vasculitis characterised by tissue deposition of IgA (it retains its somewhat misleading title for historical reasons). Nephritis with IgA deposition is a common feature of HSP and histological appearances in the kidney may be indistinguishable from IgAN[3].

Characteristics of Mesangial IgA in IgAN and HSP

Mesangial IgA in both IgAN and HSP nephritis is predominantly IgA1, and is polymeric (pIgA1)[4]. pIgA1 is derived mainly from the mucosal immune system; in conjunction with the association of disease exacerbations with mucosal infection, this has led to the supposition that IgAN is a disease of mucosal immune system overactivity[5]. IgA1 responses to common antigens are indeed exaggerated but mainly following systemic rather than mucosal immunisation[6,7]. Studies of pIgA production unexpectedly show reduced pIgA synthesis in mucosal lamina propria and increased production in the bone marrow[8,9], arguing against a mucosal origin for the deposited IgA.

Mechanism of Mesangial pIgA1 Deposition in IgAN

Although bone marrow production of pIgA1 is increased in IgAN, serum levels of IgA1 are only modestly raised[10]. In any case elevation of circulating IgA1 *per se* (whether monoclonal or polyclonal) does not result in mesangial IgA deposition - very high levels of circulating IgA occur in IgA myeloma and in AIDS, yet neither IgAN nor HSP develop[11].

Experimental models of nephritis show that mesangial immunoglobulin deposits such as occur in IgAN are most easily provoked by the infusion of pre-formed antigen-antibody complexes or by formation of immune complexes *in situ* in the kidney by infusion of antibody following fixation of antigen in the kidney[12]. However, this mechanism is not proven to be operative in human IgAN and no antigen has been consistently associated with IgA deposition in IgAN despite much study. Animal models of mesangial IgA deposition are not ideal since there are significant differences in the IgA immune system between man and other species; in particular no animal has IgA subclasses analogous to human IgA1 and IgA2[13]. Therefore the studies reviewed here are restricted to IgA in human IgAN.

Failure to identify relevant antigens has led to the suggestion that deposition may occur by mechanisms other than classical antigen-antibody interactions. Altered glycosylation of IgA has been proposed as one such mechanism.

IgA1 GLYCOSYLATION

IgA1 is a heavily glycosylated molecule. Like all immunoglobulins it has complex N-glycans, but it is unique among serum immunoglobulins in possessing a series of O-glycans. These are limited to the hinge region of the molecule, where five serine residues occur in a stretch of only 18 amino acid residues; therefore each IgA1 monomer, comprising two α1 heavy chains, carries ten closely located O-glycan moieties (Figure 1)[14,15,16]. IgA2 has no hinge region and therefore lacks O-glycosylation[16]. The IgA1 O-glycans are of the simple mucin type based on N-acetylgalactosamine (GalNAc) O-linked to serine; GalNac is usually substituted with galactose in the β1,3 configuration: **Serine-GalNAc(β1,3)Gal**[14]. In addition, a varying degree of terminal sialylation of these moieties occurs[15].

Possible Role for Altered IgA Glycosylation in IgAN

Alteration of hinge region O-glycosylation could have much functional significance in this pivotal region of the IgA1 molecule. IgA1 is cleared from the circulation by the liver, via the asialoglycoprotein receptor (ASGPR)[17], and the O-glycans of the IgA1 hinge region are the preferred ligand for this receptor[18]. Therefore, abnormalities of IgA1 O-glycosylation may well lead to inefficient interaction with the ASGPR and failure of proper IgA1 catabolism, which in turn may favour deposition of abnormal molecules in other sites such as the glomerular mesangium.

Closely located series of O-glycans are known to confer rigidity on glycoprotein domains, extending the structural formation[19,20,21]. Alterations in the core O-glycans or the sialylation of IgA1, resulting in changes in 3D structure and electrical charge, could affect both the antigen binding site and the effector sites of the Fc region. Potential consequences include influences upon Fc receptor interactions, the engagement of complement, and binding to matrix proteins. All these could modify the mesangial deposition of IgA1 as well as subsequent inflammatory reactions. Recently, it has been reported that human mesangial cells express an IgA receptor[22,23], though this has yet to be fully characterised. The possible influence of IgA1 O-glycosylation on interactions with this receptor is particularly relevant to the study of IgA-associated glomerular disease.

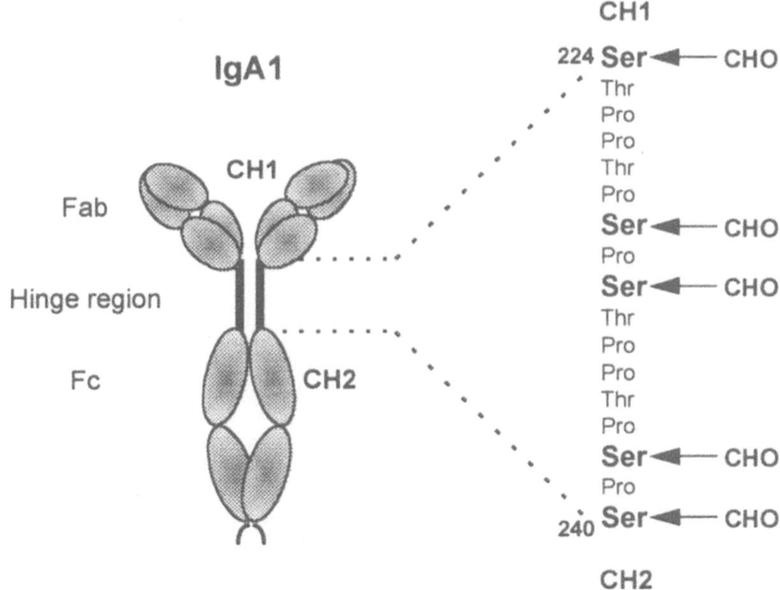

Figure 1. Structure of the human IgA1 molecule. The diagram on the left shows the position of the hinge region, between the CH1 and CH2 domains. On the right is the amino acid sequence of the hinge region, with the O-glycosylation sites (CHO) marked by arrows.

ASSESSMENT OF IgA1 HINGE REGION GLYCOSYLATION

Lectin Binding

At the present time, most of the available data concerning IgA glycosylation in IgAN is derived from lectin-binding studies. We have used a variety of lectins with affinity for O-linked sugars, and find increased binding of GalNAc-specific lectins to IgA1 in IgAN[24]. Interestingly, we find that this abnormality is shared by patients with HSP nephritis, but is not seen in patients with HSP without renal involvement (Figure 2)[25], suggesting that this alteration of IgA1 glycosylation has pathogenic significance in IgA-associated glomerular disease. The binding of lectins to the O-linked carbohydrates of serum C1 inhibitor (C1inh) is normal in IgAN, suggesting that this apparent glycosylation abnormality is restricted to IgA1, and is probably not due to altered plasma glycosidase activity[24].

Other groups using lectin-binding approaches confirm that IgA1 O-glycosylation is indeed abnormal in IgAN[26-30], though there are some inconsistencies in the exact findings. Overall, lectin binding to the galactosylated moiety (Galβ1,3GalNAc) seems reduced, while lectin binding to GalNAc is raised in IgA1 from patients with IgAN. This has been interpreted as a reduction in terminal galactosylation of O-linked carbohydrates[28], somewhat analogous to that affecting the N-glycans of IgG in rheumatoid arthritis[31] and the O-glycans of glycophorins in the Tn polyagglutinability syndrome[32]. However, the evidence for reduced galactosylation is as yet indirect, and the role of variable sialylation of the moieites on lectin binding is not clear. Lectin binding assays are rapid and convenient for assessing the glycosylation profiles of large numbers of samples, but they are subject to some uncertainties about the exact binding specificities of the lectins, and they cannot give structural information about the carbohydrate moieties involved.

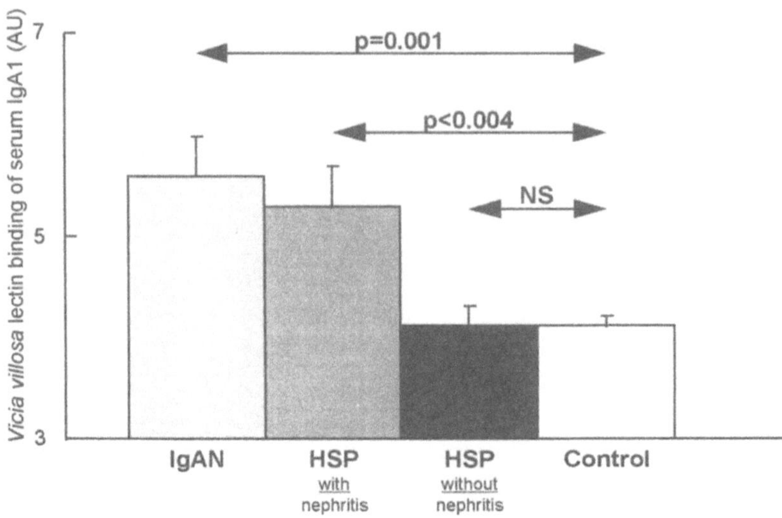

Figure 2. Binding of *Vicia villosa* lectin to O-linked GalNAc in serum IgA1. Lectin binding is significantly raised in IgAN and in HSP with nephritis, but is normal in HSP without evidence of renal involvement.

Analysis of Glycans Released from IgA1

A few attempts have been made to apply more precise methods of carbohydrate analysis to the O-glycosylation of IgA1, but these have yet to elucidate fully the abnormality found in IgAN. Gas-liquid chromatography of IgA1 carbohydrates indicates a low galactose content in IgAN[28], while HPLC analysis of O-glycans released from IgA1 by hydrazinolysis suggests reduced sialylation[33].

Recently, we have applied fluorophore-assisted carbohydrate electrophoresis (FACE) to the analysis of IgA1 in IgAN. O-glycans released from IgA1 by hydrazinolysis at 60°C are labelled with a fluorophore and separated by polyacrylamide gel electrophoresis; the bands, visualised under UV light, can be quantified by densitometry. Preliminary studies demonstrate a characteristic band pattern in IgAN (Figure 3A) , with a significant shift to a lower molecular weight band (Figure 3B). The identities of the bands are as yet unknown, but this method may be useful for screening and quantifying the putative O-glycosylation abnormality of IgA1 in IgAN with more precision than lectin binding assays can provide.

Analysis of IgA1-derived Glycopeptides

Work is currently underway to analyse IgA1 hinge region glycopeptides by matrix-assisted laser desorption/ionising mass spectrometry (MALDI-MS). This method has the potential to provide very accurate information about the structural nature of the glycosylation defect, and with the use of IgA1 protesaes which cleave specifc sites within the hinge region to obtain glycopeptides carrying individual O-glycans, it may also be possible to locate the abnormal glycans with precision.

Limitations of Serum Studies

By definition the pathogenic molecules in IgAN are in the kidney. Study of circulating IgA1 therefore has the limitation that the IgA1 of interest may be underrepresented or even

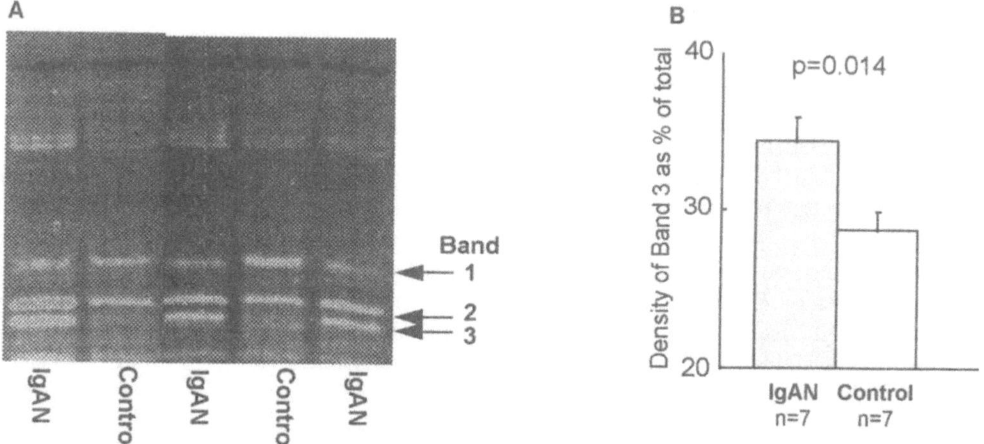

Figure 3. FACE analysis of O-glycans released from serum IgA1 by hydrazinolysis at 60°C.
   A : FACE gel visualised under UV light. The major bands are referred to as Bands 1,2 and 3. The lowest molecular weight band (Band 3) is consistently stronger in samples derived from serum IgA1 in IgAN than from matched controls.
   B : After quantification by densitometry, the relative density of Band 3 is expressed as a percentage of the total (Band 1+2+3) to allow for differences in gel loading. In IgAN, Band 3 accounted for a significantly higher percentage of the glycans released from IgA1 than in matched controls.

absent from serum. Unfortunately it is technically difficult to achieve definitive studies of mesangial IgA. Lectin binding studies on renal biopsy material would be defeated by the extensive O-glycosylation of membrane proteins which would overwhelm any variation in IgA1 signal. The small amounts of IgA1 which could be eluted from renal biopsies do not allow adequate biochemical analysis. Opportunities to study whole kidneys from IgAN are rare and rely on coincidental clinical events making kidneys available. No such opportunity has yet arisen during these studies.

UNDERLYING MECHANISM FOR ALTERED O-GLYCOSYLATION OF IgA1

Three possibilities have been considered:
- altered aminoacid sequence of the IgA1 hinge region presenting an abnormal template for glycosylation;
- enhanced glycosidase activity, resulting in increased galactose removal;
- a defect in β1,3 galactosyltransferase (β1,3GT), resulting in reduced terminal galactosylation of O-linked GalNAc.

Hinge Region Amino Acid Sequence

The hinge region is a highly conserved region of the IgA1 molecule. It is not subject to somatic mutation. Altered hinge sequence seems improbable, and an initial report suggests that it is indeed normal in IgAN[34]. Further studies are underway to confirm this.

Glycosidase Activity

No studies have addressed directly the possible role of raised plasma glycosidase activity in the generation of abnormal IgA1 glycosylation in IgAN. However, serum C1inh was found to be normally O-glycosylated in IgAN by lectin binding assay[24] This observation argues against such a mechanism being responsible.

Altered β1,3 Galactosyltransferase Activity

Glycosyltransferases are enzymes with marked specificity[35]. If the presumed defect in terminal galactose on the hinge region O-glycans were due to synthetic failure reduced β1,3GT activity would be suspected in relevant cells. The normal OLG of serum C1inh in IgAN[24] suggests that altered galactosylation is not widespread, and therefore such an enzyme abnormality could be restricted to IgA-producing cells. Such studies would best be performed on IgA plasma cells from the bone marrow, the apparent site of increased IgA1 production[9]. Thus far β1,3GT has only been investigated in peripheral blood mononuclear cell populations. A functional assay was developed using enzymatically degalactosylated IgA1 hinge region fragments as a substrate for cell-lysate derived enzyme activity, and the GalNAc-binding lectin *Vicia villosa* (VV) to detect changes in substrate glycosylation. Though this is an indirect method of assessing regalactosylation of the substrate, VV lectin proved highly specific for GalNAc in this system, and cell lysates were able to reduce its binding to the substrate in a dose dependent manner, while a commercial β1,4 galactosyltransferase preparation could not alter lectin binding (AC Allen, unpublished observations).

Cell lysates of T cells and monocytes from patients with IgAN and controls did not differ in their regalactosylating activity in this assay system. However, in IgAN B cell lysates had significantly lower activity than controls, and this had an inverse correlation with VV lectin binding of serum IgA1 from the same samples (Figures 4A and 4B)[36]. In other words, we found that in IgAN, the higher the GalNAc expression (VV binding) of serum IgA1, the lower the ability of the corresponding B cell lysate to regalactosylate the normal IgA1-derived substrate. Further elucidation of the distribution of β1,3GT activity is somewhat limited at present since the enzyme has not yet been sequenced preventing the application of molecular techniques.

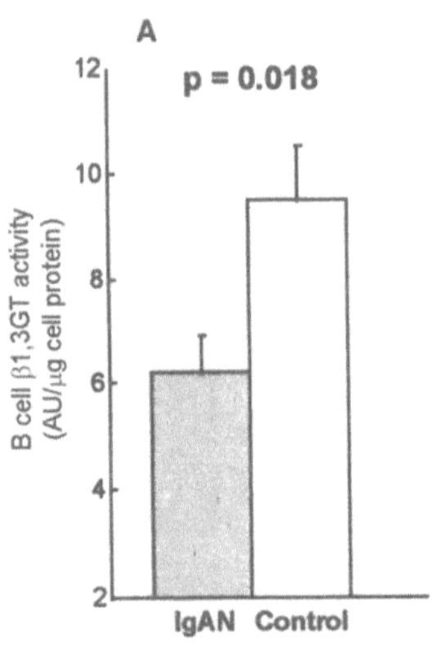

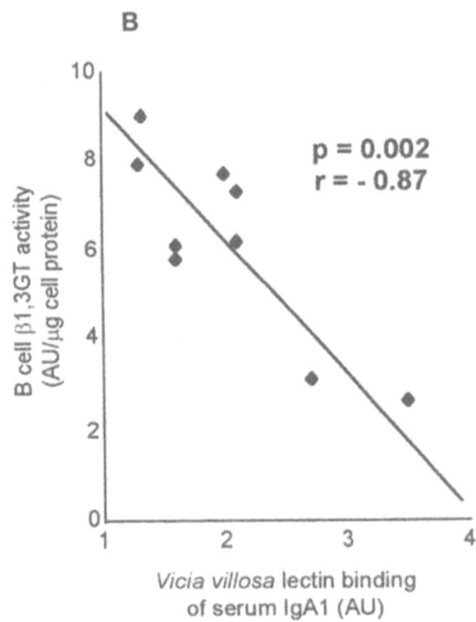

Figure 4. β1,3 galactosyltransferase activity in peripheral blood B cells.

A : B cells from patients with IgAN have significantly lower ability to O-galactosylate an IgA1-derived substrate than B cells from matched controls.

B : In IgAN, there was an inverse correlation between the B cell β1,3GT activity and O-galactosylation of serum IgA1, as assessed by binding of the GalNAc-specific lectin *Vicia villosa.*

FUNCTIONAL IMPLICATIONS OF DEFECTIVE O-LINKED GLYCOSYLATION OF IgA1

The role of defective IgA1 glycosylation in the pathogenesis of IgAN and HSP remains speculative, and functional studies are needed to confirm that this abnormalilty indeed has the potential to influence :

Efficiency of IgA1 catabolism, due to altered affinity for the asialoglycoprotein receptor;

Mesangial deposition of IgA, through interactions with

- mesangial cell receptors
- mesangial extracellular matrix
- antigen binding site promoting formation of immune complexes;

Subsequent mesangial injury, through alterations in

- mesangial cell activation via IgA- or other cell surface receptors
- engagement of complement
- recruitment of circulating inflammatory cells via Fc receptors.

CONCLUSIONS

Progress in understanding IgA has been disappointingly slow over the 25 years since it was first described. Many young people continue to develop renal failure as a result of IgAN and HSP.

The hinge region O-glycosylation defect discussed here has strong potential as an explanation for mesangial IgA deposition - the fundamental initiator of glomerular damage in this condition. Studies continue to confirm the structural abnormalities of IgA1 and to establish the functional implications.

REFERENCES

1. G d'Amico. The commonest glomerulonephritis in the world: IgA nephropathy. *Q J Med*, 245:709 (1987).
2. G d'Amico. Clinical features and natural history in adults with IgA nephropathy. *A J Kidney Dis*, 12:353 (1988).
3. FB Waldo. Is Henoch-Schönlein purpura the systemic form of IgA nephropathy?. *Am J Kidney Dis*, 12:373 (1988).
4. J Feehally. Immune mechanisms in glomerular IgA deposition. *Nephrol Dial Transplant*, 3:361 (1988).
5. M-C Béné, G Faure. Mesangial IgA in IgA nephropathy arises from the mucosa. *Am J KidneyDis*, 12:406 (1988).
6. L Layward, AC Allen, SJ Harper, JM Hattersley, J Feehally. Increased and prolonged production of specific polymeric IgA after systemic immunisation with tetanus toxoid in IgA nephropathy. *Clin Exp Immunol*, 88:394 (1992).
7. L Layward, AC Allen, JM Hattersley, SJ Harper, J Feehally. Elevation of IgA in IgA nephropathy is localised in the serum and not saliva and is restricted to the IgA1 subclass. *Nephrol Dial Transplant*, 8:25 (1993).
8. SJ Harper, JH Pringle, ACB Wicks, J Hattersley, L Layward, AC Allen, A Gillies, I Lauder, J Feehally. Expression of J chain mRNA in duodenal plasma cells in IgA nephropathy. *Kidney Int*, 45:836 (1994).
9. SJ Harper, AC Allen, JH Pringle, J Feehally. Increased dimeric IgA producing B cells in

the bone marrow in IgA nephropathy. *J Clin Pathol*, 49:38 (1996).

10. SN Emancipator, ME Lamm. IgA nephropathy: pathogenesis of the most common form of glomerulonephritis. *Lab Invest*, 60:168 (1989).

11. M-C Béné, P Canton, C Amiel, T May, G Faure. Absence of mesangial IgA in AIDS: a postmortem study. *Nephron*, 58:425 (1991).

12. V Montinaro, AR Esparza, T Cavallo, A Rifai. Antigen as mediator of glomerulr injury in experimental IgA nephropathy. *Lab Invest*, 64:508 (1991).

13. J Mestecky. Immunobiology of IgA. *Am J Kidney Dis*, 12:378 (1988).

14. J Baenziger, S Kornfeld. Structure of the carbohydrate units of IgA1 immunoglobulin. II. Structure of the O-glycosidically linked oligosaccharide units. *J Biol Chem*, 249:7270 (1974).

15. MC Field, RA Dwek, CJ Edge, TW Rademacher. O-linked oligosaccharides from human serum immunoglobulin A1. *Biochem Soc Trans*, 17:1034 (1989).

16. MA Kerr. The structure and function of human IgA. *Biochem J*, 271:285 (1990).

17. Z Moldoveanu, I Moro, J Radl, SR Thorpe, K Komiyama, J Mestecky. Site of catabolism of autologous and heterologous IgA in non-human primates. *Scand J Immunol*, 32:577 (1990).

18. RJ Stockert, MS Kressner, JC Collins, I Sternlieb, AG Morell. IgA interaction with the asialoglycoprotein receptor. *Proc Natl Acad Sci USA*, 79:6229 (1982).

19. N Jentoft. Why are proteins O-glycosylated? *Trends in Biochemical Sciences*, 15:291 (1990).

20. EF Hounsell, MJ Davies. Role of protein glycosylation in immune regulation. *Annals Rheumatic Dis*, 52:S22 (1993).

21. H Lis, N Sharon. Protein glycosylation. Structural and functional aspects. *Eur J Biochem*, 218:1 (1993).

22. C Gomez-Guerrero, E Gonzalez, J Egido. Evidence for a specific IgA receptor in rat and human mesangial cells. *J Immunol*, 151:7172 (1993).

23. AC Allen, MR Greer, C Combe, WW Stewart, SJ Harper, MA Kerr, J Feehally. Binding of human IgA to human leucocytes and mesangial cells. *J Am Soc Nephrol*, 5:739 (1994).

24. AC Allen, SJ Harper, J Feehally. Galactosylation of N- and O-linked carbohydrate moieties of IgA1 and IgG in IgA nephropathy. *Clin Exp Immunol*, 100:470 (1995).

25. AC Allen, F Willis, SJ Harper, TJ Beattie, J Feehally. Altered O-glycosylation of IgA1 in Henoch-Schönlein purpura is restricted to subjects with nephritis. *Nephrol Dial Transplant*, in press (1996).

26. PM Andre, P le Pogamp, D Chevet. Impairment of jacalin binding to serum IgA in IgA nephropathy. *J Clin Lab Anal*, 4:115 (1990).

27. Y Hiki, H Iwase, M Saitoh, Y Saitoh, A Horii, K Hotta, Y Kobayashi. Reactivity of glomerular and serum IgA1 to jacalin in IgA nephropathy. *Nephron*, 72:429 (1996).

28. J Mestecky, M Tomana, PA Crowley-Nowick, Z Moldoveanu, BA Julian, S Jackson. Defective galactosylation and clearance of IgA1 molecules as a possible etiopathogenic factor in IgA nephropathy. *Contrib Nephrol*, 104:172 (1993).

29. R Coppo, A Amore, B Gianoglio, A Reyna, L Peruzzi, D Roccatello, D Alessi, LM Sena. Serum IgA and macromolecular IgA reacting with mesangial matrix components. *Contrib Nephrol*, 104:162 (1993).

30. D Baharaki, M Dueymes, R Perrichot, C Basset, R Lecorre, J Cledes, P Youinou. Aberrant glycosylation of IgA from patients with IgA nephropathy. *Glycoconjugate J*, 13:505 (1996).

31. JS Axford, N Sumar, A Alavi, DA Isenberg, A Young, KB Bodman, IM Roitt. Changes in normal glycosylation mechanisms in autoimmune rheumatic disease. *J Clin Invest*, 89:1021 (1992).

32. LT Lee, S Frank, DS de Jongh, C Howe. Immunochemical studies on Tn erythrocyte glycoprotein. *Blood*, 58:1228 (1981).
33. Y Hiki, H Iwase, T Kokubo, A Horii, A Tanaka, J Nishikido, K Hotta, Y Kobayashi. Association of asialo-galactosylβ1-3N-acetylgalactosamine on the hinge with a conformational instability of jacalin-reactive immunoglobulin A1 in immunoglobulin A nephropathy. *J Am Soc Nephrol*, 7:955 (1996).
34. M Matsuda, Y Mishige, K Takenouchi, T Tabe, A Moriyama, T Haruta, Y Ohtsubo, Y Uchida, Y Yasumoto, T Arima. Amino acid sequence of the hinge region of IgA1 molecule can not be responsible for defective galactosylation in patients with IgA nephropathy. *J Am Soc Nephrol*, 7:1323 (1996).
35. JU Baenziger. Protein-specific glycosyltransferases: how and why they do it! *FASEB J*, 8:1019 (1994).
36. AC Allen, PS Topham, SJ Harper, J Feehally. Reduced β1,3 galactosyltransferase activity in the B cells of patients with IgA nephropathy. *J Am Soc Nephrol*, 6:820 (1995).

OLIGOSACCHARIDE PROFILING OF ACUTE-PHASE PROTEINS: A POSSIBLE STRATEGY TOWARDS BETTER MARKERS IN DISEASE

Graham A. Turner and Mohammad T. Goodarzi

Department of Clinical Biochemistry
The Medical School
Newcastle upon Tyne, NE2 4HH,U.K.

INTRODUCTION

Current diagnostic methods using serum proteins are often insensitive, non-specific and lack the ability to predict the future outcome of treatment. With the exception of albumin and C-reactive protein, most serum proteins are glycosylated. Carbohydrate structures on serum proteins are very complex and for any particular protein, there coexists in the blood the same molecule with different carbohydrate structures (glycoforms). These reflect the sources of the molecule, and the particular physiological and biochemical conditions that existed when the molecule was synthesised and released. When disease is present, the glycoform distribution changes according to the type of disease and particular protein studied. Over one hundred proteins have been identified in the blood. The carbohydrate changes on these molecules, therefore, provide a wealth of untapped diagnostic information. For reviews of glycosylation changes in disease the reader should consult previous literature[1-3].

Although three types of carbohydrate/protein linkage have been described for serum proteins, the majority bear exclusively N-glycosidic linkages, and all the changes described in this paper involve this type. Previous studies of glycosylation changes in serum proteins in disease have been mainly concerned with changes in N-glycans with the so called 'complex' structure. These units contain the neutral sugars, fucose (Fuc), galactose (Gal), N-acetylglucosamine (GlcNAc), mannose (Man), and the charged sugar, N-acetyl-5-neuraminic acid (Neu5Ac). N-glycans occur in structures with 2, 3, and 4 branches (bi-, tri- or tetra-antennary chains), with Fuc, Gal, GlcNAc, and Man present in the main molecule, and Neu5Ac on the ends of the branches. Figure 1 shows the structure of a typical biantennary chain. The boxes indicate the carbohydrate groupings that may or may not be present on the basic structure, and these groupings are the ones that are most frequently changed in disease. The situation is further complicated by the fact that oligosaccharide chains can be synthesised in an incomplete form (for example, lacking galactose on one or two arms of a biantennary), or the structure of the chain may be different at each glycosylation site. A more

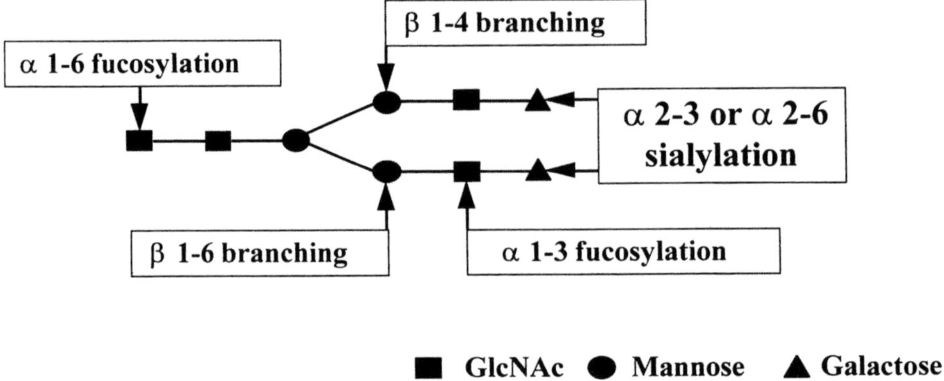

Figure 1. The structure of a typical biantennary oligosaccharide chain on N-glycans of serum glycoproteins. The groupings shown in the boxes are optional, and glycosylation changes in disease are often found at one or more of these sites.

detailed description of the carbohydrate structure of serum glycoproteins has been given previously[4].

For the last few years we have studied the carbohydrate structure of two acute-phase serum proteins, α1 proteinase inhibitor (API) and haptoglobin (Hp) in various disease states [5-10]. Both these proteins are synthesised by the liver. API is a single polypeptide chain carrying three N-linked 'complex' oligosaccharide units (~12% by weight). Hp is composed of two α subunits and two ß subunits linked by disulphide bridges, the carbohydrate content of Hp is 16-20%, and it is found exclusively as four N-linked 'complex' oligosaccharide chains on each ß subunit. The role of API is to inhibit the enzyme activity of serine proteases such as elastin and cathepsin G. This protects the body from proteolytic damage. The role of Hp, on the other hand, is to combine with any free haemoglobin (Hb) that is released by intravascular red cell destruction. The Hp/Hb complex is then broken down by the reticulo-endothelial system and this mechanism conserves iron stores and protects the renal tubules from damage by Hb. In many abnormal situations in the body (inflammation, malignancy and trauma), the concentration of API and Hp in the blood increase rapidly and substantially (2-4 fold) in the so called acute-phase reaction.

Table 1 summarises our previous findings on the changes in glycosylation of API and Hp in different inflammatory and malignant diseases as shown by monosaccharide analysis[5-10]. Increased fucosylation of Hp was found in all the diseases investigated. Increased branching was also observed for Hp in all diseases; however, for API, the branching was increased in inflammation and decreased in cancer. The magnitude of all these changes varied between diseases and in some cases the changes in fucosylation and branching were accompanied by changes in sialylation.

The aim of the current study was to obtain more detailed information on the changes in carbohydrate structure of API and Hp in disease by analysing their oligosaccharide composition. This approach became possible because we had recently developed a new procedure to completely and reproducibly remove sialylated oligosaccharides from very small amounts (~50µg) of purified serum glycoprotein using PNGase F, and analyse these components using high-pH anion-exchange chromatography with pulsed amperometric detection (HPAEC-PAD)[11]. It was hoped that this procedure would be useful for diagnosis in some clinical situations, for example, in differential diagnosis or prognosis, or that it might lead to identifying particular structures that could be developed into diagnostic markers using lectins or anti-carbohydrate antibodies.

186

Table 1. Changes in branching, fucosylation and sialylation of API and Hp in various disease states

Serum protein	Glycosylation Change				
	Increased Branching	Decreased Branching	Increased Sialylation	Decreased Sialylation	Increased Fucosylation
API	Crohn's Dis.	Breast Ca.	Crohn's Dis?	Breast Ca.	Breast Ca.
	R. Arthritis?	Ovarian Ca.			Crohn's Dis.
		Stomach Ca.			Ovarian Ca.
					R. Arthritis
					Stomach Ca.
Hp	A. cirrhosis				A. cirrhosis
	B. cirrhosis?				B. cirrhosis
	Breast Ca.		Breast Ca?		Breast Ca.
	Crohn's Dis?		Crohn's Dis.		Crohn's Dis.
	Ovarian Ca.		R. Arthritis?		Ovarian Ca.
	R. Arthritis		Stomach Ca.		R. Arthritis
	Stomach Ca.				Stomach Ca.

Conclusions given in the Table are based on the monosaccharide content of API or Hp isolated from sera in various diseases as compared with the monosaccharide content of these proteins from healthy individuals. Data taken from references 5-10 and unpublished observations. A question mark after a disease indicates that the change observed is borderline. A.=alcoholic; B.=biliary, Ca.=cancer, Dis.=disease, R. =rheumatoid

EXPERIMENTAL

Blood specimens were obtained by venupuncture from 10 healthy individuals (HI) (median age 46y); 8 patients with active rheumatoid arthritis (RA) (median age 59y); 7 patients with active Crohn's disease (CR) (median age 42); 6 patients with progressive stomach cancer (ST) (median age 75y); 14 patients with progressive breast cancer (BR) (median age 60y); and 14 patients with progressive ovarian cancer (OV) (median age 60y). Sera was obtained by low speed centrifugation. All the healthy individuals and patients were non-smokers and consumed low amounts of alcohol. None of the healthy individuals were taking medication. The majority of the patients in the disease groups were taking some form of medication.

API and Hp were purified from 250 μl serum using affinity chromatography with anti-API and anti-Hp antibodies respectively. Purity was checked by SDS/PAGE electrophoresis. The methods used have been previously described[5-10]. All specimens gave very pure preparations.

Oligosaccharides were removed from API and Hp using the following experimental procedure[11]. Purified glycoproteins (~50μg) in 50mM ammonium formate buffer, pH=8.6, containing 0.4% SDS was heated at 100°C for 3 min and an equal volume of the buffer solution containing 2% mercaptoethanol, 1.2% CHAPS and 0.1M EDTA was heated at 100°C for further 3min. After cooling, 0.2 Boehringer Unit PNGase F (Boehringer) was added and incubated at 37°C for 20h. Removal of all the oligosaccharides was checked by electrophoresis and staining for carbohydrate. Finally, the sample was dialysed overnight at 4°C in a microdialyser and lyophilised.

Sialylated oligosaccharides were separated on a Carbopac PA100 anion-exchange column using a sodium acetate gradient (20-200mM) in 100mM NaOH and the separated peaks were detected by pulsed amperometric detection (Dionex BioLC System). Samples were injected on to the column using a Spectra System AS3500 Autosampler, data from the detector was collected by an ACI interface and all data was analysed by the AI450 software[11]. An internal standard of 1,6 glucose diphosphate (GDP) was added to each sample and with every batch of samples oligosaccharide standards (monosialylated biantennary (MSB), disialylated biantennary (DSB) and trisialylated triantennary (TST), Oxford Glycosystems) were separated. The average retention times of MSB, BSB, TST and GDP were 14.9, 26.3, 36.4 and 64 min respectively. Using these standards it was possible to calculate the relative retention times of the unknown peaks, and compare separations between different samples, different diseases and different acute-phase proteins. The Neu5Ac content of the oligosaccharides is a major factor in the separation of sialylated oligosaccharides; however, the rentention time on the ion-exchange column is further modified by other factors such as fucosylation, chain branching, bisecting GlcNAc and the linkage of the Neu5NAc[12].

RESULTS

Analysis of the Oligosaccharides from Hp

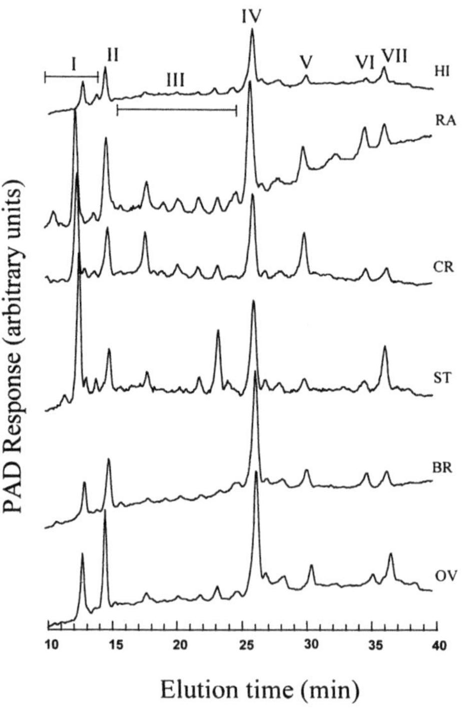

Figure 2. Typical chromatography profiles of oligosaccharides of Hp from healthy individuals and patients with different diseases.The oligosaccharides were prepared and separated as described in the text. In Figures 2, 3, 4, and 5 and Tables 2 and 3, 'HI'=healthy individuals; 'RA'=rheumatoid arthritis;'CR'=Crohn's Disease; 'ST'=stomach cancer; 'BR'=breast cancer; 'OV'=ovarian cancer; peaks 'II', 'IV' and 'VII' co-eluted with monosialylated biantennary, disialylated biantennary, and trisialylated triantennary standards respectively.

Figure 2 shows that typical oligosaccharides profiles for Hp from the different disease groups. From the retention times of known standards, peaks II, IV and VII were identified as MSB, DSB and TST. For analytical purposes other peaks or groups of peaks were also assigned roman numerals. In each disease group the profile was very consistent and all peaks had a profile that was different from the pattern given for Hp of healthy individuals. Particularly noticeable was the increase in the number of peaks between the MSB and DSB peaks (region III) in RA, CR and ST groups; an increase in peaks (region I) eluting sooner than MSB; and an increase in peak V between the DSB and TST peaks.

The oligosaccharide separation data for Hp was analysed quantitatively by calculating the area of each peak as a percentage of the total area. Initially, this was done separately for all the peaks that were detected. In order to simplify the analysis, however, some groups of the peaks were later pooled as shown in Figure 2, and other peaks were not included in the analysis because they appeared to be present in the same amount in each group. Figure 3 shows the difference between the 'healthy' group and the other groups for six regions of the chromatogram. The data are presented as a total change in the oligosaccharide content in each component compared with the amounts of oligosaccharide that are present in the 'healthy' group. The changes that are statistically significant are given in Table 2.

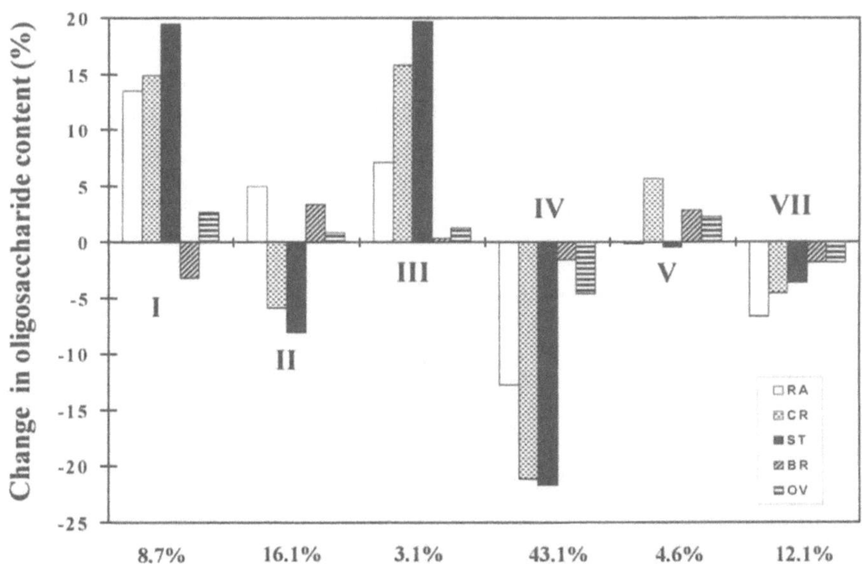

Figure 3. Changes in oligosaccharide composition of Hp in different disease states. In Figures 3 and 5, the values shown in the bar chart were obtained by subtracting the amount of oligosaccharide in healthy individuals from the amount for the same oligosaccharide obtained in disease; the coefficient of variation of the data was between 10-25%; and the oligosaccharide composition of healthy individuals is given below the figure.

In the RA, CR and ST groups there were large reductions in the identified peaks (MSB, DSB and TST). For RA this occurred for the DSB and TST peaks, for ST this occurred for the MSB and the DSB, and for CR it occurred for all three peaks. These changes were also accompanied by increases in peaks I, III and V. The only significant change observed for the BR and OV groups was an increase in peak V. The CR group showed the most changes. No change in a particular oligosaccharide was associated with any disease, but the overall changes in RA, CR and ST were unique, and also different from the change observed for the BR and OV groups.

Table 2. Summary of changes in Hp oligosaccharides in different diseases compared to the composition in healthy individuals

Disease Group	Oligosaccharide peaks					
	I	II	III	IV	V	VII
RA	↑		↑	↓		↓
CR	↑	↓	↑	↓	↑	↓
ST	↑	↓	↑	↓		
BR					↑	
OV					↑	

In Tables 2 and 3, an arrow indicates that there is a significant ($P<0.05$) increase (↑) or decrease (↓) in the amount of an oligosaccharide peak in the disease group compared to the healthy group. For Hp, the change in oligosaccharide composition was different for each disease except in breast and ovarian cancer where both show an increase in peak 'V'.

Analysis of Oligosaccharides from API

Figure 4 shows typical oligosaccharides profiles for API from the different disease groups. API gave many peaks with similar relative retention times to Hp, but the overall patterns were different. A major difference was the presence of more of the API oligosaccharides in the DSB peak. Another striking difference was the absence of peaks in the region between the MSB and the DSB peaks (region III). From visual examination of the traces in Figure 4, differences in the pattern between the 'healthy' and disease groups can be seen. These seem to be confined to increases in peak V in the disease groups.

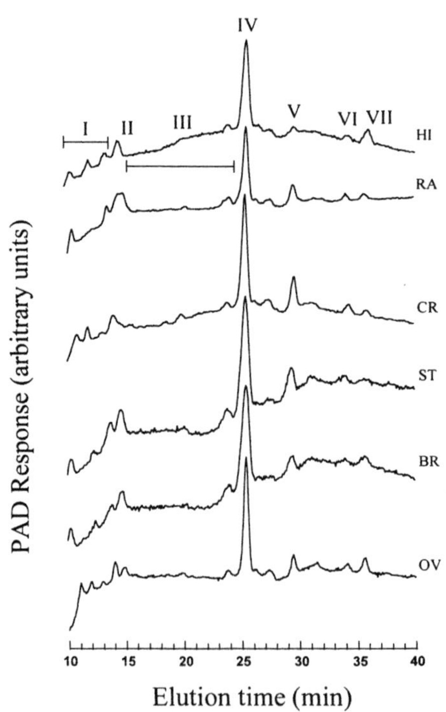

Figure 4. Typical chromatography profiles of oligosaccharides of API from healthy individuals and patients with different diseases. For further details see legend of Figure 2.

The quantitative analysis of the oligosaccharide composition of API from the different disease groups are summarised in Figure 5 and Table 3. This was carried out as described for Hp and the data is presented in a similar way. The changes in oligosaccharide composition in disease were much smaller than those observed for Hp. Peak V was significantly increased in all the disease groups, however, there were no large reductions in the MSB, DSB and TST peaks. Some of the changes were of the same type as seen for Hp, but others were not. Again, the overall profile of change was different in each disease.

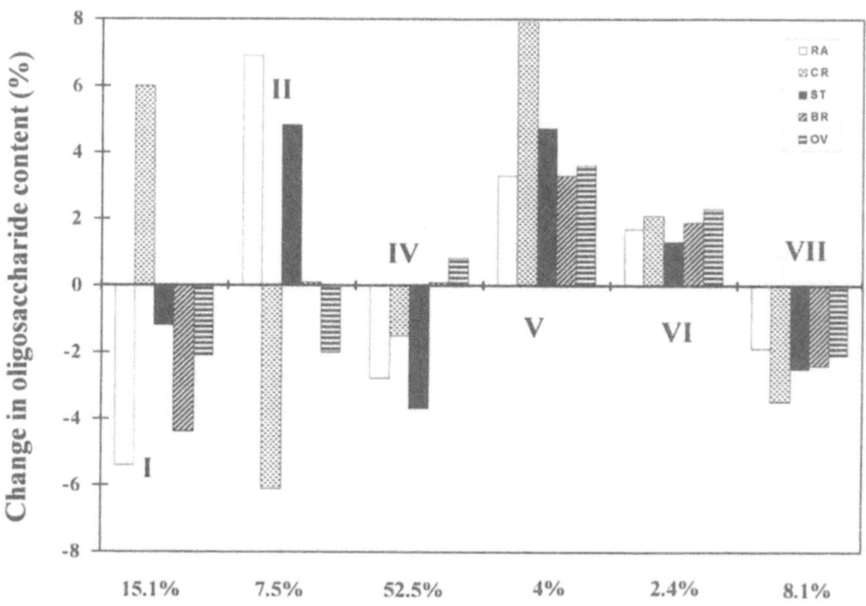

Figure 5. Changes in oligosaccharide composition of API in different disease states. For further details see legend of Figure 3.

DISCUSSION

Methodology

This study has developed a simple, reproducible and rapid method for investigating the sialylated oligosaccharide composition of API and Hp in disease. The procedure is based around a previously described PNGase F enzymatic technique for removing oligosaccharides from glycoproteins[11]. Using an approach that was slightly modified from the original procedure (the detergent CHAPS was used instead of NP40, because it was found that the latter gave some contaminant peaks that could interfere with oligosaccharide peaks that eluted early in the chromatogram) it was possible to completely deglycosylate 50µg of protein in 20h with 1/5 of the enzyme activity previously used[13]. When this deglycosylation procedure was combined with a rapid affinity chromatography method for preparing the purified glycoprotein and an automated ion-exchange method for separating the oligosaccharides, 10-20 specimens could be easily processed in 1 week. Although this timescale is too long for routine diagnostic analysis, it could be useful in specific situations.

Table 3. Summary of changes in API oligosaccharides in different diseases compared to the composition in healthy individuals

Disease	Oligosaccharide peaks				
Group	I	II	V	VI	VII
RA		↑	↑		
CR	↑	↓	↑	↑	↓
ST		↑	↑		
BR			↑		
OV			↑	↑	

The change in oligosaccharide composition of API was different for each disease. See Table 2 for more details.

Hp Oligosaccharides in Disease

Using the above approach Hp gave characteristic and reproducible profiles for different diseases, although no peak was identified as specific for any particular disease. The large reductions in the major oligosaccharides in RA, CR and ST that were accompanied by the appearance of peaks eluting sooner, may be due to the addition of fucose to these units. It is well established that the additon of fucose causes peaks to elute earlier[12,14]. The fucose content of Hp was increased in all these diseases[9]. Furthermore, we have recently demonstrated using a new lectin-binding method[15] that there is an increase in the binding of the α1-6 fucose-specific lectin, lentil culinaris agglutinin (LCA) and the α1-3 fucose-specific lectin, lotus tetragonolobus to Hp from RA, CR and ST patients (unpublished observations).

It seems likely, however, that other structural changes are contributing to the observed profiles in RA, CR and ST, because the fucose content and the fucose-specific lectin binding of Hp was also increased in the BR and OV groups (unpublished observations). The changes in the oligosaccharide profiles in the latter were small in comparison with RA, CR and ST.

One other possible feature that could be affecting the elution properties of the oligosaccharides is the position of the Neu5Ac. There was a large increase in the binding of α2-3Neu5Ac-specific lectin, maackia amurensis agglutinin, to Hp from BR and OV (unpublished observations). It may be of some relevance that it has been reported that oligosaccharides having Neu5Ac linked α2-6 to Gal rather than by a α2-3 bond are eluted earlier[12]. Therefore, earlier elution through the addition of fucose to Hp in the BR and OV groups may be partially negated by a change in the linkage of the sialic acid.

Finally, some of the extra peaks between the MSB and DSB peaks obtained for the RA, CR and ST groups may have been caused by the presence of a bisecting GlcNAc on biantennary chains or a reduction in the size of the biantennary chain. Both these modifications have been reported to lead to changes in the elution profile of oligosaccharides from an anion-exchange column[12].

API Oligosaccharides in Disease

The much smaller changes in the oligosaccharide composition of API in disease was a surprising finding. API from healthy individuals contained more DSB chains than Hp, whereas the latter contained more MSB and TST chains.Why this should be important in influencing the changes in oligosaccharide composition in disease is unclear. Previous monosaccharide and lectin-binding studies have indicated that API and Hp are glycosylated differently in ovarian cancer and it was suggested that this difference may be due to removal of API from the blood with abnormal glycosylation[8]. A similar mechanism may explain the large differences in oligosaccharide composition of API and Hp in disease in this study.

The increase in oligosaccharide peak V for API was larger and more consistent than that observed for Hp. Like Hp, all API specimens in disease showed increased fucosylation (see Table 1), but recent studies have shown that in disease the increase in binding of these proteins to LCA and LTA lectins is greater for API than Hp (unpublished observations). Therefore, it seems likely that peak V is also a fucosylated species.

Conclusions

From this study it is concluded that it is possible to reproducibly and routinely measure the oligosaccharide composition of the serum glycoproteins API and Hp in clinical specimens. Furthermore, consistent and characteristic patterns of glycosylation were observed for different diseases, although no unique oligosaccharide was found for any disease. The changes for Hp were larger than those observed for API, some involved known oligosaccharide components whereas others involved unidentified components. Lectin studies suggested that some of the latter species are due to increased fucosylation. However, many other changes in structure were probably present in these unidentified peaks. These results suggest that oligosaccharide analysis may lead to more sensitive markers of disease.

ACKNOWLEDGEMENTS

We wish to acknowledge the staff of the Royal Victoria Infirmary and Associated Hospital NHS Trust, Newcastle for help in obtaining the blood specimens and clinical details; and financial support from North of England Cancer Research Campaign, Dionex UK Ltd, The GO Fund and the Islamic Republic of Iran.

REFERENCES

1. G.A.Turner, N-glycosylation of serum proteins in disease and its investigation using lectins, *Clin. Chim. Acta.* 208:149-171 (1992).
2. W.Van Dijk , G.A.Turner, and A.Mackiewicz, Changes in glycosylation of acute-phase proteins in health and disease: occurrence, regulation and function, *Glycosylation & Disease* 1: 5-14 (1994).
3. G.A.Turner, N-glycosylation of serum proteins in disease:diagnostic potential, Proc UK NEQAS Meeting, 1:104-111.(1994).
4. J.U.Baenziger, The oligosaccharides of plasma glycoproteins: synthesis, structure and function, in: "The Plasma Proteins, vol 4," F.W.Putnam, ed. , Academic Press Inc, New York (1984) .
5. S.Thompson, E.Dargan , and G.A.Turner, Increased fucosylation and other carbohydrate changes in haptoglobin in ovarian cancer, *Cancer Lett.* 66:43-48.(1992).
6. S.Thompson, E.Dargan, I.D.Griffiths, C.A.Kelly, and G.A.Turner, The glycosylation of haptoglobin in rheumatoid arthritis. *Clin. Chim. Acta.* 220:107-114 (1993).
7. A.C.Mann, C.O.Record, C.H.Self, and G.A.Turner, Monosaccharide composition of haptoglobin in liver diseases and alcohol abuse: large changes in glycosylation associated with alcoholic liver disease. Clin. Chim. Acta. 227:69-78.(1994).
8. G.A.Turner, M.T.Goodarzi, and S.Thompson, Glycosylation of alpha-1-proteinase inhibitor and haptoglobin in ovarian cancer: evidence for two different mechanisms, *Glycoconjugate J.* 12: 211-218(1995).

9. G.A.Turner, Haptoglobin: a potential reporter molecule for glycosylation changes in disease, *Adv. Exp. Med. Biol.* 376: 231-238 (1995).

10. M.T.Goodarzi, and G.A.Turner, Decreased branching, increased fucosylation and changed sialylation of alpha-1-proteinase inhibitor in breast and ovarian cancer, *Clin. Chim. Acta.* 236: 161-171 (1995).

11. A.C.Mann , C.H.Self , G.A.Turner, A general method for the complete deglycosylation of a wide variety of serum glycoproteins using peptide-N-glycosidase-F, *Glycosylation & Disease* **1**:253-261 (1994).

12. P.Hermentin, R.Witzel , J.F.G.Vliegenthart, J.P.Kamerling, M.Nimtz, and H.S.Conradt, A strategy for the mapping of N-glycans by high-pH anion-exchange chromatography with pulsed amperometric detection. *Anal. Biochem.* 203: 281-289 (1992).

13. A. Haselbeck, and W.Hosel, Studies on the effects of the incubation conditions, various detergents on the enzyme activity of N-glycosidase F, glycopeptidase F, and endoglycosidase F *Topic in Biochemistry.* 8:1-4 (1988).

14. M.R.Hardy,and R.R.Townsend, Separation of positional isomers of oligosaccharides and glycopeptides by high-performance anion-exchange chromatography with pulsed amperometric detection, *Proc. Natl. Acad. Sci., USA.* 85:3289-3293 (1988).

15. M.T.Goodarzi, and G.A.Turner, A lectin-binding assay for the rapid characterization of the glycosylation of purified glycoproteins, in *"Protocol Proteins Handbook,"* J.M.Walker. ed., Humana Press, Totawa (1996)

THE ROLE OF N-LINKED GLYCOSYLATION IN THE SECRETION OF HEPATITIS B VIRUS

Anand Mehta[1], Timothy M. Block[2], and Raymond A. Dwek[1]

[1]The Glycobiology Institute, Department of Biochemistry, Oxford University, Oxford, United Kingdom, OX1-3QU

[2]Viral Hepatitis Group, Kimmel Cancer Center, Jefferson Medical College, Philadelphia, Pennsylvania, USA, 19107-6799

INTRODUCTION

Hepatitis B virus (HBV) is the human member of the *hepadnaviradae* family of viruses which chronically infects over 300 million people worldwide. Infection by HBV can lead to both acute and chronic liver disease. For those unfortunate to become chronic carriers of HBV, there are lifelong complication associated with the virus. Indeed up to 40% of chronic carriers may eventually develop hepatocellular carcinoma (**1**). As there is such a large pool of chronic carriers, it can be assumed that a vast number of the worlds population die from complications that arise from infection with HBV.

Hepatitis B virus is a small DNA virus which encodes for only seven proteins, three of which are the envelope proteins termed L,M,S (figure 1). The three envelope proteins are produced from a single open reading frame through the utilization of alternative translation start sites. All three proteins have a common N-linked glycosylation site at position 146 of the S domain (fig. 1) while the M protein alone contains an additional site at position 4 of the pre-S2 domain (**2**).

A peculiar feature of these envelope glycoproteins is that they are secreted in the form of smaller non-infectious, non DNA containing, lipoprotein particles. These sub-viral particles are secreted in vast numbers, often outnumbering the infectious HBV virion by 100,000 to 1 (**2**). Sub-viral particles are found in two forms: 22 nm sphere shaped particles, which consists of mainly of the S and M proteins; and long filament like particles, which vary in size and contain a much greater amount of the L protein (**3**).

The life cycle of HBV is quite complex (reviewed in 1) involving many unique steps. All three envelope glycoproteins are thought to play an important role in the viral life cycle. While it has been shown that the L and S proteins are necessary for virion secretion (**4**), the role of M is in doubt (**4,5**). The current model of virion formation involves DNA containing nucleocapsids acquiring viral envelope proteins by budding into the lumen of the endoplasmic reticulum (ER) and secretion through the trans-Golgi network (**6**).

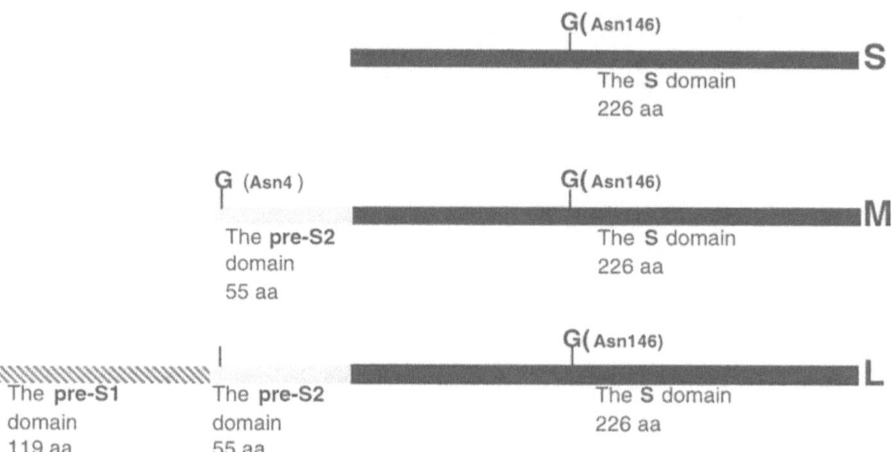

Figure 1. The HBV envelope proteins. The L protein consists of three domains: the pre-S1 domain, pre-S2 domain, and the S domain. M contains the pre-S2 and S domains while the S protein contains only the S domain. All three proteins have a common N-linked glycosylation site N at Asn-146 of the S domain with the M protein containing an additional site at Asn-4 of the pre-S2 domain. The S glycan site is partially occupied in all three envelope proteins while the pre-S2 site is fully occupied in M. The L protein, while containing the pre-S2 glycosylation sequon, only utilizes the shared S glycan site.

The HBV envelope proteins contain N-linked glycan. However what role these play in the life cycle of HBV is not fully known. Early work with tunicamycin on the secretion of sub-viral particles (**7,8**) and the finding that an avian homologue of HBV, duck hepatitis virus, lacked N-linked glycans, suggested that the N-linked glycans of HBV were not important. However, in such a small virus (3.2 kb) the conservation of the pre-S2 and S glycosylation sequons in over 20 sub-types (personal observation) tends to indicate that the glycans may play some role in HBV. To this end we have analyzed the role of N-liked glycosylation in HBV through the use of inhibitors of the N-linked glycosylation pathway and with site-directed mutagenesis of the individual glycan sites on the three envelope glycoproteins. The data presented in this chapter, together with our previous work with tunicamycin (**9**) lead to the conclusion that N-liked glycosylation of a specific viral glycoproteins is required for the secretion of the hepatitis B virus.

METHODS

Cells and Transfections: Hep G2 cells were grown in RPMI 1640 (Gibco-BRL, Rockville, MD) containing 10% fetal bovine serum (Gibco-BRL). Hep G2.2.15 cells were kindly provided by Dr. George Acs (Mt. Sinai Medical College (NY, NY, USA) and maintained as Hep G2 cells but with the addition of 200µg/ml of G418 (Gibco-BRL). DNA transfection of Hep G2 cells were carried as in (**4**).

Detection of intracellular viral DNA: Hep G2.2.15 cells were either left untreated or treated with 1000 µg/ml of the α-glucosidase inhibitor NB-DNJ (provided by Monsanto/Searle, Inc., St. Louis, MO) for the indicated times and the total DNA extracted as described (**11**). 25µg of DNA was digested with Hind III, resolved through a 1.2% agrose gel and transferred to nylon membranes (MSI). Membranes were then hybridized with a ^{32}P labeled probe containing the total HBV genome and developed as described (**11**). The relaxed circular (rc), linear (l), and closed circular (CCC) DNA were confirmed by enzymatic digestion (data not shown).

Detection of HBV proteins by ELISA. After six days, culture medium from Hep G2.2.15 cells left uninhibited or inhibited with the 1000 µg/ml NB-DNJ were resolved through a discontinuous sucrose gradient as described (**10**). Fractions were collected and subjected to antigen detection by the ELISA method as described (**10**).

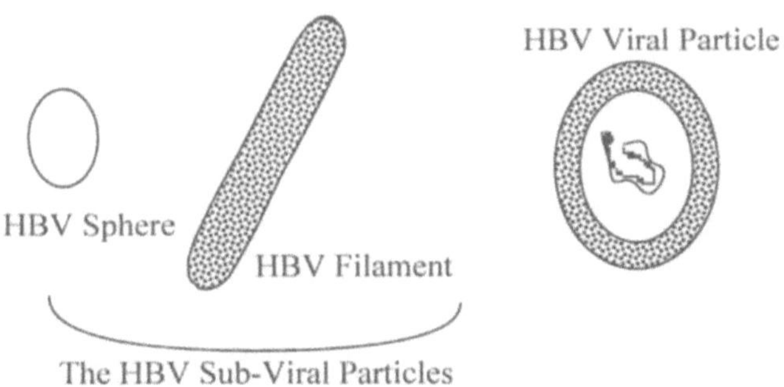

Figure 2. An infected cell secretes infectious virus along with empty lipoprotein sub-viral particles (which contain no DNA) that are composed of the HBV envelope proteins. The shape of the sub-viral particles is dependent upon the particles protein composition. See text for more details. The sub-viral particles are secreted in vast excess to the viral particle.

Release of glycans and labeling of sugars. Sub-viral particles from untreated or NB-DNJ treated cells were purified by double CsCl ultra-centrifugation (**2**) and concentrated using an Amicon Centriplus 100,000 dalton cut off protein concentrator (Amicon, Inc., Beverly, MA). Particles were further purified by P4 column chromatography (Bio-Rad, Melville, NY) before sugar release. Glycans were removed by hydrazinolysis using a Glycoprep 1000 machine (Oxford Glycosystems, Abingdon, Oxfordshire, UK) and fluorescently labeled at their reducing end with 2-aminobenzamide using the Signal Labeling Kit (Oxford Glycosystems). Glycans were subsequently analyzed by WAX chromatography (**12**) and normal phase HPLC (**13**). Glycan structures were identified using sequential exoglycosidase digestion as described (**13**).

Detection of HBV proteins by antigen capture : For detection of sub-viral particle secretion, cells transfected with the vectors PRV-HBV 1.5, Sg⁻, and Mg⁻, were incubated with 200 μci/ml ^{35}S methionine (ICN) for one hour in the presence of methionine free RPMI (Sigma Chemicals, St. Louis, MO). After one hour the radioactive media was removed, the cells washed with 1X PBS (150 mM NaCl, 10 mM Na_2HPO_4, 1.5 mM K_2HPO_4, pH 7.12) and replaced with complete RPMI (10% FBS, Gibco-BRL) for 24 hours. The media was collected and layered onto a discontinuous sucrose density gradient as described (**10**). The collected fractions were then analyzed for the presence of S antigen by an antigen capture method as described (**14**).

Mutagenesis: A 3.5 kb EcoRV to EcoRV fragment from PRV-HBV 1.5 (kindly provided by Dr. Volker Bruss) containing the entire HBV genome was cloned into the EcoRV site of sk-bluescript (Stratagene). This plasmid, referred to as sk-HBV EcoRV - EcoRV, was used for the mutagenesis experiments. Mutagenesis was performed using the Chameleon Site Directed Mutagenesis Kit (Stratagene) as per instructions. Site directed mutagenesis was performed only on the second nucleotide of the S ORF which changed with third nucleotide of the underlining POL ORF. Hence, the only amino acid change occurred in the S ORF. After mutagenesis, the 3.5kb EcoRV fragment from sk-HBV EcoRV- EcoRV was cloned back into PRV-HBV 1.5. The Sg⁻ vector contained the mutation Asn 146 to Ser 146. The Mg⁻ vector contained the change of Thr 6 to Ile 6, hence disrupting the Asn 4-xxx-Thr 6 glycosylation sequon. The Sg⁻ Mg⁻ vector contained both these mutations. No alterations within the underlining polymerase gene were created and the mutants contained only the desired mutation as confirmed by DNA sequence analysis (data not shown).

Detection of secreted viral DNA: Five days after transfection with the appropriate HBV expression vectors, the supernatant was collected and briefly centrifuged at 3000 rpm for 10 minutes followed by pelleting through 20% sucrose for 16 hours at 35,0000 RPM. As the parent vector PRV-HBV 1.5 has been shown to only secrete enveloped viral particles (**4**), PCR was then performed on the pellets using primers and PCR conditions as described (**11**). PCR products were subsequently run on 1.5% agrose gels and stained with ethidium bromide.

RESULTS

Inhibition Of α-Glucosidase Causes The Accumulation Of The Replicative Forms Of HBV DNA

The N-linked glycosylation pathway is well defined, consisting of over 13 enzymes that are involved in processing within the endoplasmic reticulum (ER) and the Golgi apparatus (**15**). The α-glucosidases are the first enzymes involved in this pathway and remove the terminal three glucose residues from the Glc3Man9NAcGlc2 glycoform after it has been transferred from the dolichol diphosphate to the backbone of the growing polypeptide chain (reviewed in **15**). Inhibition of the α-glucosidases leads to the retention and misfolding of some glycoproteins within the ER (**16**). This may be the result of tri-glucosylated glycoproteins being unable to interact with ER chaperones such as calnexin, which only recognize mono-glucosylated glycoproteins (**17,18**). Inhibitors of α-glucosidase have also been shown to have anti-viral activities against many viruses, including HBV (**10**). For example, the inhibition of α-glucosidase in Hep G2.2.15 cells, a cell line that chronically secretes HBV (**19**), prevents the secretion of enveloped HBV DNA and causes the intracellular accumulation of viral DNA (**10**). Fig. 3 shows that in the presence of the α-glucosidase inhibitor, N-butyldeoxynojirimycin (NB-DNJ), Hep G2.2.15 cells accumulate large amounts of the replicative forms of HBV viral DNA. As stated earlier, the life cycle of HBV is very complex. An infected cell contains several forms of HBV DNA which represent different stages in the HBV life cycle. For example, the covalently closed circular DNA (CCC DNA) is the nuclear form of the DNA and is thought to be the viral template (**2**). In contrast, the relaxed circular DNA (rc) is associated with the viral particle and is an indicator of encapsidation. Therefore, figure 3 shows the accumulation of HBV DNA as seen in the presence of the α-glucosidases inhibitor NB-DNJ. The upper bands (int) represent the integrated forms of the HBV DNA and their abundance remains constant in all samples. In contrast, the replicative forms of HBV DNA increase dramatically over time. At day 7 for example, there is over a 30 fold increase in the amount of relaxed circular (rc) and linear (l) HBV DNA as compared to the untreated control. This data provides information as to which stages of the HBV life cycle have been interrupted by the α-glucosidase blockade. For example, accumulation of the replicative forms of HBV DNA suggest that while encapsidation of HBV viral DNA has occurred, the envelopment process may be prevented. This is consistent with the α-glucosidase inhibitor NB-DNJ affecting the viral glycoproteins and implies that they are functionally impaired. Treatment of Hep G2.2.15 cells with tunicamycin also causes the accumulation of these replicative forms of DNA in a manner similar to figure 3 (data not shown).

Secreted sub-viral particles from cells in which α-glucosidase has been inhibited lack the M protein

Our previous study showed that sub-viral particles were still secreted in the presence of NB-DNJ (**10**). However, a more detailed analysis revealed that the individual HBV envelope antigens had varying degrees of sensitivity to α-glucosidase function (fig. 4). As shown in figure 4a-b, while the secretion of the S and L proteins within sub-viral particles was reduced by only 20-30%, the secreted sub-viral particles were virtually devoid of the M protein. That is, while sub-viral particles containing mostly the S and L glycoproteins are secreted near normally, as are many host glycoproteins from α-glucosidase inhibited cells (**16,20,21**), there was a 16 fold reduction the secretion of M containing sub-viral particles (figure 4c). Preliminary evidence suggest that the sub-viral particles containing the M protein are localized to lysosomal compartments and carry hyperglucosylated glycan structures (**14**). The lack of secretion of M containing sub-viral particles is thought to be the result of the effect of the α-glucosidase inhibitor on the M protein. The extreme sensitivity of M, compared with L and S is interesting since all three proteins are derived from alternate translation of the same open reading frame and hence have great amino acid similarity (**2**).

Sub-Viral Particles Secreted From α-Glucosidase Inhibited Cells Still Contain Fully Processed Oligosaccharides

If the failure to secrete the M protein, relative to S and L, is due to the accumulation of tri-glucosylated glycan structures on the M protein, then HBV glycoproteins secreted from α-glucosidase inhibited cells should not contain any unprocessed glycan (their N-glycans should be complex). Therefore the glycan structures from sub-viral particles secreted from untreated and α-glucosidase inhibited cells were determined and are shown in figures 5a and 5b respectively. As shown in figure 5, the S and L proteins secreted from α-glucosidase inhibited cells contain complex glycan structures. The lack of the fucosylated structure from α-glucosidase inhibited cells is assumed to be the result of the lack of M protein (and its corresponding glycans) in sub-viral particles obtained from cells treated with NB-DNJ. Consistent with this is a previous report localizing the fucosylated bi-antennary structure, lacking in the α-glucosidase treated sample, to the M protein (**22**). The complete processing of the S and L glycans in the presence of NB-DNJ arises from their ability to overcome the α-glucosidase blockade via the endomannosidase pathway (**20**). This pathway allows for the formation of complex glycan structures in the presence of inhibitors of the α-glucosidases.

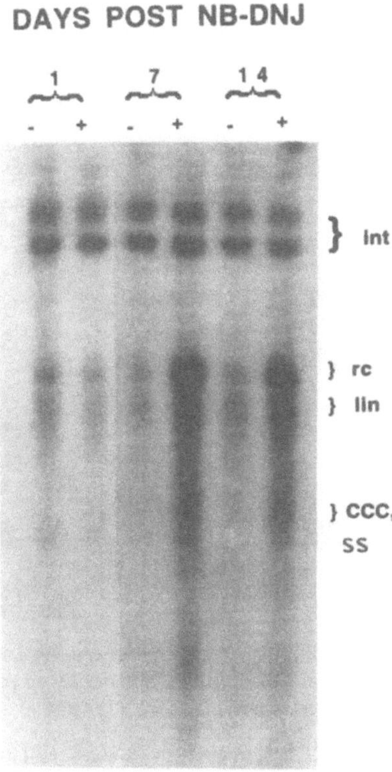

Figure 3. The intracellular accumulation of HBV viral DNA as a result of a-glucosidase inhibition. Hep G2.2.15 cells were left untreated or treated with the a-glucosidase inhibitor NB-DNJ for the indicated days and the DNA extracted as described in the materials and methods. Abbreviations are : int, integrated HBV DNA; rc, HBV relaxed circular DNA; lin, HBV linear DNA; CCC, HBV covalently closed circular DNA; SS, HBV single stranded DNA. As the SS DNA co-migrates with the CCC DNA it is difficult to determine which of these increases.

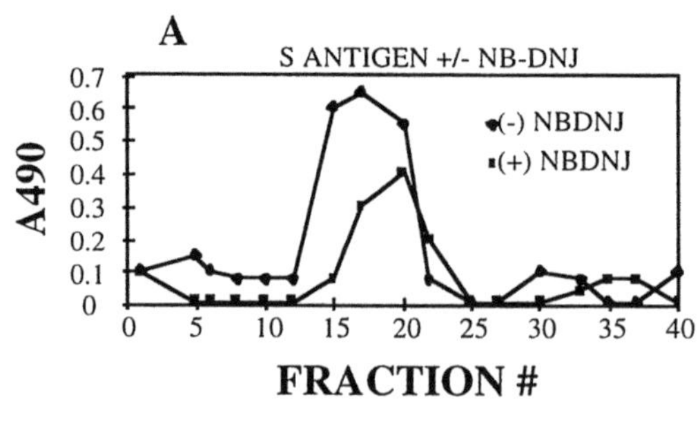

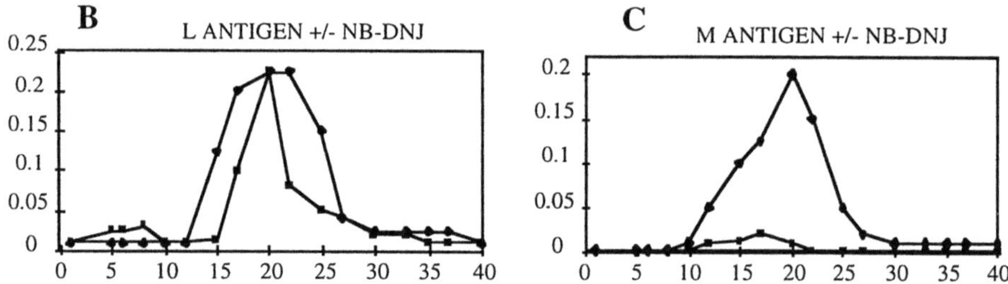

Figure 4. Detection of the **a)** HBV S protein, **b)** HBV L protein and **c)** HBV M protein in sub-viral particles secreted from NB-DNJ treated cells as described in the materials and methods.

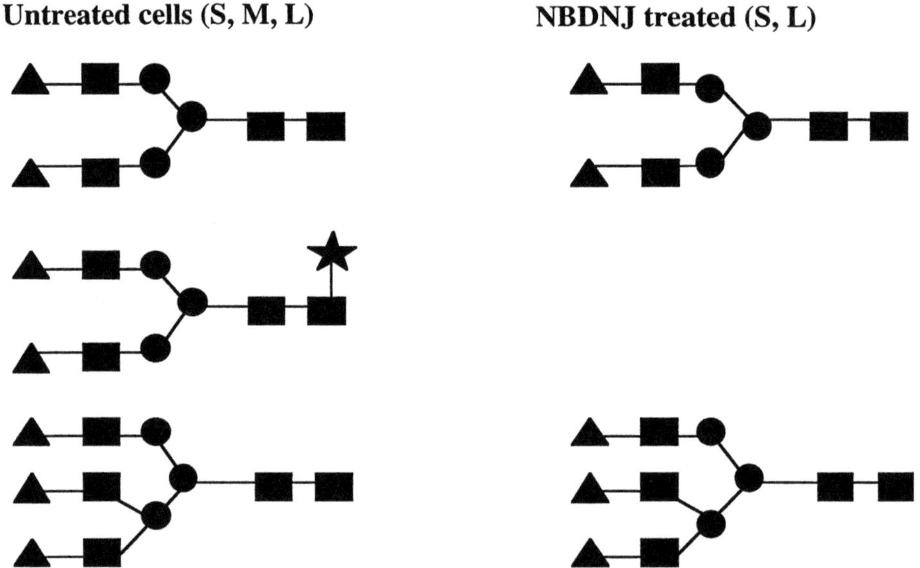

Figure 5. The Glycans associated with sub-viral particles from untreated and NB-DNJ treated cells as determined by Normal phase HPLC as described in the materials and methods. For glycan structures: triangles (galactose), squares (N-acetylglucosamine), circles (mannose), star (fucose).

HBV Glycosylation Knockout Vectors All Secrete HBV Sub-Viral Particles

To determine which, if any, glycan was necessary to mediate secretion of virus, HBV glycosylation knockout vectors were created in the background of the HBV expression vector PRV-HBV 1.5 (**4**). Three mutant vectors were created that either (*i*)

lacked the shared glycan at Asn-146 of the S domain (Sg⁻) , (*ii*) lacked both the shared glycan at Asn-146 and the additional pre-S2 glycan at Asn-4 of the M protein (Sg⁻Mg⁻), or (*iii*) lacked only the Asn-4 glycan in the pre-S2 domain of the M protein (Mg⁻).
All vectors contained amino acid changes only within the *env* open reading frame, with no alterations of the overlapping HBV DNA polymerase as determined by DNA sequence analysis (data not shown). These vectors, along with the parent vector were transfected into Hep G2 cells and assayed for the ability to induce the production of sub-viral and viral particles. Figures 6a-c show that all vectors produced sub-viral particles with sedimentation profiles similar to wild type as determined by a sucrose density gradient. The Sg⁻ Mg⁻ vector also induced the secretion of sub-viral particles as determined by UV absorbance at the correct sucrose density (data not shown). The expression vectors also maintained the ability to secrete all three envelope antigens, implying that no extreme misfolding had occurred as the result of the mutagenesis or lack of glycosylation (data not shown). This is consistent with the results obtained with tunicamycin (**7,8,9**) and further shows that the secretion of sub-viral particles is independent of N-linked glycosylation.

Expression Vectors Which Lack The Pre-S2 Glycosylation Site Do Not Secrete Viral DNA

The ability of the expression vectors to support the secretion of viral particles is shown in figure 7. Figure 7 shows that HBV DNA can be recovered from culture media of Hep G2.2.15 cells (lanes 1,2) and cells transfected with the parent vector PRV-HBV 1.5 (lanes 3,4). No detectable HBV DNA was recovered from the culture media of cells transfected with the Sg⁻Mg⁻ expression vector (lanes 5,6). This also confirms the results obtained with tunicamycin and demonstrates that the N-glycans of HBV are required for virion secretion. Surprisingly, the results obtained following the transfection of the Sg⁻ glycan vector (lanes 7-8) indicated that the common glycan in L, M and S, does not play a role in virion secretion. However the removal of the pre-S2 glycan (lanes 9-10) prevents the secretion of virus. Southern analysis of pelleted media from all HBV expression vectors confirmed the amplified DNA as from HBV viral particles (data not shown).

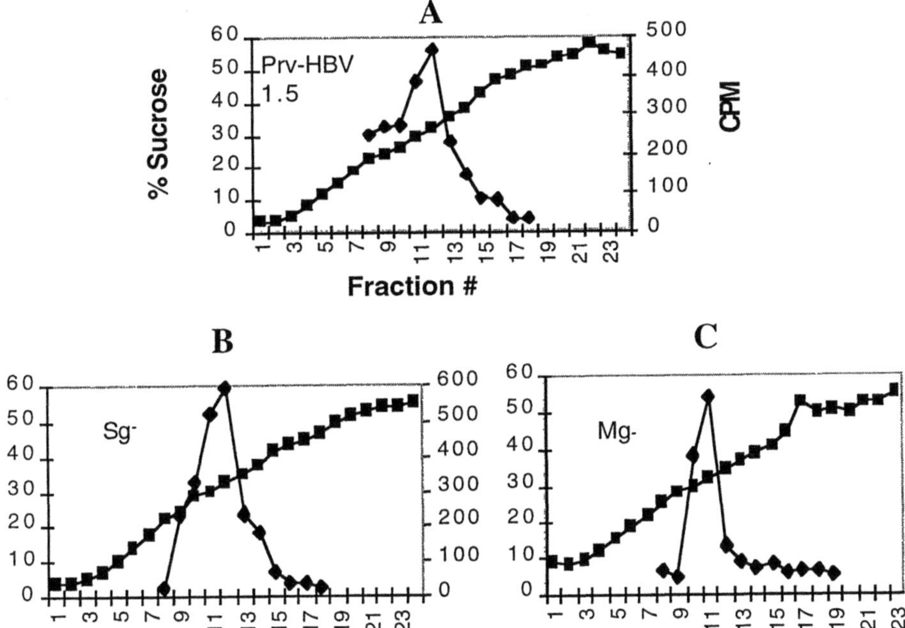

Figure 6. Secretion of sub-viral particles from the **a**) the parent vector Prv-HBV 1.5, **b**) the Sg⁻, vector and **c**) the Mg⁻ vector. Sub-viral particles were detected with a conformation dependent antibody to the S protein as described in the materials and methods.

DISCUSSION

Inhibitors of α-glucosidase are known to have different effects on different glycoproteins (**16,21**). However the differential effect of the α-glucosidase inhibitor on the three HBV envelope proteins (L,M,S) proved quite surprising. All three proteins are closely related, since they are derived from the same open reading frame. These proteins are also thought to have similar three dimensional structures (**2**). The only difference between the S and M protein is the additional 55 amino acids of the pre-S2 domain on M and importantly the additional glycan site within this region at Asn-4 (fig. 1). Although the L protein contains the pre-S2 domain as well, this glycosylation site is not utilized. The importance of glycosylation in the pre-S2 domain of M is consistent with the full utilization of this site as opposed to the variable utilization of the glycan sequon in the S domain (**2**).

The data obtained with tunicamycin and the glycan mutants demonstrate the importance of glycosylation for the secretion of viral particle (Table 1). Previous experiments with DMJ, an inhibitor of Golgi mannosidases, have shown that glycan processing outside of the ER is not important for the secretion of either viral or sub-viral particles. It is therefore the glycosylation events in the ER that must be important for the secretion of the virus. It is in this cellular compartment that protein folding, oligomerization and the viral budding process occur (**6**).
For glycoproteins, the processing of the initial oligosaccharide precursor from the Glc3Man9NAcGlc2 glycoform to the Glc1Man9NAcGlc2 glycoform in the ER can lead to an interaction with chaperones such as calnexin (**17**). Calnexin, which binds only to glycoproteins containing Glc1Man9NAcGlc2 structures, is thought to assist the folding of some glycoproteins (**18**). NB-DNJ, by inhibiting the formation of the Glc1Man9NAcGlc2 glycoform, prevents the interaction with calnexin. As the M protein is not secreted and is targeted towards lysosomes in the presence of NB-DNJ (**14**) we hypothesize that this type of interaction may be required for M, but not for the other envelope proteins. Support for this comes from the analysis that only the glycosylation state of M is altered by the presence of NB-DNJ. Presumably, sub-viral particles containing predominately S and L proteins (fig. 4) can utilize the Golgi-endomannosidase pathway, which can remove the tri-glucosylated cap and allows further processing to proceed normally. The same glycosylation profile of these proteins in the presence or absence of NB-DNJ is also indicative of their correct folding (**23**).

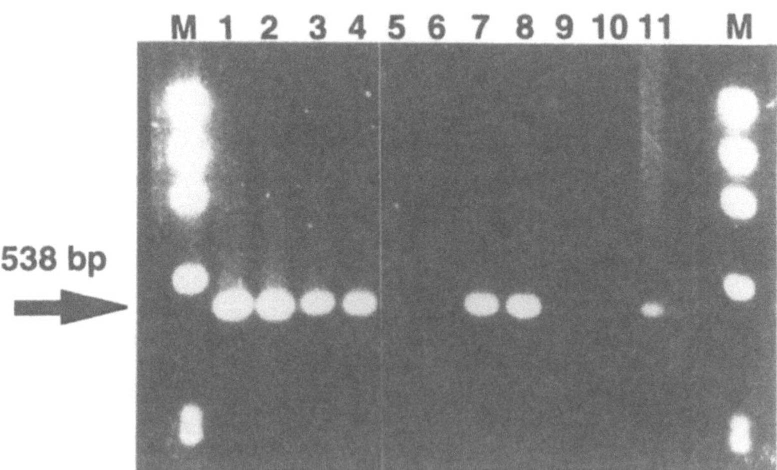

Figure 7. The secretion of the glycan mutants as detected by PCR as described in the materials and methods. First and last lanes marked "M" are DNA molecular weight standards. Lanes 1 and 2 are Hep G2.2.15 cells. Lanes 3 and 4 are from the control vector PRV-HBV 1.5. Lanes 5 and 6 are from the Sg⁻ Mg⁻ vector which secretes no virus. Lanes 7 and 8 are from the Sg⁻ vector. Lanes 9 and 10 are the Mg⁻ vector which also secretes no virus. Lane 11 is a control for the PCR reaction.

<u>**TABLE 1:**</u> Viral and sub-viral particle secretion from cells treated with inhibitors of the N-linked glycosylation pathway or transfected with the glycan mutants.

Treatment/glycan mutant	Virions	Sub-viral particles	
Tunicamycin[1]		No	Yes
NB-DNJ[2]		No	Yes - No M
DMJ[1]		Yes	Yes
Glycan mutant:	Virions	Sub-viral particles	
Sg⁻,Mg⁻		No	Yes
Sg⁻		Yes	Yes
Mg⁻		No	Yes

1) As in (**9**)

2) As in (**10**)

The site directed mutagenesis experiments highlight the important role of the M glycan in HBV. HBV expression vectors which lacked either all the glycan sites, or the additional pre-S2 glycan site, were unable to secrete enveloped virus. However, alterations of the common S glycan site had no effect on virion secretion. Since the L protein also contains the pre-S2 region, alterations in the M glycan site could affect the properties of the L protein. However the detection of the L protein as part of a sub-viral particle from the Mg⁻ vector suggest that the L protein is unaffected by the amino acid change at Asn4 (data not shown).

The results of the site directed mutagenesis experiments suggests that the lack of viral particle secretion from tunicamycin treated cells arises from the deficient glycosylation on the pre-S2 glycan on the M protein. They also imply that the effect with NBDNJ arises from the improper glycosylation of M. Thus the M glycan plays a crucial role in the formation of the HBV viral particle (Table 1).

In searching for a mechanism, it is interesting to note the anti-viral effects of NB-DNJ on HIV (**24**). Inhibition of α-glucosidase prevents the proper folding of the V1/V2 loop of HIV gp120 (**25**). However in contrast to HBV, enveloped HIV virus is formed and secreted (although it is non infectious). Another example in which the inhibition of glycan processing with NB-DNJ resulted in regional misfolding is that of tyrosinase (**26**). In this case, the hyperglucosylated enzyme is correctly transported to melanosomes but is inactive as a result of being unable to bind the copper required for activity (**26**). In the case of the HBV M protein, the most likely effect of NB-DNJ would be that the M protein is severely misfolded. This is supported by the finding of aggregated, hyperglucosylated M in the lysosomes (**14**). That the M protein is secreted in the presence of tunicamycin and from the Mg⁻ vector suggest that the degree of misfolding is less in these cases. However unlike the secretion of sub-viral particles, viral envelopment may require the precise three dimensional geometry of its assembled envelope glycoproteins. Even the slight misfolding of the individual M glycoprotein within the viral pre-envelope may be enough to prevent virion formation.

A crucial role for the M glycan is surprising as there is evidence that HBV can be secreted in the absence of the M protein (**4**, **27**). Perhaps the M protein acts in a dominant negative manner. That is, it is hypothesized that when the M protein is present and misfolded, it has the ability to destabilize the viral envelope in such a way as to hinder viral but not sub-viral particle secretion.

Overall the results presented here demonstrate that alterations within the pre-S2 glycosylation site can prevent the secretion of Hepatitis B virus. The realization that the interference of a *single* glycan site so profoundly influences HBV secretion suggests both an opportunity for anti-viral intervention and offers a simplified system to study the importance of N-linked glycosylation in the life cycle of HBV and possibly other animal viruses.

ACKNOWLEDGMENTS

Most of the text and the figures for this chapter are taken from a manuscript entitled "*HBV envelope glycoproteins vary drastically in their sensitivity to glycan processing - evidence that alteration of a single N-linked glycosylation site can regulate HBV secretion.*" that has been accepted for publication in The Proceedings of the National Academy of Sciences (USA). The authors would like to thank Dr. Volker Bruss for the expression vector PRV- HBV 1.5. Drs. P. Rudd, S. Petrescu, F. Platt, and T. Mattu are thanked for their careful reading of the manuscript and providing useful comments. This work was supported by the Hepatitis B foundation of America, Monsanto Searle Inc., and the North Atlantic Treaty Organization.

REFERENCES

1) Beasley, R. P. (1988). The Hepatitis B virus: The major etiology of hepatocellular carcinoma. Cancer. **61**: 1942-1956.

2) Heermann, K.H, Gerlich, W.H. (1992) In: Mol. Biol of HBV, A Maclachlan, ed, CRC Press, Boca Raton Fl.

3) Heerman, H-H, Goldman, I., Schwartz, W, Seyffarth, T., Baumgarten, H. Gerlich, W. H. (1984) Large surface proteins of hepatitis b virus containing preS sequence. J.Virol **52**: 396-402.

4) Bruss, V., D. Ganem. (1991). The role of envelope proteins in hepatitis b virus assembly. Proc. Natl. Acad. Sci. USA. **88**: 1059-1063.

5) Ueda, K., T. Tsurimoto and K. Matsbara (1991) Three envelope proteins of hepatitis B virus: large S, middle S and major S proteins needed for the formatioin of Dane particles. J. Virol. **65**: 3521-3529.

6) Ganem, D. (1991) Current topics in Microbiol. and Immunol. (W. Mason and C. Seeger, Eds.) **168**, 61-84.

7) Patzer, E., Nakamura, G., Yaffe, A. (1984) Intracellular transport and secretion of hepatitis B surface antigen in mammalian cells. J. Virol. **51**: 346-353

8) Sheu, S. Y. and Lo, S. J. (1994) Biogenesis of the hepatitis B viral middle (M) surface protein in a human hepatoma cell line: demonstration of an alternative secretion pathway. J. General Virol. **75**, 3031-3039.

9) Lu, X., Mehta, A., Butters,T., Dwek, R.A., Block, T.M. (1995) Evidence that N-linked glycosylation is necessary for hepatitis B virus secretion. Virol. **213**, 660-665.

10) Block, T., Platt, F., Xuanyong, L., Gerlich, W., Foster, G., Blumberg, B., Dwek, R. (1994) Secretion of human hepatitis B virus is inhibited by the imino sugar, N-butyldeoxynojirimycin. Proc. Natl. Acad. Sci. (USA) **91**: 2235-2239.

11) Lu, X., Block, T., Gerlich, W.H. (1996) Protease-induced infectivity of hepatitis B virus for a human hepatoblastoma cell line. J. Virol. **70**, 2277-2285.

12) Guile, G.R., Wong, S.Y.C., Dwek, R.A. (1994) Analytical and preparative separation of anionic oligosaccharides by weak anion-exchange high performance liquid chromatography on an inert polymer column. Anal. Biochem. **222**, 231-235.

13) Guile, G.R., Rudd, P., Wing, D., Prime, S.B., Dwek, R.A. (1996) A rapid high-resolution high-performance liquid chromatographic method for separating glycan mixtures and analyzing oligosaccharide profiles. Anal. Biochem. **240**, 210-226.

14) Lu, X., Mehta, A., Dadmarz, M., Dwek, R.A., Blumberg, B.S., Block, T.M. (1996) Aberrant trafficking and behavior of hepatitis B virus glycoproteins in cells in which glycosylation processing is inhibited. Submitted to Proc. Natl. Acad. Sci. (USA).

15) Datema, S. Olofsson, P. Romero. (1987) Inhibitors of protein glycosylation and glycoprotein processing in viral systems. Pharmacol. Thera. **33**: 221-286

16) Lodish, H., A. Kong (1984). Glucose removal from N linked oligosacchairdes is required for efficient maturation of certain secretory glycoproteins from the rough endoplasmic reticulim to the Golgi compartment. J. Cell. Biol. **98**: 1720-1729.

17) Ou, W.J., Cameron, P.H., Thomas, D.Y., Bergeron, J.M. (1993) Association of folding intermediates of glycoproteins with calnexin during protein maturation. Nature **364**, 771-776.

18) Helenius, A. (1994) How N linked oligosaccharides affect glycoprotein folding in the endoplasmic reticulum. Mol. Biol. Cell. **5**: 253-265.

19) Sells, M.A., M. L. Chen, G. Acs, (1987). Production of hepatitis b virus particles in Hep G2 cells transfected with cloned hepatitis b virus DNA. Proc. Natl. Acad. Sci. USA. **84**: 1005-1009.

20) Moore, S. E., Spiro, R.G. (1990) Demonstration that Golgi endo-α-D-mannosidase provides a glucosidase independent pathway for the formation of complex N-linked oligosaccharides of glycoproteins. J. Biol. Chem. **265**, 13104-13112.

21) Karlsson, G.B., Butters, T. D., Dwek, R.A., Platt, F.M., (1993) Effects of the imino sugar N butyldeoxynojirimycin on the N glycosylation of recombinant gp120 J. Biol. Chem.**268**: 570-576.

22) Poulter, L., Burlingame, A.L. (1990) Methods in Enzymology **193**, 661-689.

23) Dwek, R.A. (1996) Glycobiology: Towards understanding the function of sugars. Chemical Reviews, **96**, 683-720.

24) Fisher, P.B., Karlsson, G.B., Dwek, R.A., Platt, F.M. (1996) N-Butyldeoxynojirimycin- mediated inhibitionof human immunodeficiency virus entry correlates with impaired gp120 shedding and gp41 exposure. J. Virol **70**, 7153-7160.

25) Fisher, P.B., Karlsson, G.B., Butters, T.D., Dwek, R.A., Platt, F.M. (1996) N-Butyldeoxynojirimycin- mediated inhibitionof human immunodeficiency virus entry correlates with changes in antibody recognition of the V1/V2 region of gp120. J. Virol. **70**, 7143-7152.

26) Petrescu, S.M., Petrescu, A.J., Dwek, R.A., Platt, F.M. (1996) Inhibition of N-glycan processing in B16 melanoma cells results in inactivation of tyrosinase but does not prevent its transport to the melanosome. Accepted J. Biochemistry.

27) Fernholz, D., Galle, P.R., Stemler, M., Brunetto, M., Bonino, F., Will, H. (1993) Infectious hepatitis B virus variant defective in pre-S2 protein expression in a chronic carrier. Virology **194**, 137-148.

ROLE OF GLYCAN PROCESSING IN HEPATITIS B VIRUS ENVELOPE PROTEIN TRAFFICKING

Timothy M. Block[1], Xuanyong Lu[1], Anand Mehta[1,2],
Jason Park[1], Baruch S. Blumberg[2,3] and Raymond Dwek[3]

[1]Viral Hepatitis Group, Kimel Cancer Center of Jefferson Medical College,
Philadelphia, Pa., 19107-6799, USA
[2]Glycobiology Intsitute, Oxford University, Oxford, OX1-3QU, UK
[3]Fox Chase Cancer Center, Philadelphia, Pa., 19111-2497, USA

INTRODUCTION

We have previously shown that the glucosidase inhibitor, NBDNJ, prevents HBV virion but not subviral particle secretion from stably transfected HepG2 cells (Block et al, 1994). Moreover, the inhibition of virion secretion was due to the inhibition of the ER glucosidase (Lu et al, 1995). To determine which envelope protein functions require glycan processing and to understand the consequences to a glycoprotein of bearing unprocessed glycan, it was of interest to know the specific steps in the virus life cycle which were upset in cells in which ER glucosidase was inhibited

Background

Viral envelope proteins and virions themselves have been useful tools in the study of secretory and glycosylation pathways (Datema, et al, 1987). Often, the virus will exploit host secretory pathways to its own advantage, and the study of virus particle assembly and secretion will lead to valuable insights about cell biology. In addition, there is basic virological justification for studying the secretion of viruses, since knowledge about this process may help identify useful antiviral agents. We have been studying the role of glycan processing in the fate and function of glycoproteins from the perspective of hepatitis B virus (HBV) secretion. Since (i) HBV morphogenesis requires viral envelope protein function (ii) the viral glycoproteins are a family of homologous proteins, carrying only one or two N-glycans at common sites, and (iv) secretion of the viral glycoproteins varies dramatically in their sensitivity to glycan processing; the HBV system holds the promise of providing a simple system to understand the role of N-glycosylation and trimming in the function of glycoproteins.

Hepatitis B virus (HBV)

Hepatitis B Virus (HBV) is a causative agent of acute and chronic liver disease (Beasley, 1988). Although protective immunity is conferred by the use of both human serum derived and recombinant vaccines (reviewed in Peterson, et al, 1987), there are more than 300,000,000 people in the world who are chronically infected with the virus. For them, the vaccine has no therapeutic value. Between 25 to 40% of those who are chronically infected eventually progress to serious liver disease (cirrhosis and hepatocellular carcinoma)(Beasley, 1988). It is therefore important to find effective antiviral therapies. Alpha interferon has been used with promising results in a selective minority of patients (Perrillo, et al, 1990). Recent reports of compounds with anti-HBV activity suggest that the virus encoded DNA polymerase, which functions as a reverse transcriptase (Summers & Mason, 1982), is an attractive target. Other virus mediated processes have not been targeted for antiviral intervention. Effective antiviral therapy for HBV is likely to involve multiple strategies, including agents that influence the host immune system as well as those that interfere with different steps in the life cycle of the virus. It is therefore of interest to explore the possibility that other, non polymerase mediated steps in the virus life cycle are vulnerable targets for intervention. Since glycan processing inhibitors have been safely used in primates and people (as in Fischl, et al, 1994), we have been exploring the possibility of their use in the antiviral intervention for HBV.

HBV Life cycle

The envelope of HBV contains three related proteins called L, M and S (large, medium and small, respectively) (See Figure 1 and ref. Gerlich 1993) . These result from the alternate translation initiation of a single open reading frame (reviewed in Gerlich 1993 and Heermann & Gerlich, 1992). S occurs as both a 24 kilodaltons (kd) unglycosylated and 27 kd protein containing a single N linked glycan (Gerlich & Bruss, 1993, Peterson, 1982; 1987). L and M contain the S domain in addition to additional amino acids (see Heermann & Gerlich, 1992; Gerlich, 1993). M occurs as a singly N-glycosylated (33kd) and doubly N glycosylated (36 kd) species. L is found as a 39 kd and 42 kd N-glycosylated protein. Mature virions possess all forms of the envelope antigens.

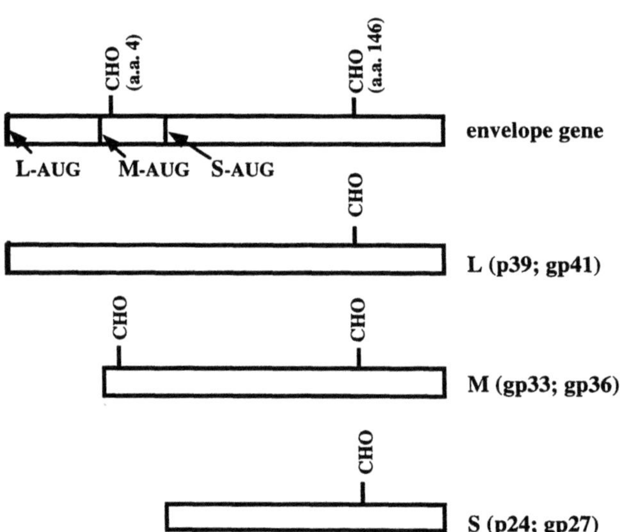

Figure 1.   The HBV envelope proteins.   The HBV envelope glycoprotein gene, which extends from nucleotides 2854 to 832 (where the EcoR1 site is nucleotide number 1) of the circular viral chromosome is shown (see Gerlich 1993). The three envelope proteins (L, M and S) are derived from alternate translation initiation from the same reading frame. The translation initiator AUGs are shown for each glycoprotein. The location of the N-glycosylation sites (CHO) and the molecular weights of the glycosylated (prefix "gp") and unglycosylated ("p"), as they appear in the virion, are also provided.

The morphogenesis of HBV is also noteworthy. Infected cells produce and secrete a vast excess, relative to intact 42 nM diameter infectious virus (*virions*), of "empty" (DNA negative) 20 nM diameter filament and spherical polymers of the envelope antigens (*subviral particles*) (Ganem, 1991) . The empty particles contain mostly S, with significantly less M and L then do virions. The role of empty particles in pathogenesis is not known.   This relative lack of L in subviral particles is significant, since polymers containing only L are not secreted and L, itself, may actually inhibit the secretion process (Bruss & Ganem, 1991). Infectious virions therefore possess an enrichment of L and M. In infected cells, by analogy

208

to the situation with sub viral particles, HBV is thought to acquire its envelope by the budding of nucleo-capsids through intracellular membranes (possibly of the endoplasmic reticulum or intermediate compartment), which have been modified with the various viral envelope proteins (Huovila, et al, 1992; Patzer, et al, 1984; 86). The envelope proteins are transported through the endoplasmic reticulim (ER), where terminal glucose residues of N linked glycans are removed. Further processing of the carbohydrate chain occurs (probably in the Golgi) and assembled viral particles are secreted via conventional "constitutive" vesicular pathways (see Ganem, 1991, Eble et al, 1987).

Secretion of liver glycoproteins and glycosylation

The mechanisms whereby liver cells sort and secrete proteins is of fundamental interest and represents one of the central questions in cell biology (reviewed in Barinaga, 1993 and Rothman & Orci, 1992). Since many, if not most, secreted proteins bear N-linked carbohydrates, it has been speculated for many years that N-glycans are involved in this process. Attention has recently turned to an intriguing peculiarity of N-glycan maturation: the trimming process. In the ER, mammalian cells begin to degrade (process) the 14 residue oligosaccharide immediately after its addition to the nascent, polypeptide backbone at specific amino acids, called "sequons" (see Elbein, 1991). This trimming reaction is temporally and spatially regulated, beginning with the removal of terminal and penultimate glucose residues by ER glucosidases I and II (see Elbein et al, 1991 or Lu et al, 1995). After additonal trimming, the glycoprotein is transferred to the Golgi coimplex, where the oligosaccharide is reconstructed in to its characteristic glycoform. Although the enzymology of processing is well understood, the reason that the N-linked glycan is degraded soon after its transfer only to be rebuilt in the Golgi is an enigma (see Helenius, 1994).

The function of N-glycan modification of polypeptides is the subject of debate but includes assisting folding, domain stabilization, spécial solubilzation, conferral of longevity, recognition of cell surface receptors, and other activities (Dwek, 1995; Helenius, 1994; Leavit, et al, 1979). The reasons for and the role of the tremendous cellular investment in *glycan processing* are even more surprising when considering that the process, in the ER, is non essential [for tissue culture growth] (Stanley, 1991; Ray, et al, 1991). Removal of the terminal glucose is apparently necessary for the transport of some, but not all, glycoproteins from the ER to the Golgi (Kong & Lodish, 1984; Moore & Spiro, 1993; Parent et al, 1991; Platt et al, 1991, Block et al, 1994). It has been proposed that, after dissociation from BiP, the mono-glucosylated glycoprotein (the product of glucosidase) binds calnexin (Hammond, et al, 1994; Hammond & Helenius, 1994). Calnexin is hypothesized to help in protein folding and subsequent protein-protein interactions (oligomerization) and transport from the ER (Helenius, 1994; Ou et al, 1993). Misfolded glycoprotein is either retained in the ER or continuously cycles between the ER and Golgi, possibly in association with BiP (Hannond & Helenius, 1994a,b; Haas & Wabl, 1983; Hammond & Helenius, 1994). Calnexin may recognize mono-glucosylated glycoproteins via the glucose residue (or) or via another domain on the protein (Herbert et al, 1995; Kim & Arvin, 1995).

On one hand, the vast diversity of secreted glycoproteins and lipoproteins suggests that common mechanisms of chaperoning and transport must be used. Transport of proteins from the ER to the Golgi has been hypothesized to occur by simple concentration driven "bulk flow" (see Rothman & Ochi, 1992). This theory may need to be modified, in light of the fact that retrograde transport vesicles can leave the Golgi for the ER without returning proteins which are in high concentration in the Golgi (see Pelham, 1993). Moreover, different glycoproteins are secreted at different and characteristic rates, in spite of similar synthesis rates (Lodish & Kong, 1985; Parent et al, 1986). In hepatic parenchymal cells, albumin and lipoprotein particles are localized to different terminal dilations of the rough ER and cis Golgi (Yokota,S. and D. Fahimi, 1981). We have shown, for example, that secretion of HBV virions, but not sub viral particles, are highly sensitive to glycan trimming inhibitors (Block et al, 1994). Taken together, suggests that there are different transport mechanisms for even similar classes of macromolecules. In this regard, it is already clear that many glycoproteins do use the Calnexin pathway (e.g. Ottechen & Moss, 1996) and, even for glycoproteins which interact with Calnexin, the role of glycan trimming is not entirely clear (Kim & Arvin, 1995). Certainly, although Calnexin and, in some cases, calreticulin (a lumenal ER chaperon) are important in the protein folding process, there are likely to be other host factors involved. The study of a variety of different glycoproteins will be necessary to fully understand the process.

Moreover, glycoproteins which do not experience proper trimming may not fold correctly and are usually not secreted. To the extent that *retention* of a glycoprotein with *unprocessed* glycan occurs, understanding how the cell recognizes and deals with misfolded glycoproteins is a very important part of the glycan trimming / host chaperon story. Although binding to BiP, Calnexin or calreticulin is usually very transient, recent evidence suggests that misfolded vesicular stomatitis virus (VSV) G protein remains bound to BiP/GRP 78, cycling indefinitely between the ER and the Golgi (Hammond & Helenius, 1994a). However, misfolding in that system was the result of a temperature sensitive mutation and not inhibition of glycan processing. Whether or not a similar consequence would result from inhibiting glycan trimming is not known. In addition, the ultimate fate of the misfolded protein is uncertain, since this was studied in productively and lethally infected cells. Finally, from the perspective of liver secretion, that work was mostly limited to cells in which there is virtually no endomannosidase, a Golgi enzyme which could serve as a shunt pathway for glycoproteins with unprocessed glycan (Moore & Spiro, 1990). That point underscores the importance of studying the role of glycan processing to glycoproteins in liver cells, where endomannosidase is abundant. In any event, it is obvious that since glycoproteins fall in to different categories with respect to their sensitivities to glycan processing, it will be necessary to study a range of different proteins before any general principles are proved.

HBV envelope proteins as valuable tools in the study of secretion and glycan trimming

HBV glycoproteins (envelope proteins) offer a powerful means of studying the role of glycan processing in the fate and function of glycoproteins. Although HBV glycoproteins all possess common glycosylation sequons, occupancy of the sequons differ, with M doubly glycosylated and S and L monoglycosylated in only 50% of the molecules (Gerlich & Bruss, 1992). This offers an attractive system to study the role of protein structure upon glycan sequon occupancy and visa versa. There is clearly a selection for secretion of subviral particles containing glycosylated M. Intracellular (22 nM) sub viral particles contain unglycosylated as well as doubly glycosylated M, whereas secreted (22 nM) sub viral particles contain only the glycosylated form. Moreover, our prel. evid. suggests that the secretion of L, M and S is exquisitely sensitive to glycan trimming. That is, in cells in which glycan trimming is inhibited, only sub viral particles bearing fully processed, complex, carbohydrates is secreted (not shown, see Mehta et al, this volume). Thus, even though ER glucosidase is completely inhibited in those cells, processed glycoproteins which presumably were trimmed by the Golgi endomannosidase, were selected for secretion. The unprocessed envelope glycoproteins were ostensibly retained within the cell (with viral DNA and nucleocapsid. This extreme requirement for processing to acheive secretion is not the situation with many other glycoproteins (such as HIV gp120: Kaarlson et al, 1992) which are secreted in unprocessed form from cells in which processing is inhibited. Thus, HBV envelope proteins represent an extreme case of dependency upon glycan trimming for secretion, since the cell clearly discriminates against particles containing unprocessed glycan.

HBV glycoproteins are highly cross-linked by intramolecular and inter-protein interactions (Gerlich & Bruss, 1992) and play an essential role in the morphogenesis and secretion of the virion lipoprotein (Bruss and Ganem, 1991). Being physically and functionally well characterized, they offer an ideal opportunity to study the functional consequences of preventing glycan trimming.

Exploring why glycan processing is necessary for virion secretion from liver cells will, in itself, have big pay offs. For example, secretion of HBV virions (as opposed to sub viral particles) may have a great degree of tissue specificity. Although transfected liver HepG2 cells secrete both virions and sub viral particles, transfected kidney CV-1 cells appear to only secrete sub viral particles (in spite of the fact that viral nucleocapsids are made) (Lu, unpublished, Bruss, personal comm.). The HBV system thus provides an example of tissue dependent secretory elements.

Also, different glycoproteins mature and are secreted at different rates (Lodish & Kong, 1985). However, morphogenesis of lipoprotein particles, in general, and HBV lipoproteins, in particular, require a high degree of spatial and temporal coordination between several different glycoprotein, core protein and glycolipid species. The realization that HBV glycoproteins bearing unprocessed glycan are sorted, with other non viral proteins, from secreted materials, dramatizes the tremendous organizational demands placed upon the host cell. Our prelim. evid. suggests that, in glucosidase inhibited cells, while L and S are secreted, M is fated for intracellular retention and probably lysosomal degradation. The preliminary pulse chase experiemnts further *suggest* that L, M and S can traffic from the ER to the Golgi and then, return to the ER! This is consistent with the results of Hammond and Helenius (1994a) in suggesting that abberrant proteins can recycle between the Golgi and ER. Our data would be the first to suggest that this can happen as a consequence of unprocessed glycan and that similar glycoproteins can have radically different sorting fates. For example: a significant fraction of M is discriminated from L and S and then fated for possible lysosomal compartments, whereas L and S are mostly secreted! How this "sorting"and differential sending of L, M and S is accomplished and how the cell preferentially sends M to the lysosomes raises fundemental questions. We will ultimately need to determine what host proteins manage these events. Answering that question will be helped by determining whether or not any of the discrimination involves the glycan status. Answering these questions should provide significant insights in to cellular mechanisms of glycoprotein and lipoprotein transport. Perturbing the system by preventing glycan trimming will help precisely determine the role that glycan processing has in mediating these events.

EVIDENCE THAT GLUCOSIDASE INHIBITED LIVER CELLS DIFFERENTIALLY SORT L, M AND S ENVELOPE PROTEINS

Secretion of HBV from chronically infected cells, in culture, is prevented by inhibitors of ER glucosidase

The fate and function of HBV envelope proteins was evaluated in 2.2.15 cells in the absence and presence of an inhibitor of ER glucosidase. 2.2.15 cells are derived from HepG2 cells, following the stable transfection of a plasmid bearing a dimer of the HBV genome, and chronically secrete infectious HBV (Sells et al, 1987). DNJ, and its alkyl derivatives, are potent inhibitors of ER glucosidase (Saunier et al, 1982). 2.2.15 cells treated with 1000 ugs/ml NBDNJ for 6 days remain completely viable, in spite of the inhibition of glucosidase (Lu et al, 1995). Consistent with the inhibition of glucosidase, NBDNJ treated 2.2.15 cells fail to secrete enveloped HBV (Block et al, 1994) and accumulate alpha1 antitrypsin polypeptide bearing unprocessed (glucosidase I sensitive) N-glycan (data not shown). This confirmed that NBDNJ had the expected effect upon 2.2.15 cells: low toxicity with a high effectiveness in inhibiting ER glucosidase.

Sub-cellular distribution of HBV DNA

It was next of interest to determine into which intracellular compartments HBV DNA

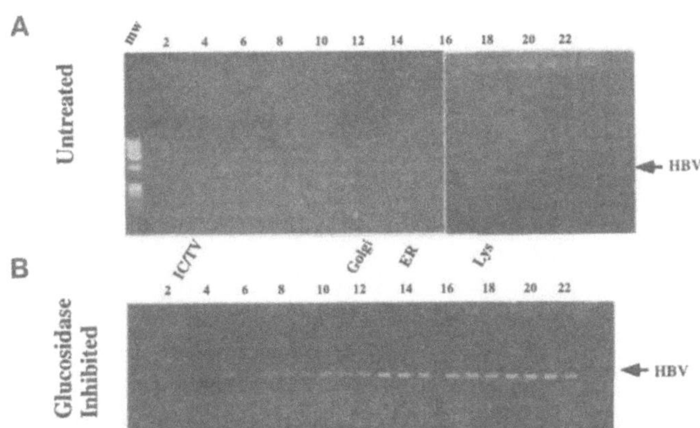

Fig. 2. Sub cellular location of HBV DNA in glucosidase inhibited cells.
The presence of HBV DNA in each fraction from percol gradients was determined by PCR amplification. Peaks of the enzyme and antigenic markers corresponding different sub cellular fractions are provided: IC/TV (Intermediate compartment/ Transport vesicles; Golgi; ER and Lys (Lysosomes).

had accumulated in the ER glucosidase inhibited cells. Previous work had shown that the accumulated DNA was located in the cytoplasm. Therefore, after 5 days of mock treatment or glucosidase inhibition, 2.2.15 cell cytoplasms were resolved through percoll gradients. These gradients permit a reasonable separation of ER and Golgi compartments, as determined by the location of characteristic markers (Fig. 2). The presence of HBV DNA in each fraction of the gradient was detected by a quantitative PCR (Lu etal, 1995) and the results are shown in Fig.2. PCR is used because, in these experiments, the amount of material is so little. In the cytoplasms from untreated 2.2.15 cells (Fig 2A), most of the HBV DNA migrates with the ER. Lesser amounts co-migrate with the Golgi. Since HBV nucleocapsids are presumed to acquire envelopes in the ER (or post ER) and are subsequently modified in the Golgi (Ganem, 1991), these are the expected locations of viral DNA during the normal secretory route. Some HBV DNA is detected at the bottom of the gradient and presumably represents free nucleocapsids (under investigation).

The distribution and abundance of steady state HBV DNA in the cytoplasms of ER glucosidase inhibited cells is completely different. In glucosidase inhibited cells (Fig. 2B), there is, as expected, considerably more HBV DNA than in untreated cells. However, the DNA predominantly co migrates with the ER and more dense fractions. There is little DNA co-migrating with the Golgi, suggesting a block in the maturation process at an ER and pre-ER step. The DNA in the dense regions (possibly in lysosomes as nucleocapsids) of the gradient are assumed to be in nucleocapsids, based upon work in progress.

Sub cellular distribution of envelope proteins in glucosidase inhibited cells

Since, in glucosidase inhibited cells, the amount and distribution of HBV DNA was aberrant and proper morphogenesis requires functional viral glycoprotein, it was of interest to know the relative abundance and distribution of the viral glycoproteins. Therefore, after 5 days of incubation in the presence or absence of glucosidase inhibitor, 2.2.15 cells were

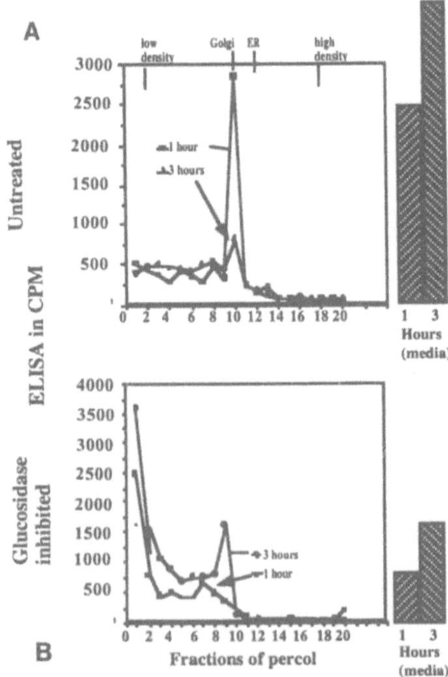

Fig. 3. Sub cellular location of envelope proteins in glucosidase uninibited and inhibited cells as a function of time following synthesis. 2.2.15 cells which had been left untreated (A) or were glucosidase inhibited (B) for 3 days with NBDNJ, were labelled and chased for either 1 or 3 hours (as indicated), homegenized and resolved through percol. The locations of the peaks of the Golgi/IC and ER markers are shown. Envelope protein was detected in each fraction with a mAb that recognizes S, M and L (does not distinguish). Note the rapid secretion in untreated cells (A) and the aberrant retention and location of envelope protein in glucosidase inhibted cells (B).

labeled for 20 minutes with 35-S-methionine, and cytoplasms were resolved through percoll gradients after 1, 3 or 24 hours of labeling. Each fraction from the gradient was tested for the presence of HBV envelope antigen by a quantitative solid state assay (as in Block et al, 1994) and the results are shown in Fig. 3.   The kinetics of synthesis and secretion of envelope antigen from control (untreated) 2.2.15 cells is straightforward, with most of the radioactivity corresponding to envelope proteins having left the ER and co-migrating with the Golgi/intermediate compartment (IC) apparatus by 1 hour after labeling (Fig. 3A). By 3 hours, 95% of the radioactivity corresponding to the envelope antigens is no longer present within the cell and is almost entirely recovered in the culture medium (not shown). Therefore, the period of time from synthesis to secretion is nearly complete within 3 hours, as observed by others (Patzer et al, 1984,6).

On the other hand, the kinetics of intracellular trafficking of HBV antigens in glucosidase inhibited cells is completely different. At 1 hour after labeling, there are very few HBV envelope proteins in the ER or Golgi/IC, with most material at the top of the gradient (Fig. 3B). It is not clear if this low density location represents a position in between the ER and the Golgi (possibly in transport vesicles) or if it is a "dead end" pathway, leading to degradation. To distinguish between S, M and L, conformation specific mAb was used on samples from glucosidase inhibited cells at 3 and 24 hours after labelingIt appears that the peaks of S, M and L occur in different fractions (Fig. 4A), suggesting they may be independent of each other.

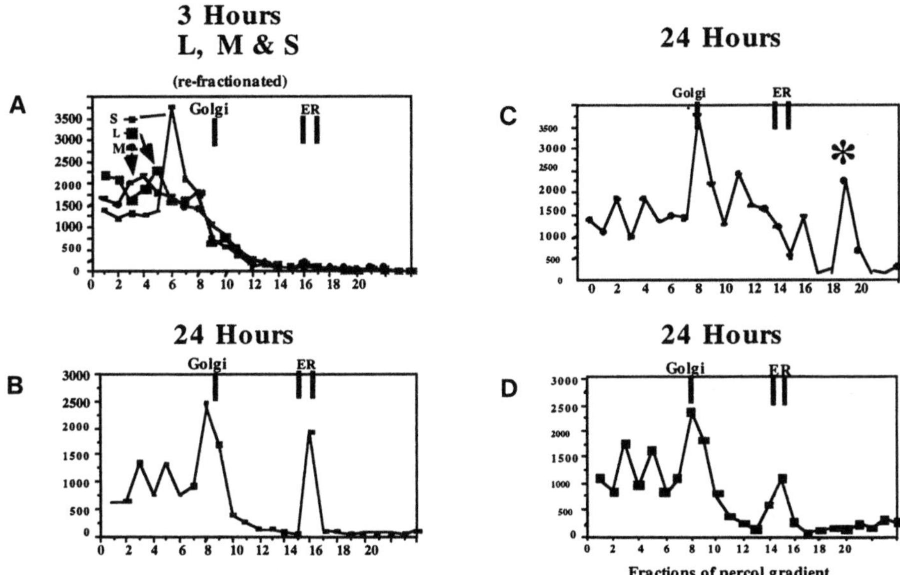

Fig. 4.   In glucosidase inhibited 2.2.15 cells, envelope protein synthesized in the ER, moves from low density regions to the Golgi and then back to the ER. 2.2.15 cells, incubated with NBDNJ , were resolved through percoll gradients after the indicated times following labelling. Peaks of the Golgi/IC and ER markers are shown, based upon assays described above. The star in (C) indicates the expected position of lysosomes. Note the appearence of envelope protein in positions associated with the ER at 24 hours, long after they would have initially left the ER.

Figs. 4B, C and D show fractionation of glucosidase treated cells 24 hours after labelling. By 24 hours in glucosidase inhibited cells, approximately half of the counts per minutes associated with S have been secreted in to the medium (not shown). Of the S protein remaining within the cell, approximately 60- 70% co-purified with the Golgi/IC and low density markers. The other 30% appears to have returned to the ER (Fig. 4B)! The pattern is similar for L (Fig. 4C). This recycling is consistent with the results of Hammond and Helenius (1994a) and represents the first example of recycling of proteins in cells unable to process N-glycan. Whether or not the recycling is the result of unprocessed glycan on the glycoprotein is a matter of speculation.   Also of special note is the fate of the "M" glycoprotein. At 24 hours after synthesis in glucosidase inhibited cells, more than half of the "M" is recovered in a very dense region of the gradient, suggesting containment in lysosomes. This region is virtually free of "S", suggesting that the cell has distinguished between M and S.

Conclusion

This work thus dramatizes the consequences of inhibition of glucosidase to the function of a set of "reporter" glycoproteins. Glucosidase inhibited cells treat L, M and S very differently, in spite of the fact that they share 66 and 77% of their amino acid sequence and usually assemble together as lipoprotein particles. Thus, *the results are very exciting and underscore how the HBV system can be used to help understand how cells sort very similar proteins.* Clearly, in glucosidase inhibited cells, HBV envelope antigens do not function properly. They do not traffic properly through the cell and, unlike in untreated cells, they are not associated with viral DNA. It is not clear if this is a direct result of unprocessed glycan on the HBV envelope glycoproteins themselves as opposed to an affect upon a cell protein. Others have shown that some glycoproteins must be properly trimmed to bind cell chaperons in the ER assume proper folding conformations (Hebert et al, 1995). Unfolded or misfolded glycoproteins might be expected to be retained within the ER, not move to the Golgi and eventually become targets for degradation. However, HBV proteins in glucosidase inhibited cells appear to be relatively stable. We have recently shown that HBsAgs, retained within glucosidase inhibited cells are greatly enriched in "M" and "L", relative to "S":, have long (not short) half lives, and travel throughout the cell as large particles contained within intracellular vesicles (not shown). As shown in the model in Fig. 5, they do not migrate to the proper compartments with proper kinetics. It may be that the upset in trafficking kinetics prevents coordination of the envelopment process. Envelope proteinsand nucleocapsids may not arrive in the same place at the same time. It is not clear if the failure of the HBV envelope proteins to properly route and associate with viral DNA, in glucosidase inhibited cells, is a consequence of improper folding, inability to bind necessary transporter or chaperon proteins, or both. It is clear, however, that proper host cell glucosidase function is necessary for HBV envelope protein function, as defined by the mediation of envelopment. Determination of the exact reasons for dysfunction will likely provide a new perspective upon the role of glycan trimming in glycoprotein behavior.

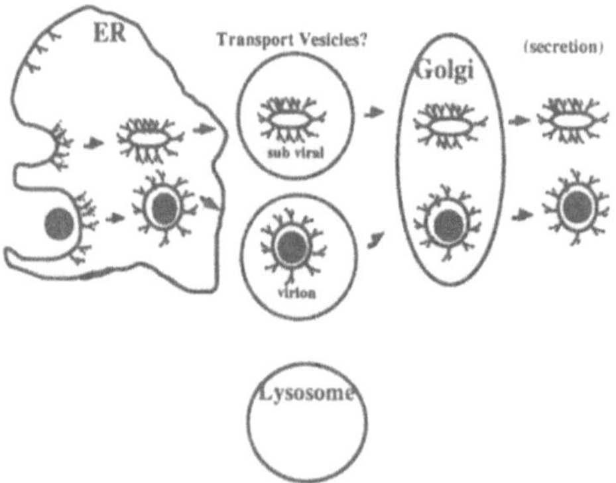

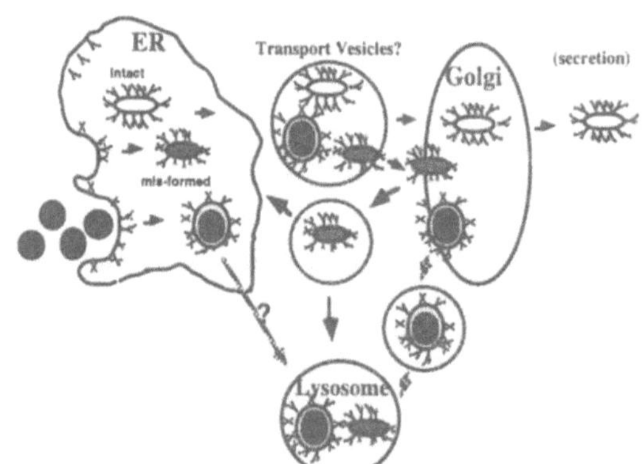

Figure 5. Model of how HBV particles containing unprocessed glycan are rejected from the Golgi, not secreted, recycled to the ER and eventually degraded in the lysosomes.

ACKNOWLEDGEMENTS

This work was supported by The Hepatitis B Foundation with an award from the Blanche and Irving Laurie Foundation, NATO and Searle, Inc.

LITERATURE CITED

Barinaga, 1993, Secrets of secretion revealed, Science, 260: 488-9

Beasley, R. P. (1988). The Hepatitis B virus. The major etiology of hepatocellular carcinoma. Cancer. 61: 1942-1956..

Block, T., Platt, F., Xuanyong, L., Gerlich, W., Foster, G., Blumberg, B., Dwek, R. Secretion of human hepatitis B virus is inhibited by the imino sugar, N-butyldeoxynojirimycin. Proc. Natl. Acad. Sci. (USA) 91: 2235-2239.

Bruss, V., D. Ganem. (1991). The role of envelope proteins in hepatitis b virus assembly. Proc. Natl. Acad. Sci. USA. 88: 1059-1063.

Chen, W., J. Helenius, I. Braakman and A. Helenius (1995). Cotranslational folding and calnexin binding during glycoprotein synthesis. Proc. Natl. Scad. Sci. (USA). 92: 6229-6233.

Datema, S. Olofsson, P. Romero. (1987), Inhibitors of protein glycosylation and glycoprotein processing in viral systems. Pharmacol. Thera., 33: 221-286

Dwek, R. A. (1995). Glycobiology: more functions for oligosaccharides. Science. 269: 1234-1235.

Elbein, A. D. (1991) Amino sugar compounds as antiviral and anti tumor agents, Sem. cell Biol., 2; 309-317;

Fischl, M, L. Resnick, R. Coombs, A. Kremer, J. Potage et al (1994), The safety and efficacy of combination therapy N-butyl deoxynojirimycin (SC-48334) and Zidovudine in patients with HIV infection and 200-500 CD4 cells / mm3. J.Acq. Imm. Def. S., 7: 139-147

Ganem, D. (1991) Assembly of hepadnavirus virions and subviral particles. In: Hepadnaviruses, Current topics in Microbiol. and Immunol. (W. Mason and C. Seeger, Eds.) 168: 61-84

Gerlich, W. (1993) Structure and molecular biology (of HBV). In: Zuckerman, A. & H. C. Thomas, Viral hepatitis, pp. 83-113. Chruchill Livingston Publishers, London, UK.

Hammand and Helenius, 1994a, Quality control in the secretory pathway: retention of a misfolded viral membrane glycoprotein involves cycling between the ER, Intermediate compartment and Golgi Apparatus, JCB, 126: 41-52).

Hammond, C., and A. Helenius, 1994b, Folding of VSV G protein: sequential interaction with BIP and calnexin. Science. 266: 456458.

Hammond, C., I. Braakman and A. Helenius, 1994, Role of N linked oligosaccharide recognition, glucose trimming and calnexin in glycoprotein folding and quality control. Proc. Natl Acad. Sci. . 91: 913-917

Haas, I. G. and Wabl (1983). Immunoglobulin heavy chain binding protein. Nature. 306: 387-389.

Heerman, H-H, Goldman, I., Schwartz, W, Seyffarth, T., Baumgarten, H. Gerlich, W. H. Large surface proteins of hepatitis b virus containing preS sequence. (1984) J.Virol 52: 396-402)

Heermann, K-H, Gerlich, W.H. (1992), In: Mol. Biol of HBV, A Maclachlan, ed, CRC Press, Boca R raton Fl

Helenius, A. (1994) How N linked oligosaccharides affect glycoprotein folding in the endoplasmic reticulum. Mol. Biol. Cell. 5: 253-265.

Herbert, D. N., B. Foellmer and A. Helenius (1995). Glucose trimming and reglucosylation determine glycoprotein association with calnexin in the endoplasmic reticulum. Cell. 81: 425-433.

Huovila, A. P., A. Eder and S. Fuller (1992) Hepatitis B surface antigen assembles in a post ER, pre golgi compartment. J. Cell Biol. 118: 1305-1320.

Karlsson, G.B., Butters, T. D., Dwek, R.A., Platt, F.M., (1993) Effects of the imino sugar N butyldeoxynojirimycin on the N glycosylation of recombinant gp120 J. Biol. Chem.,268: 570-576.

Kim, P. S. and P. Arvin (1995) Calnexin and BiP as sequential molecular chaperones during thyroglobulin folding in the endoplasmic reticulum. J. Cell Biol. 128: 29-38.

Leavitt, R., Schlesinger, S., Kornfeld, S. 1979. Impaired intracellular migration and altered solubility of of non glycosylated glycoproteins of vesicular stomatitis virus. J. Biol. Chem. 252: 9018-9023

Lenter, M. and D. Vestweber (1994). The integrin chains b1 and a6 associate with the chaperon calnexin prior to integrin assembly. J. Biol. Chem. 269: 12263-12268.

Lodish, H., A. Kong (1984). Glucose removal from N linked oligosacchairdes is required for efficient maturation of certain secretory glycoproteins from the rough endoplasmic reticulim to the Golgi compartment. J. Cell. Biol. 98: 1720-1729.

Lu, X., A. Mehta, T. Butters, R. Dwek and T. Block (1995) Evidence that N-linked glycosylation is necessary for hepatitis B virus secretion. Virol. 213: 660-665.

Lu, X., T. Block and W. Gerlich (1996) Protease induced infectivity of human hepatitis B virus for continuous cultures of human liver cells. J. Virol., in press.

Lu, X., A. Mehta, M. Dadmarz, B. Blumberg, R. Dwek and T. Block (1997) Aberrant trafficking of hepatitis B virus glycoproteins in cells in which N-glycan processing is inhibited. Proc. Natl. Acad. Sci. USA, vol 94: 2380-2385.

Lubas, W.A., Spiro, R.G. (1988). Evaluation of the role of rat liver golgi endo alpha mannosidase in processing N linked oligosaccharides. J. Biol. Chem. 263: 3990-3998.

Mehta, A., X. Lu, T. Block, B. Blumberg, R. Dwek (1996) Similar glycoproteins with drastically different sensities to glycosylation processing (Submitted).

Moore, S. and R. Spiro (1993). Inhibiton of glucopse triming by castaospermine results in rapid degradation od unassembled major histocompatribility complex class I molecules. J. Biol. Chem. 268: 3809-3812.

Moore, S. E. and R. G. Spiro (1990) Demonstration that Golgi Endo-a-mannosidase provides a glucosidase-independent pathway for the formation of complex N-linkied oligosaccharides og glycoproteins. J. Biol. Chem. 265: 1304-13112.

Otteken, A. and B. Moss (1996) Calreticulin interacts with newly synthesized human immunodeficiency virus type I envelope glycoproetin, suggesting a chaperone function similar to that of calnexin. J. Biol. Chem. 271: 97-103.

Ou, W. J., P. H. Cameron, D. Y. Thomas and J. J. M. Bergeron (1993) Association of folding intermediates of glycoproteins with calnexin. Nature. 364: 771-776.

Parent, B., T-K. Yeo and K. Olden (1986). Differential effects of 1-deoxynojirimycin on the intracellular transfport of sceretory glycoprotiens in human hepatoma cells in culture. Mol. Cell. Biochem. 72: 21-33.

Patzer, E., Nakamura, G., Yaffe, A., Intracellular transport and secretion of hepatitis B surface antigen in mammalian cells. (1984), J. Virol., 51: 346-353

Patzer, E.J., Nakamura, G.R., Simonsen, G.C., Levinson, A.L., Brands, R.(1986), Intracellular packaging and assembly of hepatitis b surface antigen particles occurs in the endoplasmic reticulim. J. Virol. 58: 884-892

Pelham, H. R. B. (1991) Recycling of proteins between the endoplasmic reticulum and the Golgi complex. Curr. Opin. Cell Biol. 3: 585-591.

Perrillo, R.P, Schift, ER, Davis, GL, Bodenheimer, HC, Lindsay, K, Payne, J., Dienstag, JL,O'Brien, C., Tamburro, C., Jacobson, IM, Sampliner, R., Feit, D., Lefkowitch, J., Kuhns, M., Meschievitz, C., Sanghvi, B., Albrecht, J., Gibas, A. (1990), A randomized controlled trial of interferon alpha 2b alone and after prednisone withdrawal for the treatment of chronic hepatitis b. New Eng. J. Med. 323: 295-301..

Peterson, , DL, (1981), Isolation and characterization of the major protein and glycoprotein of hepatitis B surface antigen. J. Biol. Chem., 256: 6975-6983

Platt F. G. Nieses, R. Dwek, T. Butters, (1994), NBDNJ is a novel inhibitor of glycolipid biosynthesis, J. Biol. Chem., 269: 8362-8365

Platt, F., G. B. Karlsson & G. S. Jacob, (1992), Modulation of cell surface transferrin receptor by the imino sugar N butyldeoxynojirimycin. Eur. J. Biochem. 208: 187-193.

Rothman, J. E. and L. Orci (1992) Molecular dissection of the secretory pathway. Nature. 355: 409-415.

Saunier, B., Kilker, R.D., Tkacz, J.S., Quaroni, A., Herscovics, A.C. Inhibition of N linked complex oligosaccharide formation by 1 deoxynojirimycin. (1982). J. Biol. Chem. 257: 14155-14161.

Sells, M.A., M. L. Chen, G. Acs, (1987). Production of hepatitis b virus particles in Hep G2 cells transfected with cloned hepatitis b virus DNA. Proc. Natl. Acad. Sci. USA. 84: 1005-1009.

Summers, J., Mason, W.H. (1982). Replication of the genome of hepatitis b virus by reverse transcription of an RNA intermediate. Cell. 29: 403-415.

GLYCOTHERAPEUTICS

COMBINATORIAL CARBOHYDRATE CHEMISTRY

Zhi-Guang Wang and Ole Hindsgaul

Department of Chemistry
University of Alberta
Edmonton, Alberta, Canada T6G 2G2

INTRODUCTION

Modern organic/medicinal chemistry is undergoing a "cultural revolution" in the way new drugs are discovered and developed. Instead of discrete synthess and biological screening of individual compounds, which often takes years to identify and optimize leads, the currently developing technology called "combinatorial chemistry" can rapidly provide large numbers of chemicals (libraries) in a short time. In conjunction with these new synthetic methods, high-throughput screening (HTS) can rapidly screen the libraries produced, and in so doing, can provide information for optimizing lead compounds. This is going to lighten the increasing burden traditional drug development places on the pharmaceutical industry. For this reason, the generation of chemical libraries through combinatorial chemistry, including parallel synthesis, is making explosive progress both in academic and industrial areas in the last two years, especially for creating peptide, nucleotide and small molecule libraries. A number of excellent reviews[1] on this field have appeared.

In contrast to the remarkable progress on small molecule libraries and biopolymers, the development of combinatorial carbohydrate chemistry has been slow. Reasons for this arise mostly from the inherent chemical difficulties presented by this class of compounds[2]. For one, there is no general glycosylation methodology available comparable to the coupling approach of peptide and phosphodiester formation, where the linkage yields are essentially quantitative. Secondly, the process of glycosylation creates a new stereocentre at the anomeric carbon (which can be either α or β), and there is no reaction which can give a glycosidic bond with absolute stereospecificity, a problem not encountered with the amide or phosphate linkages in nucleotide and peptide chemistry. Moreover, each of the naturally occurring mammalian sugar monomers (all pyranoses: Glc, Gal, Man, Xyl, GlcNAc, GalNAc, Fuc, GlcA, IdoA and NANA, Fig. 1) carry at least three hydroxyl groups of various activities towards glycosylation. Extensive branching or other functionalization such as sulfation or phosphorylation therefore further complicates the chemistry. All of the above difficulties render the preparation of oligosaccharide libraries a formidable challenge.

Cell-surface oligosaccharides are involved in many important physiological processes and inhibiting such processes with small-molecule drugs is gaining interest. Recently, some

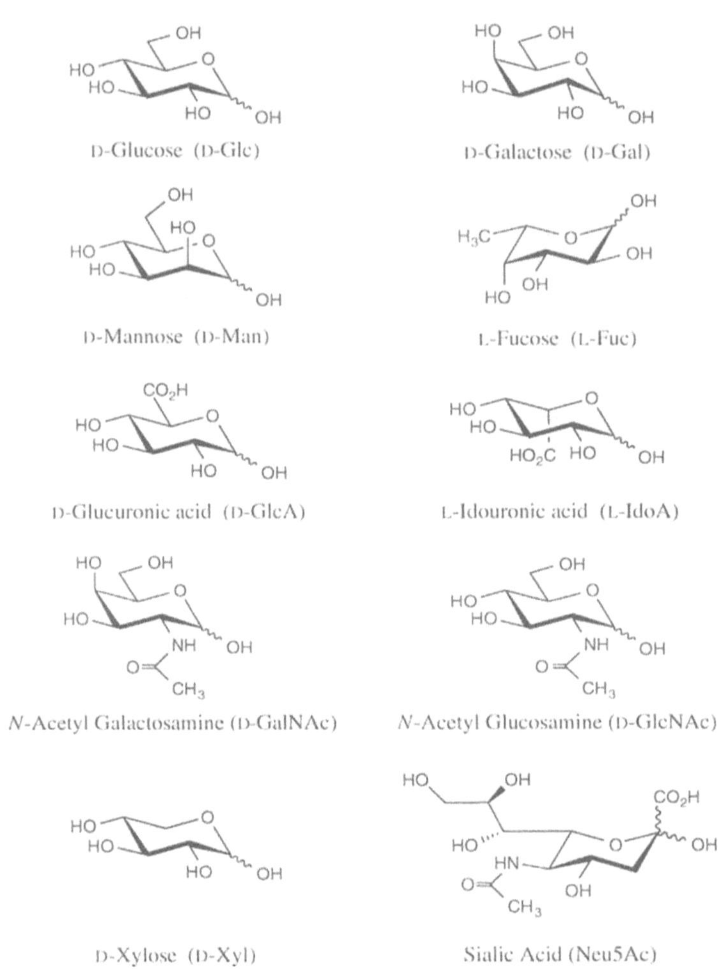

D-Glucose (D-Glc)   D-Galactose (D-Gal)

D-Mannose (D-Man)   L-Fucose (L-Fuc)

D-Glucuronic acid (D-GlcA)   L-Iduronic acid (L-IdoA)

N-Acetyl Galactosamine (D-GalNAc)   N-Acetyl Glucosamine (D-GlcNAc)

D-Xylose (D-Xyl)   Sialic Acid (Neu5Ac)

Figure 1. Monosaccharides found in mammalian glycoconjugates.

biologically-active carbohydrate-based drugs have reached phase I, II and III clinical trials as anti-inflammatories, cancer chemotherapeutic agents, anti-thrombotics, anti-diabetic therapies, anti-infectives, and anti-virals[3]. It can be anticipated that combinatorial chemistry, an efficient tool for finding leads, will play an important role in the future of carbohydrate drug research. In this mini review, we summarize the strategies that have been reported to date.

1. SOLUTION-PHASE SYNTHESIS OF OLIGOSACCHARIDE LIBRARIES

a. Random-Glycosylation Strategy

The first approach to oligosaccharide libraries was reported by Kanie *et al.*.[4] Rather than applying a more systematic "Orthogonal Protection/Glycosylation" strategy, they investigated "random-glycosylation" (Figure 2a) for the preparation of oligosaccharide libraries. In this strategy, less than 1 eq. of protected donor per free acceptor hydroxyl group, is reacted with an unprotected acceptor attached to a hydrophobic aglycon such as 8-(*p*-methoxyphenoxy)-octyl at its reducing end. The increased solubility of the unprotected sugar provided by the

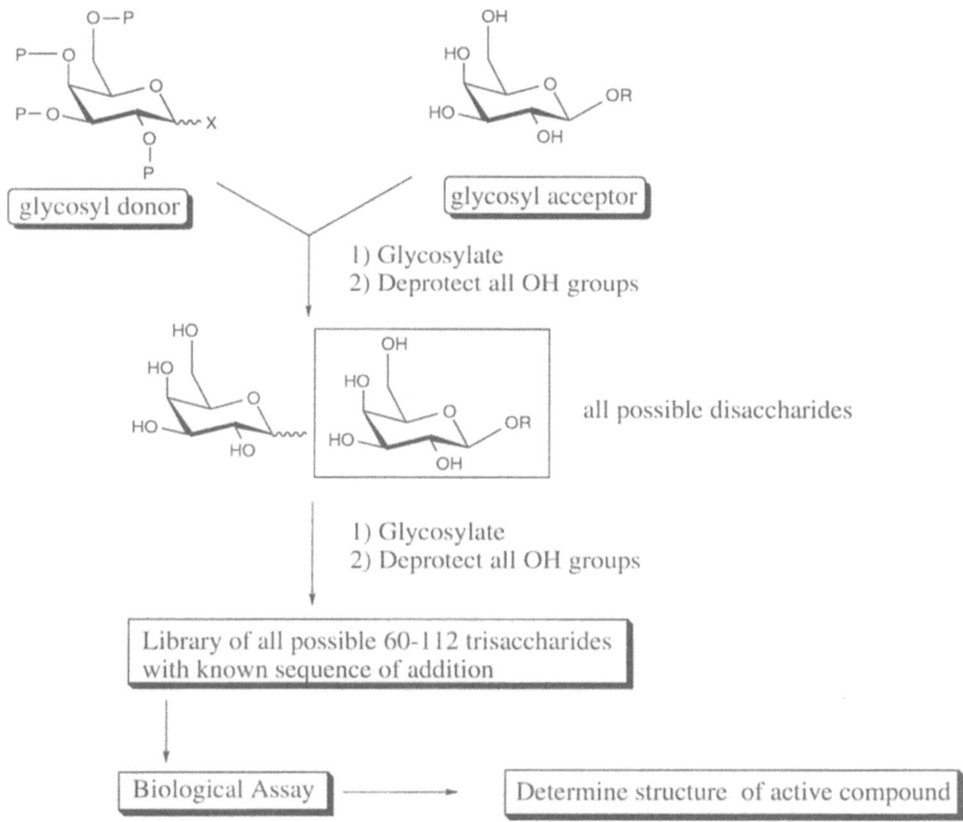

Figure 2a. Strategy for oligosaccharide library synthesis using random glycosylation. "P" denotes a protecting group.

hydrophobic aglycon in organic solvents facilitated the glycosylation and product isolation. Fig 2b demonstrates two sets of glycosylations using very similar acceptors and protected D-galactose trichloroacetimidate as the donor, with the only difference being the protecting groups on the donor moiety. Under appropriate conditions of promoter, solvent and reaction temperature, all possible disaccharides, including positional isomers and anomers (six compounds) could be obtained in a single reaction. As acceptor size increases, the glycosylation products obtained in the library increase exponentially, as is demonstrated in Figure 2c, showing an α-fucosylation with a disaccharide acceptor. Six α-fucosylated trisaccharides are generated in a single glycosylation (with minor amounts of β-fucosylated compounds also produced).

As protected donors were utilized in this strategy, the generated compounds have significant differences in their properties from both acceptor and donor in terms of polarity and molecular weight, which facilitates the purification of the library by size-exclusion and silica gel chromatography. Notably, reaction of the generated library (Fig. 2c) with radioactive UDP-GalNAc in the presence of GalNAc-transferase results in the blood group A type 1 tetrasaccharide where GalNAc has been added to O-3 of the Gal residue in α-Fuc $(1{\rightarrow}2)$ β-Gal $(1{\rightarrow}3)$ β-GlcNAc-OR (unpublished data). This implies that proteins or enzymes can recognize specific ligands or acceptors from among other similar compounds, and direct screening of the library is therefore of value.

While the random-glycosylation strategy is the most direct, avoiding extensive protecting group manipulation, the low conversion of acceptor and need for extensive product purification remain serious impediments to its widespread use.

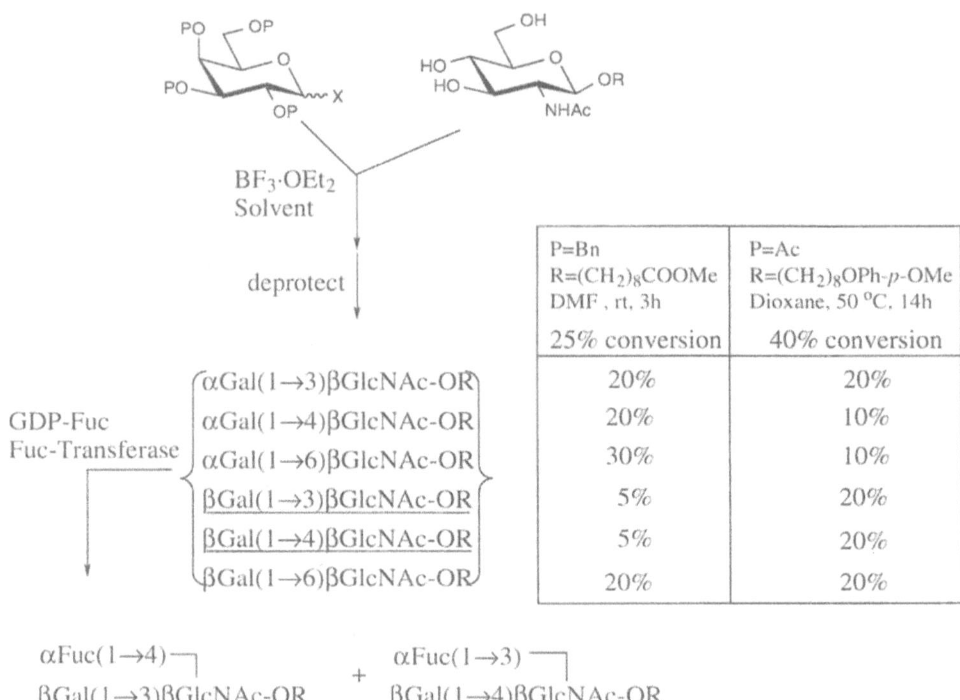

	P=Bn R=(CH₂)₈COOMe DMF, rt, 3h 25% conversion	P=Ac R=(CH₂)₈OPh-*p*-OMe Dioxane, 50 °C, 14h 40% conversion
αGal(1→3)βGlcNAc-OR	20%	20%
αGal(1→4)βGlcNAc-OR	20%	10%
αGal(1→6)βGlcNAc-OR	30%	10%
βGal(1→3)βGlcNAc-OR	5%	20%
βGal(1→4)βGlcNAc-OR	5%	20%
βGal(1→6)βGlcNAc-OR	20%	20%

Figure 2b. Examples of disaccharide Gal→βGlcNAc library preparation by random glycosylation. In both case, six desired disaccharides were obtained. Reaction of the library with human milk fucosyltransferase in the presence of labeled GDP-Fuc results in the formation of radioactive products. Only the disaccharides underlined are substrates for this enzyme, producing trisaccharides.

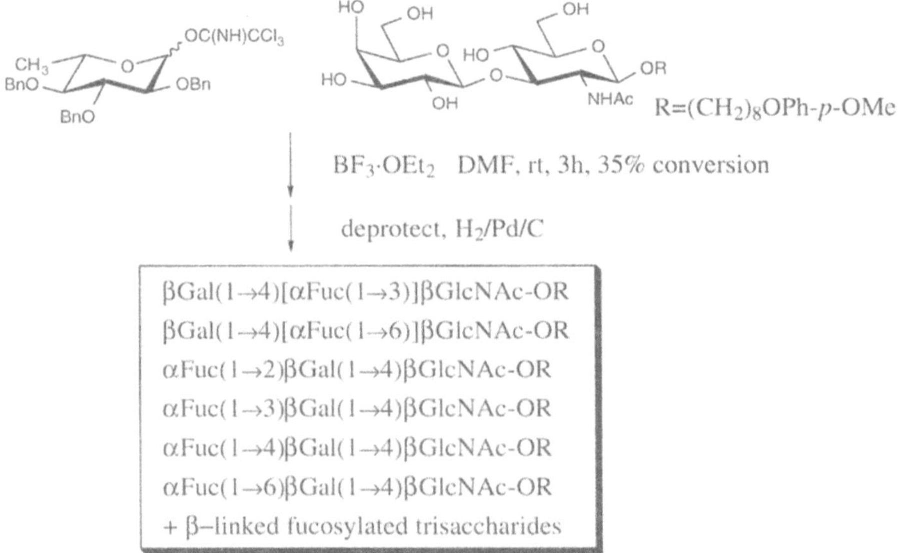

Figure 2c. Random fucosylation of a βGal(1→3)βGlcNAc derivative. Each of the six α-fucosylated trisaccharides were formed in yields ranging from 8% to 23%.

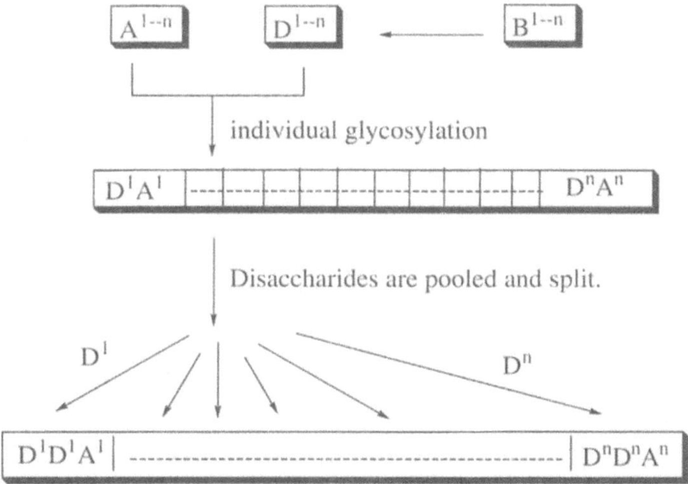

Figure 3a. Strategy for the controlled synthesis of an oligosaccharide library. Both donor (D) and acceptor (A) can be accessed from same intermediate B. Each acceptor can react with each donor individually and the formed disaccharide can be purified . The pooled disaccharides are glycosylated again with donor(D^1--D^n) forming a trisaccharide library.

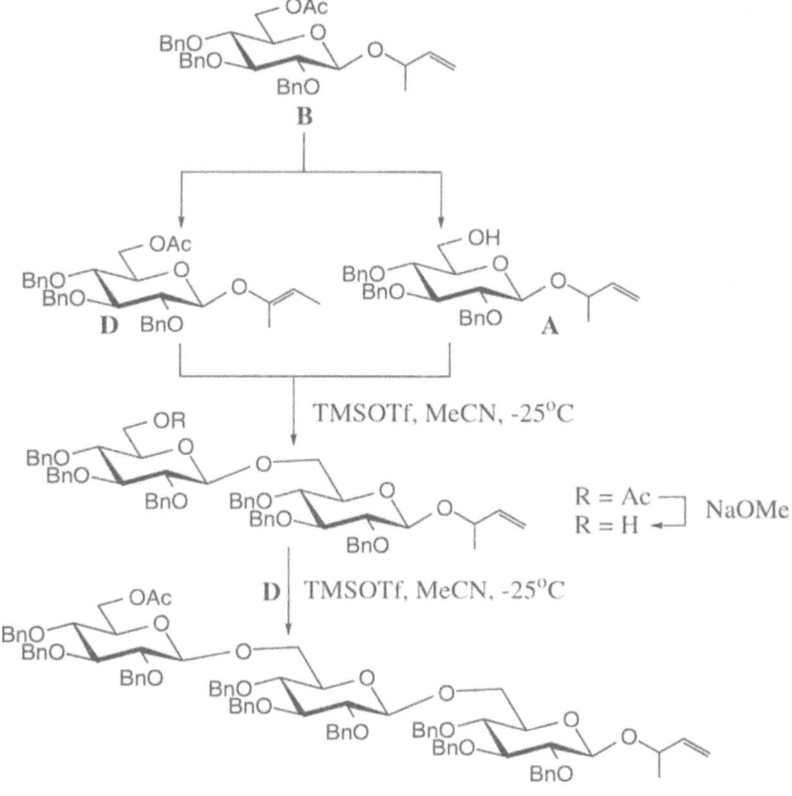

Figure 3b. A representive synthesis of a single trisaccharide by the strategy described in Figure 3a.

b. Controlled Synthesis of Oligosaccharide Libraries

While deconvolution of libraries and resynthesis of their active compounds (if found) present challenges to the random glycosylation strategy, so-called controlled synthesis of oligosaccharides[5] may overcome this difficulty to some extent. Figure 3a depicts the strategy for generating trisaccharide libraries by controlled synthesis. In this strategy, monosaccharide donors and acceptors react with each other individually and the disaccharide intermediates produced are purified (and a portion is saved, partly for future deconvolution). The products are pooled (after de-O-acetylation), and again glycosylated using individual excess donor D^1--D^n to provide trisaccharides. Since the product trisaccharides have higher molecular weights, size-exclusion chromatography readily removes excess reagents. The preparation of the trisaccharide in Figure 3b can exploit the strategy shown there. Vinyll glycosides (several other types of glycosides can in principle perform the same function, such as pentenyl glycosides, thioglycosides, etc.), using simple chemistry, can be converted into vinyl glycosyl donors and acceptors with ease. After glycosylation, the newly-formed disaccharide is again able to be converted into a new acceptor. Since the purification of the mixture obtained from each glycosylation is straightforward, pooling of the products can afford various sublibraries. Further manipulations of functionality and glycosylation of the sublibrary can give a trisaccharide library.

2. SOLUBLE POLYMER-SUPPORTED (POLYETHYLENE GLYCOL) SYNTHESIS OF OLIGOSACCHARIDE LIBRARIES

As described above, the solution-phase generation of oligosaccharide libraries requires some time-consuming purification regardless of whether it is based on the strategy of random glycosylation or that of controlled synthesis. For large libraries, this will become very problematic. Polyethylene glycol (PEG), is a polymer soluble in water and many organic solvents which has been extensively utilized in the combinatorial chemistry of peptides and

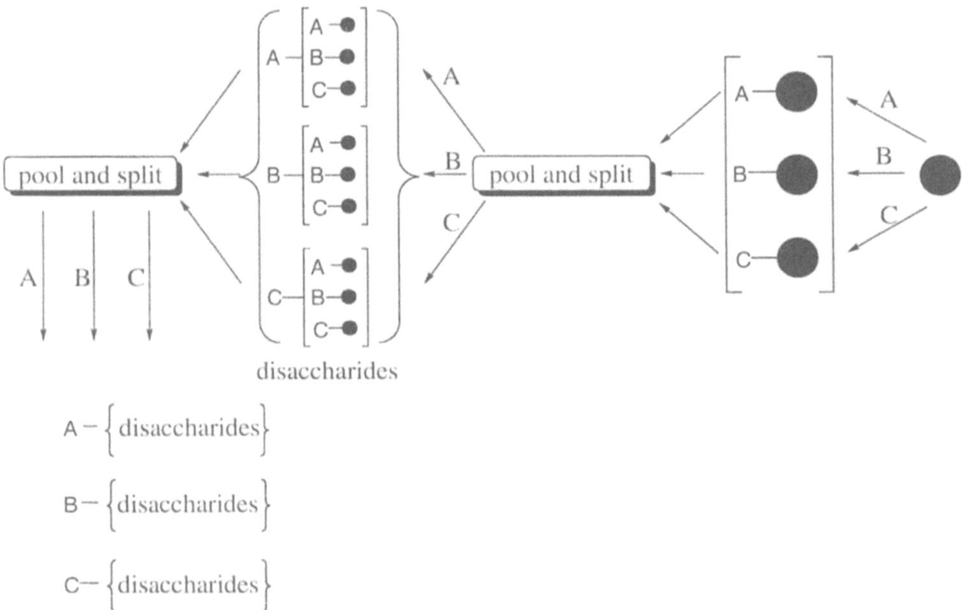

Figure 4a. Schematic of the pooling and splitting method of library synthesis.

nucleotides. PEG-supported chemistry can take advantage of most glycosylation methodologies currently available. Furthermore, carbohydrate-derivatized polymer is easily precipitated out when the reaction mixture is poured into *tert*-butyl methyl ether or diethyl ether. Unreacted reagents and by-products, not bound to the polymer, are thus easily removed. Oligosaccharide library synthesis can therefore adopt the general strategy of most peptide or small molecular library syntheses, as showed in Figure 4a. The synthesis[6a] of αMan(1→4)βGlcNAc(1→4)GlcNAc derivatives with difficult 1→4 linkages on PEG demonstrates the utility of this where the phthalimido group is used to direct complete β-glycosylation. The synthesis[6b] of a library of αMan(1→N)βGlcNAcβ(1→N)GlcNAc trisaccharides employing a pooling/splitting strategy is illustrated in Figure 4b. Three trichloroacetimidates (A, B, C) equipped with one temporary functionality (an *O*-acetyl group) and other persistent protecting groups (*O*-benzyl) can be easily immobilized to methyl PEG *via* an α,α'-*xylo* linker in the presence of triethylsilyl triflate. After de-*O*-acetylation and pooling (or the reverse), the monosaccharide attached to the PEG can be glycosylated individually with A, B and C to obtain a disaccharide library. The sublibrary resulting from pooling and de-*O*-acetylation is mannosylated again using tetra-*O*-acetyl mannosyl imidate, yielding a trisaccharide library. The composition of this library can indirectly be verified by detaching the oligosaccharides from the polymer by hydrogenation and acetylation of its three component sublibraries. The anomeric proton at the reducing end of each library component gives a characteristic signal in proton NMR, thus allowing the integration of this signal to reveal the ratio of component sugars in the library. It is difficult, however, to assign signals to specific individual compounds.

PEG-supported chemistry is particularly useful because of its similar reactivity to the solution phase while retaining the advantages of a solid-phase workup. Furthermore, the solid form of PEG makes pooling/splitting technology operationally simple. The ability to apply routine NMR techniques to the monitoring of reaction progress is a another advantage. Two major detractions, however, are the instability of PEG to repeated exposure to glycosylation conditions and PEG's inability to be encoded with currently-available coding tags.

3. POLYETHYLENE GLYCOL-GRAFTED POLYSTYRENE (TENTAGEL)-SUPPORTED SYNTHESIS OF OLIGOSACCHARIDE DERIVATIVES

When polyethylene glycol is grafted onto polystyrene, the solid part of the polymer is surrounded by soluble material, which not only facilitates reaction in a solution-like medium, but also makes coding techniques possible on the polymer. Through the combination of carbohydrate and small molecule chemistry, Liang *et al.*[7] were able to produce 1300 compounds in very few steps, as shown in Figure 5. First, six glycosides with carboxylic acid aglycons were attached to TentaGel separately, and the resulting beads were encoded individually. After deacetylation, the resulting intermediates were pooled and then split into twelve parts and then glycosylated with twelve donors individually. Once more, the products were separately encoded with 12 new tags. Since the acceptor was immobilized on Tenta-Gel, excess donor can be utilized to assure that the reaction goes to completion, without causing any difficulties in purification as in the PEG-supported chemistry. After the construction of a disaccharide or trisaccharide library, the azide function was transformed into an amine, which could be further acylated with 20 different acyl groups into a large number of oligosaccharide derivatives. Coding of those derivatives attached on the TentaGel and subsequent removal of protecting groups provided libraries on the polymer for direct bioassay with carbohydrate-binding proteins such as *Bauhinia purpurea* lectin. Importantly, the protein can discriminate very well between beads containing the hit ligand and beads containing related ligands in spite of the inhibitory concentrations being so similar.

Since TentaGel contains polyethylene glycol, the strategy faces the same potential problem with respect to the stability of PEG to the necessary conditions, but coding and deconvolution can be realized on this polymer.

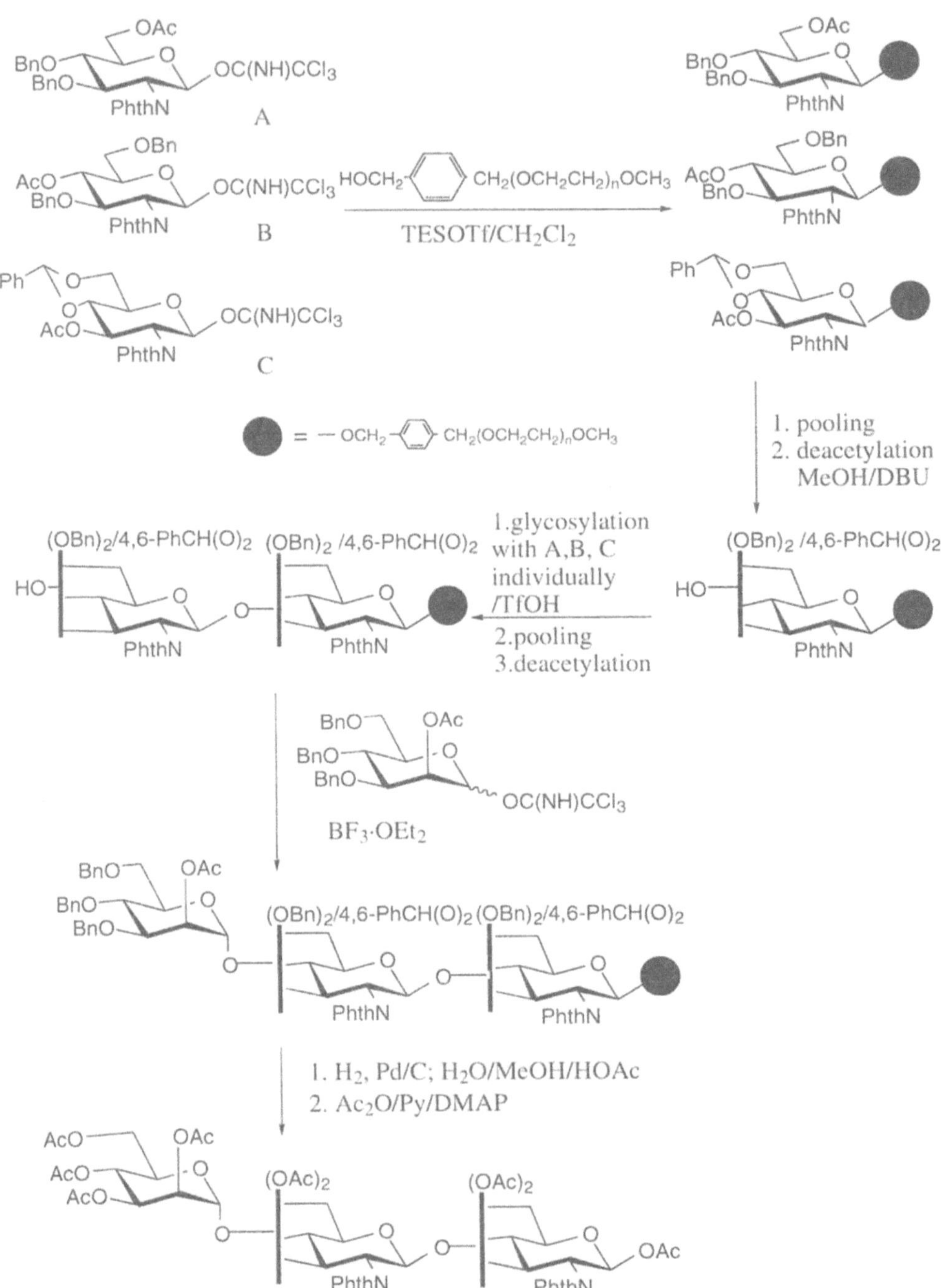

Figure 4b. Soluble-polymer (PEG) -supported synthesis of a αMan(1→N)βGlcNAc(1→N)GlcNAc library.

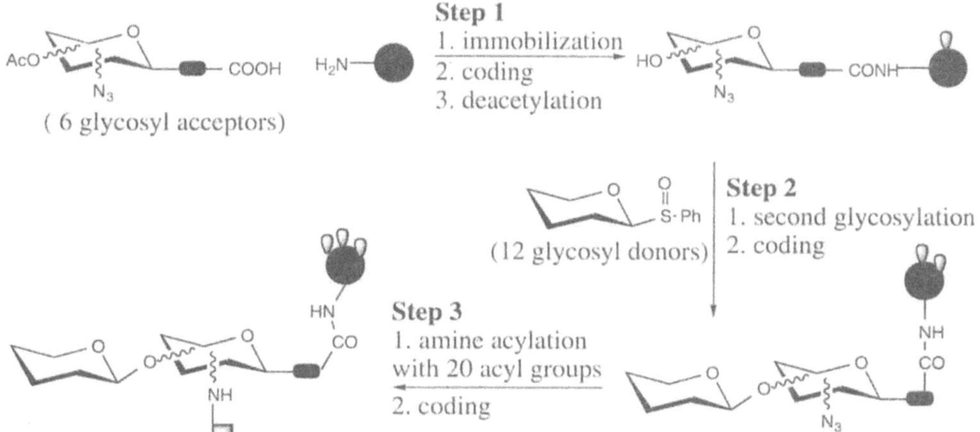

Figure 5. Library synthesis using Tenta Gel *via* glycosylation and amino derivativation. The coding tag is represented by ♀ .

4. INSOLUBLE POLYMER-SUPPORTED SYNTHESIS OF OLIGOSACCHARIDES

To the best of our knowledge, there are no reports yet concerning oligosaccharide library preparation on traditional insoluble polymers such as the Merrifield resin (polystyrene). A possible reason for this is a chemical obstacle, that is, most glycosylation methods are extremely sensitive to structural variations in glycosyl donor and acceptor pairs. An excellent set of conditions for one donor-acceptor pair may give virtually no product for another donor-acceptor pair. This problem will likely be magnified on the insoluble polymer. Tenta-Gel (PEG-grafted polystyrene) appears to be better than insoluble polymers since it contains soluble moieties for facilitating reactions, but using completely insoluble polymers as supports has great potential advantages. These include the "fishhook" function, pseudo-dilution, the stabilization of reactive species and steric and bulk effects of the polymer in synthesis.[8] We therefore choose to include some papers dealing with solid-phase-supported oligosaccharide synthesis, which can be considered as a precursor of library generation and may be of value in designing a combinatorial strategy. The earliest work reported is the synthesis of isomaltose on a modified Merrifield resin or on a light-sensitive solid support.[9] A strategy where the oligosaccharide is elongated from the reducing end to the non-reducing end is shown in Figure 6a (A), and has recently been extended to higher oligosaccharide synthesis on a linker-modified Merrifield resin. Examples (see Figure 6b) include van Boom's synthesis of a β(1→5)-linked D-galactofuranosyl heptamer *via* a classical halide donor (pivolate as a participating function at C-2), using an L-homoserine linker[10]; Schmidt's synthesis[11] of an α,β(1→6)glucose pentamer employing a trichloroimidate donor attached to a thio linker; Roush's synthesis[12] of a 6-deoxy trisaccharide using *O*-acetyl and trichloroimidate reagents and a sulfonate linker, with phenylthio as a participating group; Nicolaou's synthesis[13] of a heptasaccharide phytoalexin elicitor using thioglycosides and a

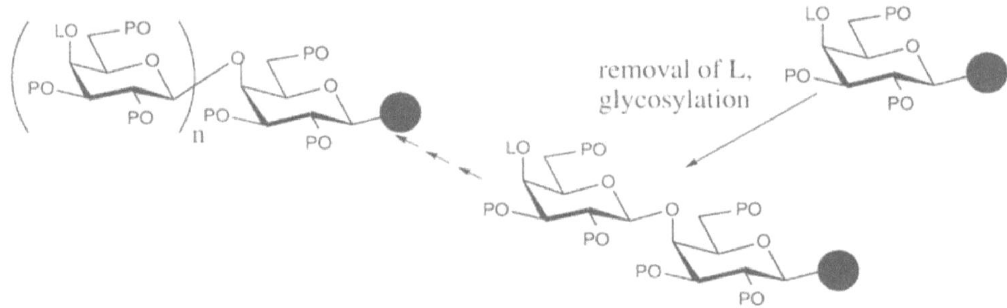

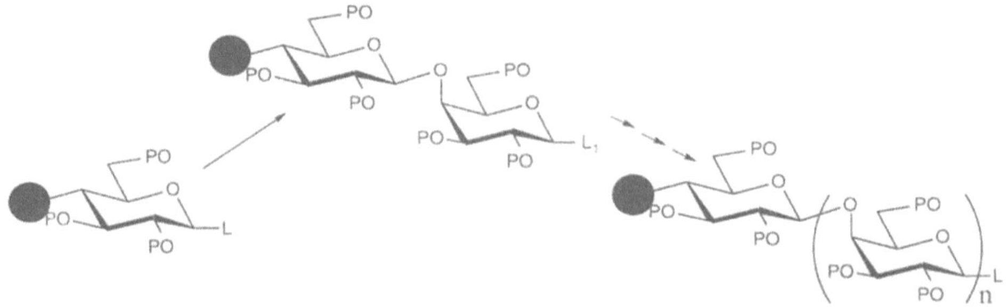

Figure 6a. Two general strategies for oligosaccharide synthesis on a solid-phase polymer support. **A:** The polymer is attached to the reducing end of the sugar, then the sugar chain is extended from the reducing end to non-reducing end. **B:** The polymer is attached to a non-anomeric position of the sugar, and the chain is extended from non-reducing end to reducing end.

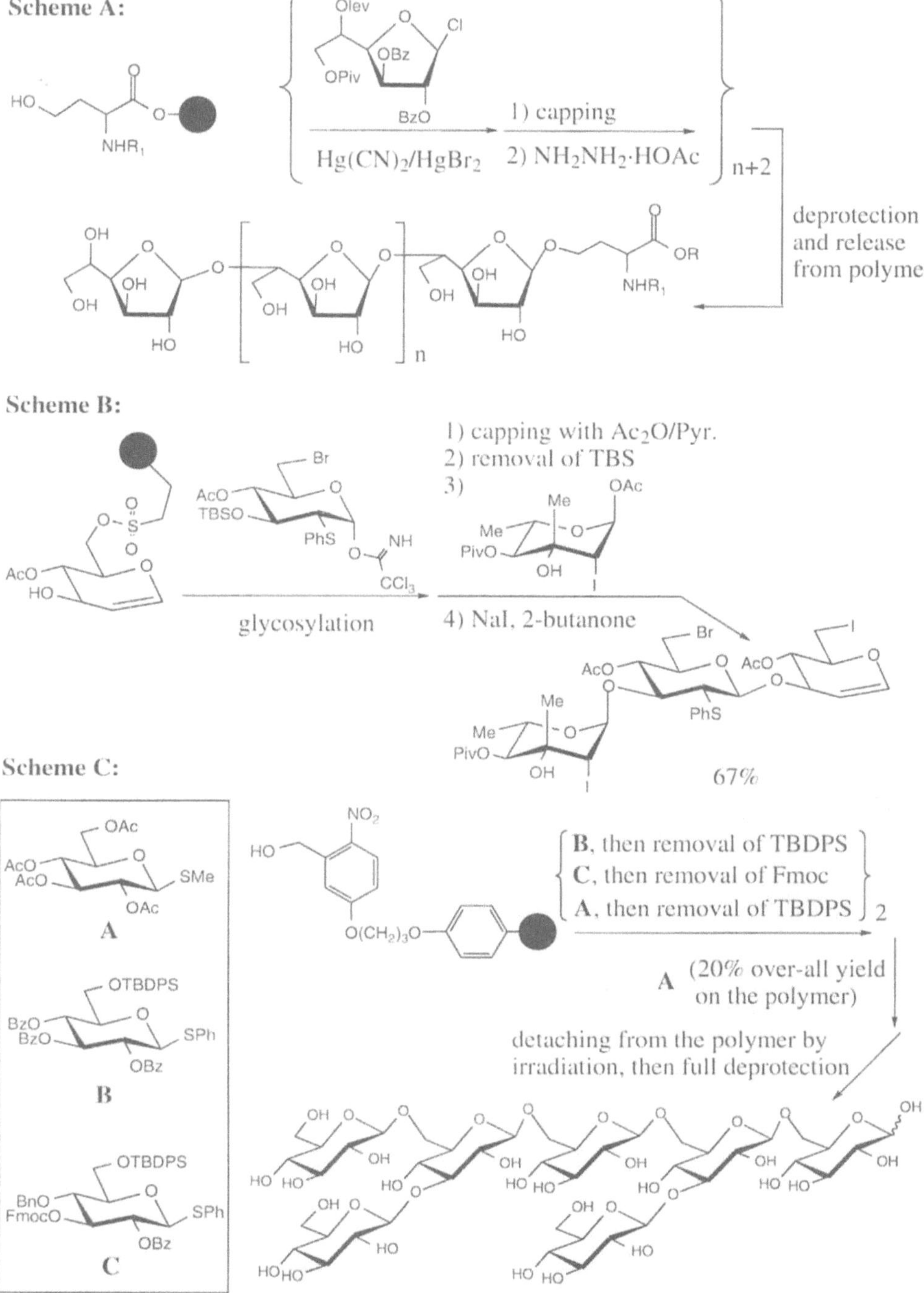

Figure 6b. Three recent examples of oligosaccharide synthesis on solid-phase polymers from the reducing end to the non-reducing end. **Scheme A**: Synthesis of a galactofuranosyl heptamer by linear iterative addition using glycosyl chlorides, pivolates as participating groups for β-glycosidic bond formation, and levulinoyl as a selectively removable functionality in the presence of esters. **Scheme B**: Synthesis of a 2-deoxy trisaccharide using a sulfonate linker, and empolying phenylthio and iodine as neighboring participating groups for β-glycosidic bond formation. **Scheme C**: Synthesis of the branched heptasaccharide phytoalexin elicitor using thioglycosides as donors.

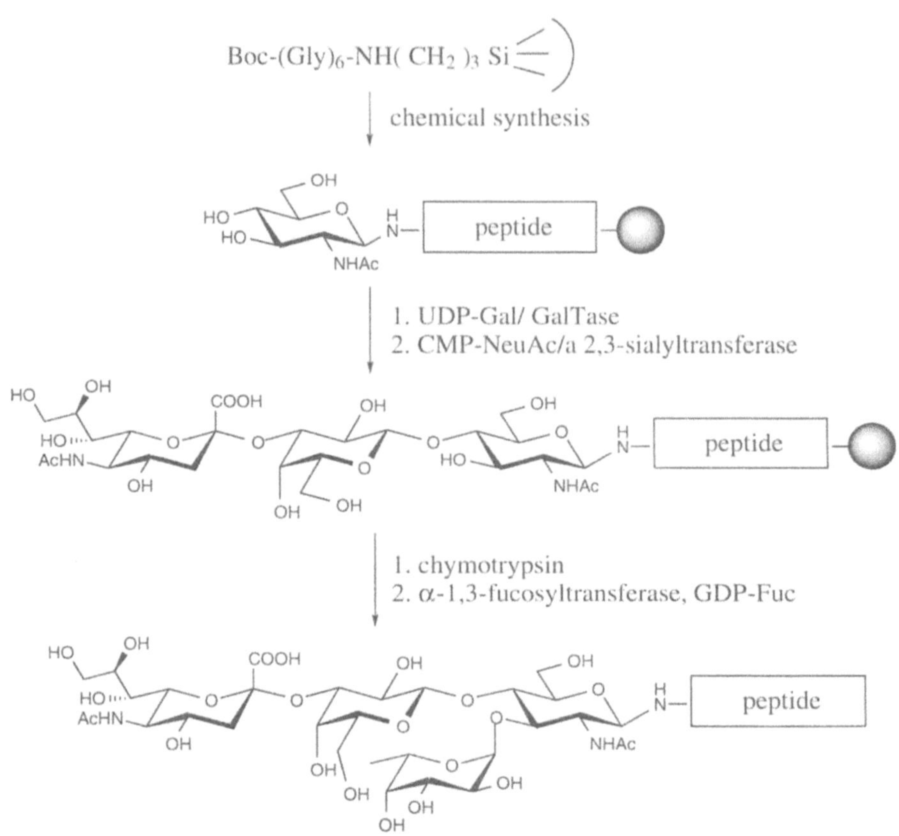

Figure 6c. Chemo-enzymatic synthesis of a glycopeptide containing SLe^x on a solid polymer support (aminopropylsilica). The hexapeptide spacer (Gly6), chymotrypsin-sensitive ester linkage, and oligopeptide-containing GlcNAc were attached to the support by chemical synthesis. The additional sugar resdiue was then added by glycosyltransferases.

light-sensitive linker, among others[14]. The elegant chemo-enzymatic synthesis[15] of a sialyl Lewis X glycopeptide has also been reported. Those results show that most current glycosylation donors, such as halide, trichloroimidate, thioglycoside, sulfoxide and some others can be used for solid-phase synthesis of oligosaccharides once optimal conditions are established.

Instead of linear iterative addition of donor to acceptor immobilized on the polymer where, in principle, excess donor reagents and promoter should drive the reaction to completion, the reverse strategy (from non-reducing end to reducing end) in Figure 6a(B) has also been demonstrated to work well. Provided that orthogonal glycosylation is applied[16], the synthesis of oligosaccharides needs fewer synthetic steps, and can avoid some blocking and deblocking manipulations. Danshifsky's glycal-strategy-based convergent synthesis of *N*-glycopeptides on a solid support [17] (see Figure 6d) is successful in this respect. In this work, the polymer seems to stabilize the 1,2-epoxide, usually somewhat unstable in solution. A large excess of acceptor (10 to 20 eq.) drives the conversion of epoxide into glycoside so that the whole sequence of reactions are very high-yielding.

5. GLYCOMIMETIC LIBRARIES

While "natural" oligosaccharide library synthesis remains under investigation, carbohydrate chemists are beginning to explore carbohydrate mimics (glycomimetics) as

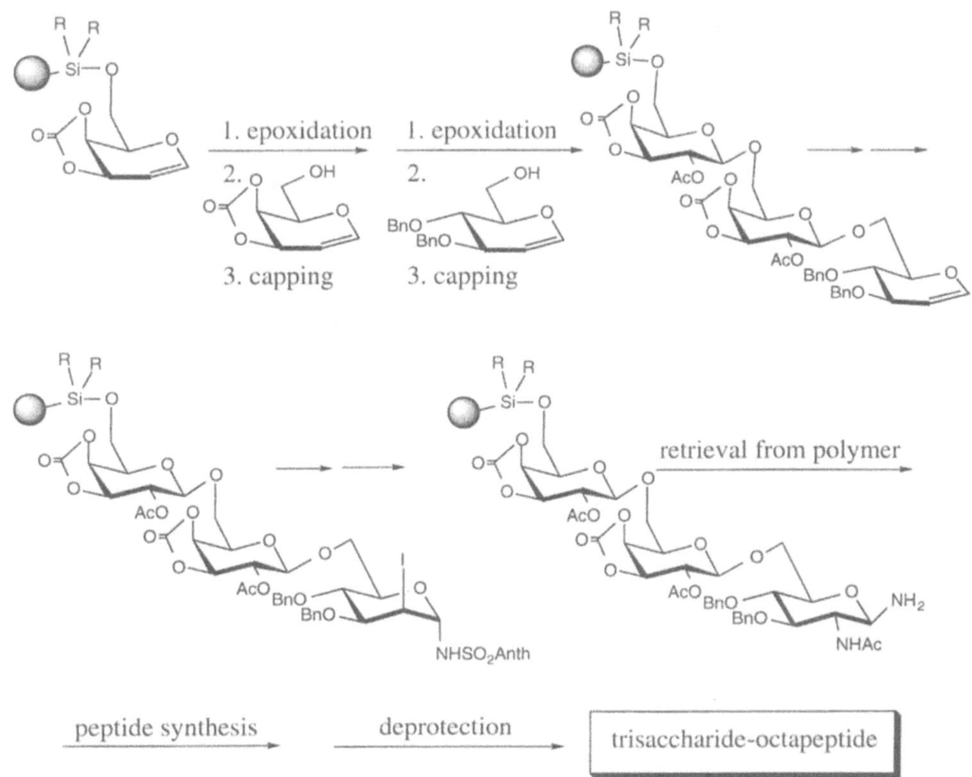

Figure 6d. Synthesis of an *N*-linked glycopeptide. The glycal moiety was initially immobilized through a silyl linker; subsequent epoxidation and glycosylation with incoming glycal stereoselectively gave the β-glycosylation product. Iteration extended the sugar chain. The product is retrieved by treatment with Bu₄NF and can be coupled with amino acids resulting in a glycopeptide.

simpler replacements for the natural ligands. Among such sugar analogs, *C*-glycosides, *S*-glycosides , *N*-glycosides and aza-sugars are being investigated in order to enhance sugar binding activity to carbohydrate-binding proteins.

Figure 7a illustrates the preparation of an aglycon-functionalized β-Gal library[18]. It utilizes a 1-thio-β-D-Gal derivative in Michael addition reactions of carbonyl compounds, followed by reductive amination of the carbonyl group. Each hydroxyl group in the galactose residue was present as its lauroyl ester, which facilitated the isolation of products by reverse-phase chromatography.

Saccharide-peptide hybrids, or so-called saccharide-aminoacids (oligoglycotides) such as the structure presented in Figure 7b, have amide linkages between units, instead of naturally occurring *O*-glycosidic linkages. A "tetraglycotide" could be conveniently prepared from a single intermediate. Varying the carboxylic acid and amine positions in the sugar can potentially create a large family of monosaccharide building blocks.[19] Clearly, a large number of oligoglycotide libraries can be generated by employing this type of chemistry .

C-glycosides form another class of oligosaccharide substitutes which are resistant to glycosidases. Their preparative chemistry is well reviewed.[20] The Ugi four-component condensation[21] (aldehyde, amine, isocynate, carboxylic acid) in Figure 7c appears to be a powerful approach for accessing a wide diversity of libraries, for each component can contain a carbohydrate moiety, depending on how the target library is going to be constructed. Armstrong applied this condensation to generate a library of 192 compounds bearing resemblance in structure to sialyl Lewis X, clustering *C*-fucosides with peptides on solid polymer supports after the demonstration of the general application of the Ugi condensation for preparing libraries.[22] Details are shown in Figure 7c.

Another application of the Ugi condensation is the combinatorial synthesis of aminoglycoside antibiotic mimetics of neomycin. The aldehyde derived from a neomycin disaccharide was reacted with an amine attached to a soluble polymer (polyethylene glycol), an isocynate and a carboxylic acid, rapidly generating a neomycin-mimetic library. The library, after deblocking, was screened, and four compounds (listed in Figure 7d) have been found to be of equal or better activity than neomycin.[23]

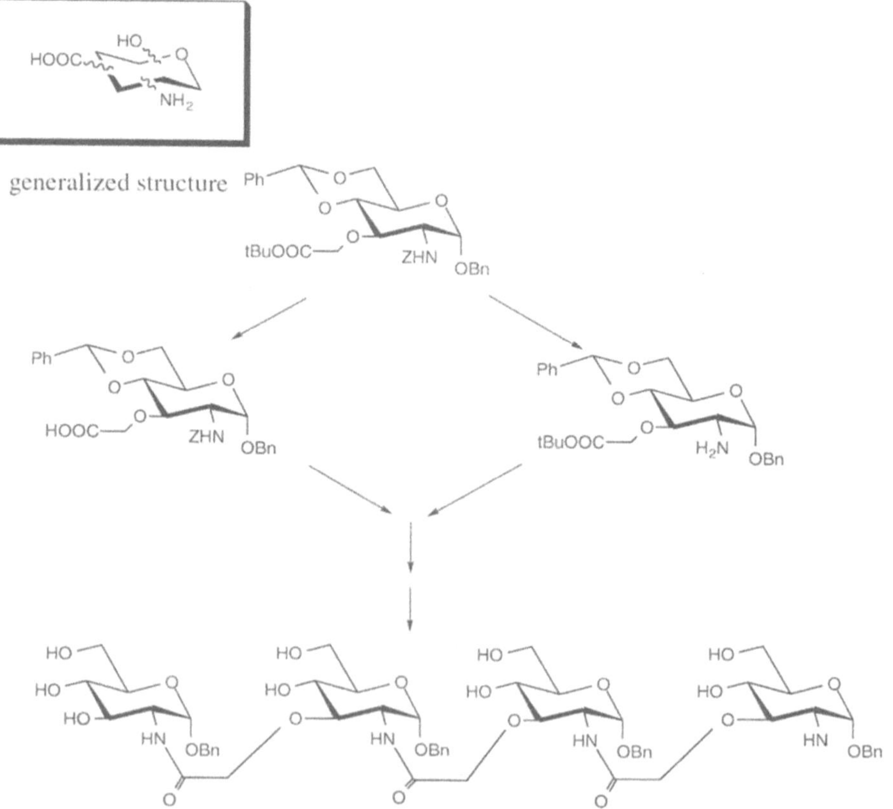

Figure 7a. Library synthesis assisted by multiple hydrophobic functionalities. The four laurate groups faciliate the rapid separation of products from the reaction mixture.

Figure 7b. Saccharide-peptide hybrids as novel glycomimetics. From the generalized structure as outlined, a number of saccharide-peptide structure libraries can be generated. As an example, a glycotide containing four sugar units and three amide bonds can be obtained from one common intermediate using well-established chemistry.

CONCLUSION

The synthesis of both oligosaccharide libraries and glycomimetic libraries is a very new area, but one that holds great potential for drug discovery. The next few years should see a large increase in activity. The final justification for this elegant chemical research, however, must be the discovery and optimization of small-molecule inhibitors of carbohydrate-protein interactions. Clever synthetic chemistry, for its own sake, will therefore become increasingly difficult to publish in the absence of biological data.

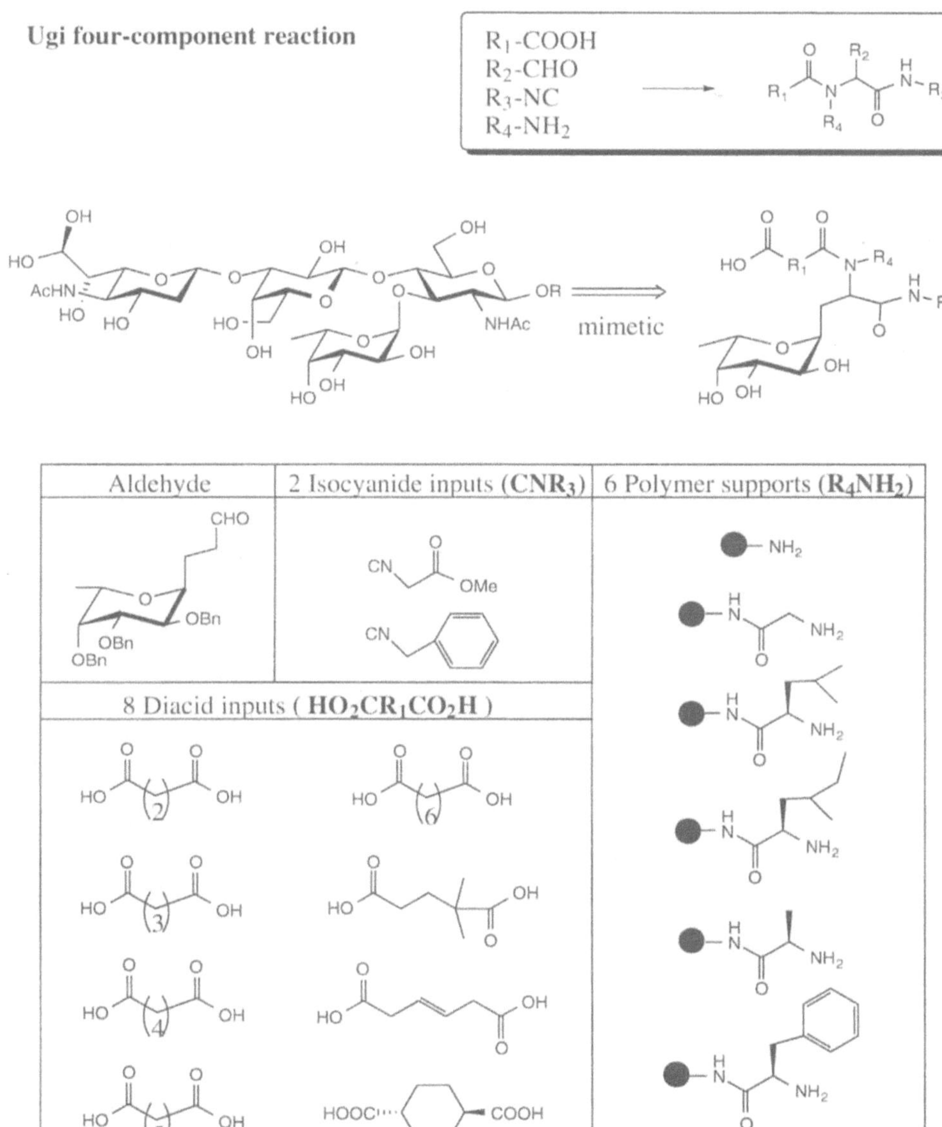

Figure 7c. Library synthesis on a solid polymer directed towards sialyl Lewis X mimetics. The mimetic library was generated from a single aldehyde, 2 isocyanides and 6 amines attached to the solid polymer and the 8 diacids listed. Individual wells contain a single reagent from each box, and all wells contain the acetylated C-fucose as the aldehyde component.

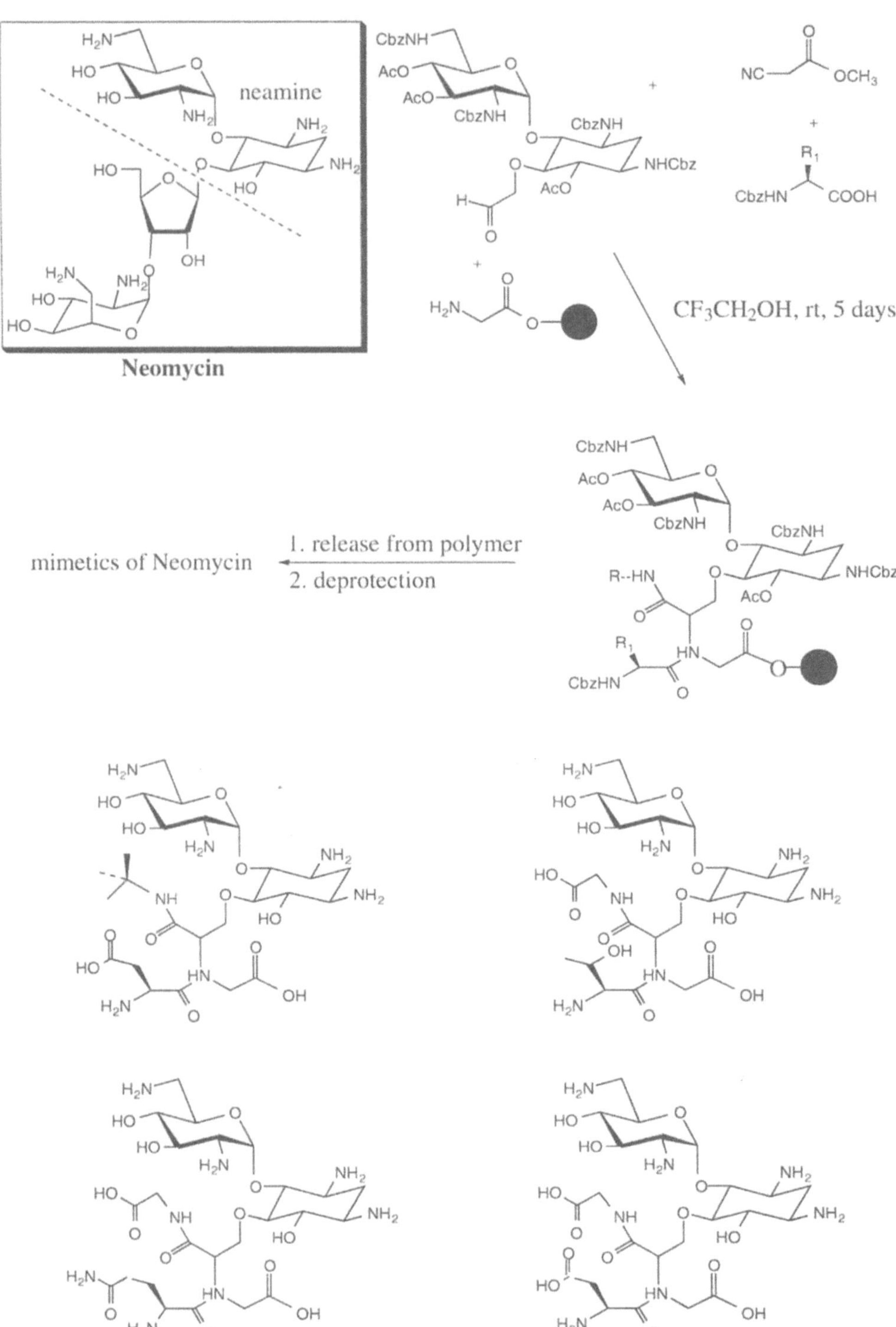

Figure 7d. Generation of a Neomycin B mimetics library using the Ugi four-component condensation on PEG by parallel experiments. The four compounds listed have the same or better activity than neomycin B.

REFERENCES

1. (a) M.A. Gallop, R.W. Barrett, W.J. Dower, S.P.A. Fodor, and E.M. Gordon, Application of combinatorial technologies to drug discovery. 1. Background and peptide combinatorial libraries, *J. Med. Chem.* 37:1233 (1994). 2. Combinatorial organic synthesis, library screening strategies and future directions, 37:1385 (1994). (b) L.A. Thompson and J.A. Ellman. Synthesis ands applications of small molecular libraries. *Chem. Rev.* 96:555(1996). (c) P.H.H. Hermkens, H.C.J. Ottenheijm, and D. Rees , Solid phase organic reactions: a review of the recent literature, *Tetrahedron* 52:4527 (1996) (d) F. Balkenhohl, C. Bussche-Hunnefeld, A. Lansky, and C. Zechel, Combinatorial synthesis of small organic molecules, *Angew. Chem. Int. Ed. Engl.* 35:2288 (1996).
2. A recent general review about glycosylation: K. Toshima, and K. Tatsuta, Recent progress in *O*-glycosylation methods and its application to natural products synthesis, *Chem. Rev.* 93:1503 (1993).
3. See a recent review: J.C. McAuliffe, and O. Hindsgaul, Carbohydrate drugs—an ongoing challenge, *Chemistry and Industry* 170 (1997).
4. (a) O. Kanie, F. Barresi, Y. Ding, J. Labbe, A. Otter, L. S. Forsberg, B. Ernst, and O. Hindsgaul. A strategy of " random glycosylation" for the production of oligosaccharide libraries, *Angew. Chem. Int. Ed. Engl.* 34:2720 (1995). (b) Y. Ding, J. Labbe, O. Kanie, and O. Hindsgaul, Towards oligosaccharide libraries: a study of the random galactosylation of unprotected *N*-acetylglucosamine, *Bioorg. Biomed. Chem.* 4:683 (1996).
5. G.J. Boons, B. Heskamp, and F. Hout, Vinyl glycosides in oligosaccharide synthesis: a strategy for the preparation of trisaccharide libraries based on latent-active glycosylation, *Angew. Chem. Int. Ed. Engl* . 35:2845 (1996).
6. (a) Z.G. Wang, S. Douglas and J.J. Krepinsky, Polymer-supported synthesis of oligosaccharides: using dibutylboron triflate to promote glycosylation with glycosyl trichloroacetimidates, *Tetrahedron Lett.* 37:6985 (1996). (b) J.J. Krepinsky, 212th Amercian Chemical Society National Meeting at Orlando, August 25-29, Division of Organic Chemistry, 003 (1996). Previous works on soluble-phase polymer-supported synthesis See: S. Douglas, D.M. Whitfiled, and J.J. Krepinsky, Polymer-supported synthesis of oligosaccharides using a novel versatile linker for the synthesis of D-mannopentaose, a structural unit of D-Mannans of pathogenic yeasts, *J. Am. Chem. Soc.* 117:2116 (1995).
7. R. Liang, L. Yan, J. Loebach, M. Ge, Y. Uozumi, K. Sekanina, N. Horan, J. Gildersleeve, C. Thompsom, A. Smith, K. Biswas, W. C. Still, and D. Kahne, Parallel synthesis and screening of a solid phase carbohydrate library, *Science.* 274:1520 (1996).
8. C.C. Lenzof, The use of insoluble polymer supports in general organic synthesis, *Acc. Chem. Res..* 11:327 (1978).
9. For a review of solid-phase synthesis of oligosaccharides up to 1980, see J. M. Frechet, In polymer-supported synthesis of oligosaccharides, Hodge, P., Sherrington, D.C., eds., Wiley: Chichester, 407 (1980) and reference cited there. (a) U. Zehavi, and A. Patchornik, Oligosaccharide synthesis on a light-sensitive solid support. the polymer and synthesis of isomaltose(6-O-a-D-glucopyranosyl-D-glucose). *J. Am. Chem. Soc.* 95: 5673 (1973) (b) J. M. Frechet, and C. Schuerch, Solid-phase synthesis of oligosaccharides. Preparation of the solid support. poly[p-(1-propen-3-ol-1-yl) styrene], *J. Am. Chem. Soc.* 93: 492 (1971).
10. G. H. Veeneman, S. Notermans, R. M. Liskamp, G. A. van der Marel, and J. H. van Boom, Solid-phase synthesis of a natually occuring β-(1→5)-linked D-galactofuranosyl

heptamer containing the artificial linkage arm L-homoserine, *Tetrahedron Lettt.* 28 :6695 (1987)

11. J. Rademann, and R. R. Schmidt, A new method of the solid-phase synthesis of oligosaccharides, *Tetrahedron Lett.* 37:3989 (1996).

12. J. A. Hunt, and W. R. Roush, Solid-phase synthesis of 6-deoxyoligosaccharides, *J. Am. Chem. Soc.* 118:9998 (1996).

13. K.C. Nicolaou, N. Winssinger, J. Pastor, and F. DeRoose, A general and highly efficient solid phase synthesis of oligosaccharides: total synthesis of a heptasaccharide phytoalexin elicitor (HPE), *J. Am. Chem. Soc.* 119:449 (1997).

14. (a) L. Yan, C.M. Taylor, R. Goodnow, and D. Kahne, Glycosylation on the Merrifield resin using anomeric sulfoxides, *J. Am. Chem . Soc.* 116:6953 (1994). (b) S. Nilsson, M. Bengtsson, and T. Norberg, Solid-phase synthesis of a fragment of the capsular polysaccharide of haemophilus influenzae type B using H-phosphonate intermediates, *J. Carbohydrate chemistry* 11:265(1992).

15. M. Schuster, P. Wang, J.C. Paulson, and C.H. Wong, Solid-phase chemical-enzymatic synthesis of glycopeptides and oligosaccharide, *J. Am. Chem. Soc.* 116:1135(1994).

16. Y. Ito, O. Kanie, and T. Ogawa, Orthogonal glycosylation strategy for rapid assembly of oligosaccharides on a polymer support, *Angew .Chem. Int. Ed. Engl.* 35:2510 (1996).

17. (a) J. Y. Roberge, X. Beebe, and S.J. Danishefsky, A strategy for a convergent synthesis of *N*-linked glycopeptides on a solid support, *Science* 269: 202 (1995). (b) S.J. Danishefky, K.F. McClure, J.T. Randolph, and R.B. Ruggeri, A strategy for the solidphase synthesis of oligosaccharides, *Science* 260:1307(1993). (c) J.T. Randolph, K.F.McClure,and S.J. Danishefky, Major simplifications in oligosaccharide syntheses arising from a solid-phase based method: an application to the synthesis of the Lewis b antigen, *J. Am. Chem. Soc.* 117:5712 (1995).

18. U.J. Nilsson, and O. Hindsgaul, Combinatorial michael additions and reductive aminations towards carbohydrate libraries designed to inhibit galactose-binding protein. XVIII International Carbohydrate Symposium. Milano, Italy, Abstract P 212 (1996).

19. (a) H.P. Wessel, C.M. Mitchell, C.M. Lobato, and G. Schmid, Saccharide-peptide hybrids as novel oligosaccharide mimetics. *Angew. Chem. Int. Ed. Engl.* 34:2712(1995). (b) E.G. von Roedern, and A. Kessler, A sugar amino acid as a novel peptidomimetic, *Angew. Chem. Int. Ed. Engl..* 33:687 (1994). (c) J.P. McDevitt, and P.T. Lansbury, Glycosamino acid: new building blocks for combinatorial synthesis, *J. Am. Chem. Soc.* 118:3818(1996).

20. M.H.D. Postema, Recent development in the synthesis of *C*-glycosides, *Tetrahedron* 48:8545 (1992).

21. I. Ugi, From isocyanides *via* four-component condensations to antibiotic synthesis, *Angew. Chem. Int. Ed. Engl.* 21:810 (1982).

22. D.P. Sutherlin, T.M. Stark, R. Hughes, and R.W. Armstrong, Generation of C-glycoside peptide ligands for cell surface carbohydrate receptors using a four-component condensation on solid support, *J. Org. Chem.* 61:8350 (1996).

23. W.K. Park, M. Auer, H. Jaksche, and C.H. Wong, Rapid combinatorial synthesis of aminoglycoside antibiotic mimetics: use of a polyethylene glycol-linked amine and a neamine-derived aldehyde in multiple component condensation as a strategy for the discovery of new inhibitors of the HIV RNA rev responsive element, *J. Am. Chem. Soc.* 118:10150 (1996).

BACTERIAL LIPOPOLYSACCHARIDES: CANDIDATE VACCINES TO PREVENT *NEISSERIA MENINGITIDIS* AND *HAEMOPHILUS INFLUENZAE* INFECTIONS

E. Richard Moxon[1], Derek Hood[1] and Jim Richards[2]

[1]Molecular Infectious Diseases Group
University Department of Paediatrics
John Radcliffe Hospital
Oxford OX3 9DU
U.K.
[2]National Research Council, Canada
Institute for Biological Sciences
100 Sussex Drive
Ottawa, Ontario
Canada   K1A 0R6

INTRODUCTION

An exciting area of microbiological research over the past decade has been the progress made in the molecular characterisation of the cell surface carbohydrates of pathogenic bacteria. Many of these are important virulence factors, for example capsular polysaccharides, so that an understanding of their structure and function is crucial to investigations on pathogenesis of infectious diseases. Following from this, many carbohydrate structures on the cell surface of bacteria are ideal targets for therapy and immunoprophylaxis. Thus, the diverse and sometimes highly complex polysaccharides located in the bacterial cell wall or envelope that surround the fragile cytoplasmic membrane of the bacterial cell have been investigated in depth applying state-of-the-art technologies such as those available through molecular genetics and fine structural analyses. The bacterial cell wall, which is rigid, protects the membrane and the cytoplasm within from the adverse effects of the environment and is thus responsible for the resistance of the bacterial cells to mechanical and osmotic injury. It is the structure that provides the bacterial cell with its characteristic shape, whether spherical or rod-like. The cell wall is just one of the features distinguishing bacteria from animal cells which are enclosed only by a plasma membrane. There are marked differences in the composition of the cell walls of gram positive and gram negative bacteria. Both may have exopolysaccharides, often referred to as capsules, which, owing to their external location on the surface, often confer antigenic specificity which is useful in their classification. Walls of both classes of bacteria also contain peptidoglycan. Distinguishing facts are that gram positive organisms contain teichoic acids whereas lipopolysaccharides are typical of gram negative bacteria.

As with other gram negative bacteria, lipopolysaccharide (LPS) is present in the cell wall of *H. influenzae* (Hi), but there are a number of significant differences in the LPS of Hi as compared to those of the best studied gram-negative bacteria, *enterobactericaeae*.[1] The major difference is that the LPS of Hi lacks the O antigens (repeating side chains) which are characteristic of many pathogenic enteric bacteria. This is probably related to the fact that gram negative bacteria living in the hydrophobic environment of the gastrointestinal tract have relatively hydrophilic surfaces as a result of the repetitive side chains which render them resistant to intestinal enzymes and intestinal secretions such as bile. On the other hand, the original concept of a molecule comprising lipid A attached to an inner core (R-core) and an outer core including side chains (S-core) is preserved.

In Hi, the S core is highly variable owing to the considerable variation found between strains and a number of molecular mechanisms which generate phase variation of terminal epitopes [2]. A further factor is the microheterogeneity that is characteristic of all gram negative bacteria reflecting the variable extent to which complete biosynthesis of the LPS molecule occurs. The implications of these structural features in relationship to function are that the antigenic variation of the terminal saccharides of Hi LPS facilitates escape from host clearance mechanisms but at the same time provides the organism with a repertoire of molecular structures which can engage in host interactions. However, these characteristics of phenotypic variation are strongly unfavourable to their use as vaccine candidates. The lipid A component of the LPS of all gram negative bacteria is responsible for the mediation of inflammation; this potential for cytotoxicity to animal tissues is the reason why LPS has been classically referred to as "endotoxin". Any consideration of vaccine development must take into account the potentially potent toxicity of the molecule. However, the inner or rough core is highly conserved, does not form structures which are cross-reactive with human tissues and has no intrinsic toxic properties. It could therefore be considered as a potential vaccine candidate.

This presentation will focus on the lipopolysaccharide (LPS) of *Haemophilus influenzae*. LPS of Hi provides a good example of how genomics, molecular genetics, fine structural analysis and animal models of infection have been recruited to investigate the role of these molecules in pathogenesis and their candidacy as vaccines.

EPIDEMIOLOGY AND PATHOGENESIS OF *HAEMOPHILUS INFLUENZAE* AND *NEISSERIA MENINGITIDIS* INFECTIONS

Haemophilus influenzae (Hi) is indigenous to humans; no other natural hosts are known. It is one of the genera of bacteria normally found in the pharynx, but not the oral cavity. Hi also colonise the mucosae of other sites such as the conjunctivae. Spread from one individual to another occurs by airborne droplets or by direct contagion within secretion. Exposure to these organisms begins shortly after birth so that, from infancy, carriage of one or more strains for periods of days to months is common and these organisms are often not eliminated by antibiotic therapy. Capsular polysaccharides are major virulence factors since they are able to inhibit the host clearance mechanisms of opsonophagocytosis and complement mediated killing. Hi is responsible for a wide spectrum of diseases [reviewed in [3]] which can be considered to be of two kinds. First, encapsulated pathogenic strains (especially those of serotype b) are responsible for invasive infections of the bloodstream including septicaemia, meningitis, epiglottitis, cellulitis, pneumonia and septic arthritis. These infections arise when type b organisms translocate from the nasopharynx to the bloodstream from where they are disseminated to distant sites. However, a number of diseases result from contiguous spread of Hi within the respiratory tract. These are mostly caused by organisms which lack a capsule (often referred to as nontypeable Hi). These diseases include otitis media, conjunctivitis and sinusitis in the upper respiratory tract. Hi is also a major cause of lower respiratory tract infections, particularly young infants in socio-economically deprived parts of the world but, in addition, in individuals in whom clearance of organisms from the lung is compromised, such as those with cystic fibrosis, immunodeficiency diseases and chronic obstructive lung disease.

USE OF THE COMPLETE GENOME SEQUENCE INFORMATION OF *HAEMOPHILUS INFLUENZAE* TO INVESTIGATE LIPOPOLYSACCHARIDE BIOSYNTHESIS

In order to investigate the candidacy of inner core structures of LPS as vaccine candidates, our laboratory has taken a molecular genetic approach. By identifying key enzymes involved in the biosynthetic pathway, especially glycosyl transferases, it is possible to construct mutants in which the biosynthesis of LPS molecules is arrested after the lipid A and core structures have been assembled but prior to the steps leading to the formation of the highly variable outer core structure. This provides a source of LPS which can be used to raise monoclonal or polyclonal antibodies which recognise only epitopes on inner core structures. This in turn provides a means for being able to map the epitopes of the inner core structures and thereby to define potentially suitable targets for protective immune responses. The feasibility of this approach is dependent on identifying epitopes which are accessible to host immune responses, conserved across the genus and which elicit bactericidal antibodies.

The availability of the complete 1.83Mb pair sequence of the Hi strain Rd genome[4] has facilitated significant progress in investigating the biology of Hi LPS. Prior to the availability of the complete genome sequence of Hi, progress was dictated by classical molecular genetic analysis and relatively few genes had been characterised[1]. These included the *lic* genes which were found to be involved with LPS variation of the terminal outer core structures, the *lsg* locus, *rfaE* and *rfaD*[5]. Finally, a gene, *isn*, involved with heptose biosynthesis, mutations of which resulted in a severely truncated LPS had been identified[6].

The approach was to use the sequences of a wide range of LPS biosynthetic genes from gram negative bacteria which were to be found in the general databases. These "data base probes" were used to screen for homologous genes by both DNA or amino acid similarity in the Hi strain Rd genome. In practice, searches with amino acid sequences against translated Rd genome sequence proved the most useful for identification of putative LPS biosynthetic genes. DNA sequences identified from the Rd genome as potential homologues were translated and then searched against the GenBank/EMBO data banks to confirm likely function. Given the tendency for clustering of genes in bacterial genomes where functions are related, an obvious strategy was to explore stretches of DNA sequences upstream and downstream when loci of interest were isolated[7].

Following the identification of candidate LPS genes, the relevant sequences were amplified by the polymerase chain reaction from Hi strain Rd. In addition, we also amplified the equivalent DNA tracts from two epidemiologically distinct type b strains. In each of these cases, the oligonucleotide primers designed from the Rd genome sequence allowed us to amplify the relevant loci. PCR amplified products were cloned, mutated by insertional activation and transformed into the appropriate Hi strains to obtain mutants by allelic replacement. LPS molecules from wild-type and mutant strains were analysed by colony immunoblotting using a panel of monoclonal antibodies, each specific for the particular LPS epitopes. These data have allowed us to assemble a much more complete picture of the biosynthesis of Hi oligosaccharides and include at least one of the several possible genes involved in each step of the pathway (Figure 1).

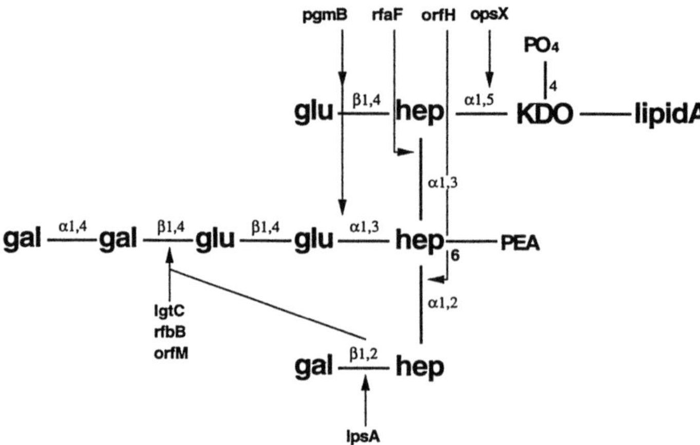

Figure 1. A schematic representation of the structure of LPS from *H. influenzae* strain Eagan (RM153). Proposed sites of action of genes characterised through application of data from the complete genome sequence of strain Rd are indicated. Represented in the LPS structure are: KDO, 2-keto-3-deoxyoctulosonic acid; hep, L-glycero-D-manno-heptose; glu, D-glucose; gal, D-galactose; PEA, phosphoethanolamine; and PO$_4$, phosphate. The heptose residues (hep) are listed from top to bottom as heptose I, heptose II then heptose III. Adapted from [7].

The complete genome sequence of *H. influenzae* strain Rd has allowed us to identify 25 candidate LPS genes. Detection of homologues of LPS biosynthetic genes in the *H. influenzae* genome data base using DNA sequences as probes was much more rapid and reliable than could have been achieved by hybridisation experiments. However, the major advantage of the genome sequence information was to identify candidate loci found only, or most reliably, by amino acid homology. This is particularly important when studying biological systems such as LPS where it is not uncommon for proteins of related function, such as the sugar transferases, to be encoded by genes of divergent sequence. 60% of the genes could not have reliably been identified by DNA sequence alone and therefore would not have been found by hybridisation experiment using the relevant heterologous probe[7].

The 25 LPS related genes identified in this study were dispersed around the genome of strain Rd either singly or in small groups (Figure 2). The largest cluster of novel genes were the eight contiguous orfs with an apparent bias of function directed towards O-antigen biosynthesis. LPS from *H. influenzae* lacks an O-antigen specific side chain but it is found in some related species, for example *Actinobacillus pleuropneumoniae*. This region of DNA may have some function in other macromolecular synthesis, such as elaboration or modification of capsule polysaccharide, but as yet we have found no change in capsule production in the type b mutant strains. The alteration of LPS seen in some of these mutants might indicate that less typical sugars, *rfbB* is rhamnose specific, or modifications of existing sugars can be incorporated into the LPS as minor components under certain, as yet uncharacterised, conditions. Rhamnose is a component of the O antigen of many pathogenic *Salmonella* strains and is synthesised from glucose by the *rfbA-D* gene products. Homologues of the other genes are not known to be present in *H. influenzae*. There may be changes in the pattern of LPS expressed *in vivo* when compared to that analysed after repeated laboratory culture. Transcription studies must confirm whether adjacent genes are functionally linked and co-regulated.

USE OF LPS MUTANTS TO INVESTIGATE VIRULENCE OF *H. INFLUENZAE* IN AN ANIMAL MODEL

The range of LPS mutants constructed in the study has allowed us also for the first time to undertake a comprehensive study to correlate LPS structure with the virulence of

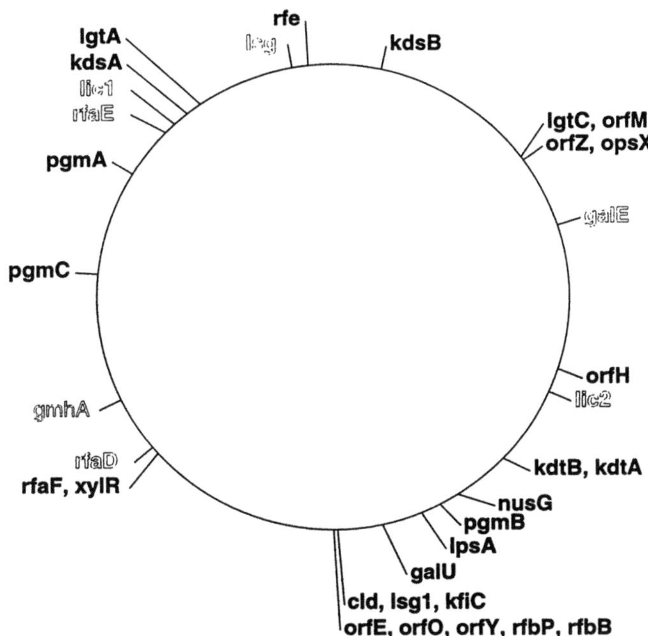

Figure 2. Identification of genes for lipopolysaccharide biosynthesis in *H. influenzae*. The gene names shown in white outlined with black indicate those which had been identified and characterized as *H. influenzae* LPS-related genes prior to this study and which are present in the strain Rd genome database. The remaining genes were identified in this study and are shown at their relative positions on a circular 1.83 Mbp strain Rd genome map. *kdsA* and *kdsB* are KDO biosynthetic genes[7].

Hi using an infant rat model[7]. Results from our experiments on virulence combined with data from other previously identified LPS related mutants allow us to make some predictions as to the minimal LPS structure required for efficient intravascular dissemination of *H. influenzae* in the infant rat (Figure 3). Three heptose molecules and at least two hexose sugars are required for maintenance of high levels of bacteraemia in the infant rat after intraperitoneal inoculation. Under these conditions there is good correlation of structure with function, with mutants elaborating a majority of LPS molecules with less than two hexose sugars being severely attenuated and strains with higher molecular weight LPS being only mildly, or not at all, attenuated. The exception is RM153*rfbB* where the minor change in the LPS, but dramatic attenuation in the infant rats, is unexplained. Possibilities include a role for the gene in some more subtle modifications which may affect only a portion of the LPS molecules or may be an *in vivo* specific effect not detected after laboratory culture. *rfbB* is involved with the production of rhamnose, a sugar not routinely associated with LPS from *Haemophilus* but which has been identified in one type b strain as a minor component. It is therefore possible that some subtle additions or modifications to LPS structure are host specific and are not detected after laboratory culture. Mutants with truncated LPS so far tested, are not defective in other virulence determinants, nor are impaired for growth *in vitro*, and it is reasonable to conclude that this attenuation is LPS specific. It is evident that factors influencing maintenance of bacteria in the bloodstream are very different to those affecting colonisation and invasion and a further important correlation between LPS size, structure and virulence may be evident after experiments following intra-nasal inoculation of the infant rats have been completed.

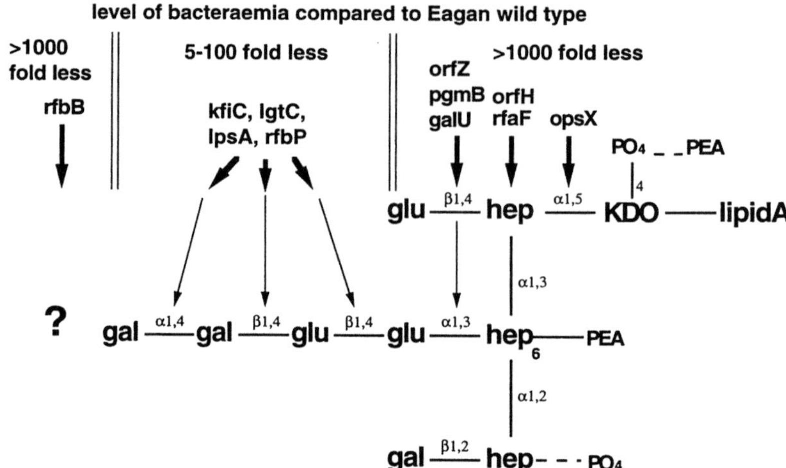

Figure 3. The grouping of mutant strains is from our virulence experiments and corresponds to the schematic representation in figure 1.

USE OF LPS MUTANTS TO INVESTIGATE FINE STRUCTURE AND TO CONSTRUCT A 3-DIMENSIONAL MODEL

Preliminary comparative analysis by mass-spectrometry of the LPS from selected mutants has helped to confirm some of our predictions of gene function. Samples of LPS were examined by negative iron electrospray mass-spectrometry (ES-MS). For example, initial structural analysis confirmed that the LPS of an *opsX* mutant contained no heptose and that the *rfa* mutant contained one heptose molecule linked to KDO[7]. This agrees with the proposed functions of the gene products as heptose-I and heptose-II transferases respectively.

Molecular modelling was used to generate a three-dimensional model and to calculate conformational flexibility of the core oligosaccharide region of LPS from strain Eagan[7]. The LPS consists of a conserved heptose containing an inner core trisaccharide unit attached by a phosphorylated KDO residue to a lipid A component. In the major LPS population group, each heptose is substituted with further chain elongation from the central unit where molecules containing 4, 5 and 6 residues are most prevalent. Space filling and ball and stick models of the core oligosaccharide have been generated. As noted previously for LPS from *Eschericiae coli*, calculations indicated the heptose-containing inner core region forms a compact structural unit. The galabiose-containing side chain attached to heptose-II adopts an elongated shape and it shows considerable conformational flexibility. It was found that even though these outer core regions of the molecule are flexible, the heptose-III and the β-glucose in the side chain were facing on the same side of the molecule. This modelling is being extended to investigate the conformation of the LPS from other mutants. The conformational flexibility implied in this model of the core oligosaccharide does not fully explain the orientation of the oligosaccharide in relation to the bacterial cell membrane, but provides a starting point for understanding the accessibility of oligosaccharide epitopes and their reactivities with monoclonal antibodies. The contribution of individual structures within LPS populations is being investigated by Western blot studies and further structural analysis and modelling of LPS derived from mutant strains should allow a better insight into mAb binding specificity.

PERSPECTIVE

In summary, this study has allowed us to advance rapidly our understanding of *H. influenzae* LPS biology. We have identified genes involved in the crucial stages of LPS biosynthesis; precursor supply, sugar transferase, export and potential regulatory functions. Mutations in these genes have allowed us to fulfill our main objective and to compile a bank of mutant strains elaborating a range of LPS molecules. These have helped to elucidate some of the steps in LPS biosynthesis and to estimate a minimal structure required for invasive infection of the infant rat. It has demonstrated the potential use of whole genome sequencing to extend biological knowledge. The speed and ease of detection of genes using the whole genome sequence of Hi strain Rd are significantly greater than that by classical molecular genetic analysis and in particular allow the identification of genes found even under circumstances of weak amino acid homology. In other organisms, the availability of sequence from random clones giving almost complete coverage of the genome, should still allow successful gene identification. The accumulated information has allowed us to identify many key LPS genes and to extend our study to the LPS structure in other strains. Studies are underway to confirm the extent of conservation of LPS related genes and the core LPS structure across all *H. influenzae* types and should help evaluate the use of LPS core structure as a potential candidate for broad range vaccine development.

REFERENCES

1.   Moxon ER and Maskell D. *Haemophilus influenzae* lipopolysaccharide: the biochemistry and biology of a virulence factor. In: Hormaeche C, Penn CW and Smyth CJ (eds). Molecular biology of bacterial infection: current status and future perspectives. SGM Symposium **49**:75-96, 1992.

2.   Moxon ER, Rainey PB, Nowak MA and Lenski RE. Adaptive evolution of highly mutable loci in pathogenic bacteria. Current Biology **4**:24-33, 1994.

3.   Moxon ER. *Haemophilus influenzae*. In: Mandell GL, Bennett JE and Dolin R, eds; Principles and Practice of Infectious Diseases (Fourth Edition), Vol 2, Churchill Livingstone Inc, NY, p2039-2045, 1995.

4.   Fleischmann RD, Adams MD, *et al*. The genome of *Haemophilus influenzae* Rd. *Science* **269**: 496-512, 1995.

5.   Lee N, Sunshine MG and Apicella MA. Molecular cloning and characterisation of the nontypeable *Haemophilus influenzae* 2019 *rfaE* gene required for lipopolysaccharide biosynthesis. *Infect Immun* **63**:818-824, 1995.

6.   Preston A, Maskell D, Johnson A and Moxon ER. Altered lipopolysaccharide characteristic of the I69 Phenotype in *Haemophilus influenzae* results from mutations in a novel gene, *isn*. J Bacteriol **178**:396-402, 1996.

7.   Hood DW, Deadman ME, Allen T, Martin A, Brisson JR, Fleischmann R, Venter JC, Richards JC and Moxon ER. Use of the complete genome sequence information of *Haemophilus influenzae* strain Rd to investigate lipopolysaccharide biosynthesis. *Molec Microbiol* **22**:951-965, 1996.

DEVELOPMENT OF DOUBLE COPY DICISTRONIC RETROVIRAL VECTORS FOR TRANSFER AND EXPRESSION OF GLYCOSYLTRANSFERASE GENES

Dariusz Iżycki, Maciej Wiznerowicz,
Maria Łaciak, Artur Słupianek,
and Andrzej Mackiewicz

Department of Cancer Immunology, Chair of Oncology
University School of Medical Sciences at GreatPoland Cancer Center
Poznań, Poland

INTRODUCTION

Glycosylation of secretory and cell surface proteins is a multistep enzymatic process which takes place on the endoplasmic reticulum and in the Golgi complex, and involves series of highly specific glycosyltransferases and glycosidases (Kornfeld and Kornfeld, 1985). Control of the relative activities of these enzymes is one of the mechanisms regulating pattern of glycosylation of end-products of protein biosynthesis. Number of groups have demonstrated that cell clones lacking or displaying high activity of particular enzymes glycosytate the same protein differently. Indeed number of hereditary diseases linked to the impaired glycosylation mechanisms due to the lack or low expression of glycosylating enzyme genes was described (Jacken et al. 1994). Moreover, overexpression or knockout of genes encoding number of glycosyltransferases caused altered protein glycosylation in various experimental models (Gorelik et al. 1995, Yoshimura et al. 1995, 1996). Genetic modification of cells leading to changes in the glycan structures is more often referred to as genetic sugar engineering. On the other hand cells displaying defects in particular enzyme gene expression may be corrected by insertion of functional gene. Such procedure is referred to as gene therapy (Friedmann, 1992).

Basic tools for gene therapy form gene delivery systems. The advances in recombinant DNA-based technologies led to the development of powerful viral vectors and non-viral systems. The choice of vector system for delivery and expression of genes into cells depends on: (i) the target cell or tissue; (ii) a requirement for integration with the genome; (iii) level of expression required; (iv) means of delivery; and (v) safety issues. Retroviral vectors are currently the most often used systems in clinical trials. They infect (transduce) dividing cells and integrate proviral DNA into genom of the target cell, thus are often the vectors of choice for use in *ex-vivo* therapies. Their limits are capacity for caring foreign genes or propensity to mutation or recombination. Retroviral vectors are in the third and fourth generation of development (Mulligan 1993).

In the present studies we constructed double-copy dicistronic retroviral vector containing strong CMV-IE promoter as a tool for delivery of human ß1,4 Galactosyltransferase (1,4GT) gene. In order to assess the effect of overexpression of 1,4GT gene on growth of tumor cells *in vivo*, murine melanoma cells (B-78-H1) were transduced with 1,4GT and injected into mice.

MATERIALS AND METHODS

Cell Lines

Murine melanoma cell line: B-78-H1 was maintained in DMEM supplemented with 5% FCS in 5% CO_2/95% air in 37°C. PA-317 amphotropic packaging cells obtained from Dr D. Miller (Miller et al. 1986) were cultured in DMEM supplemented with 10% FCS.

Construction of Double-Copy Retroviral Vector (DCCMV)

DCCMV was constructed on the basis of murine stem cell virus retroviral vector (MSCV) as described in detail (Wiznerowicz et al. 1997). Briefly, into U3 region of MSCV's 3'LTR downstream of NheI site artificial polylinker containing ClaI, NruI, SacII and MluI unique restriction sites was cloned. At the same time pgk-Neo cassette was removed from the transcription region of the vector. Next, PstI fragment from pKEX plasmid containing human CMV-IE promotor/enhancer was blunt ligated into NruI site and then SalI-BamHI fragment containing IRES-Neo cassette was blunt ligated into filled MluI site. DCCMV-1,4GT was generated by insertion into Sal I site of DCCMV human 1,4GT cDNA (obtained from Dr M. Fukuda, La Jola, CA, USA) using EcoRI. Fig 1. shows principles of construction of the DCCMV-1,4GT vector.

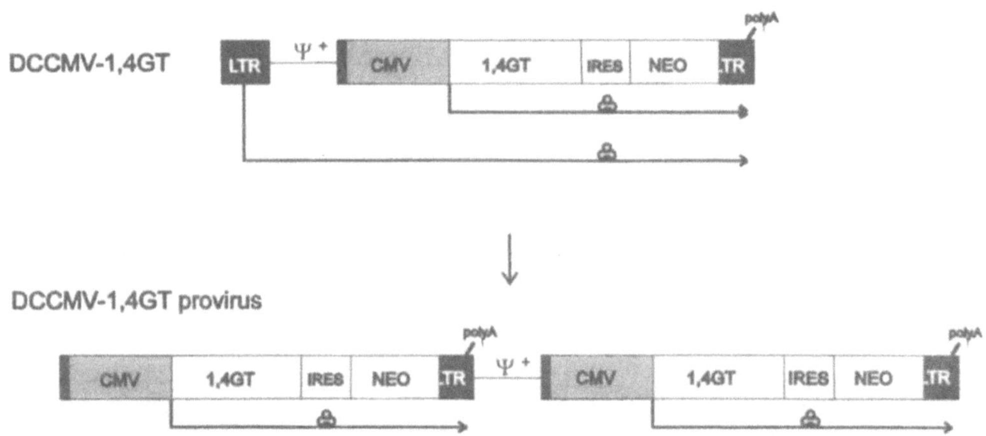

Figure 1. Concept of the double copy dicistronic retroviral vector caring 1,4GT cDNA. DCCMV-1,4GT vector electroporated into PA317 packaging cells. DCCMV-1,4GT provirus in the B-78-H1 cells (target cells).

Transduction of Murine Melanoma Cell Line

PA-317 cells were electroporated (300V/100ms) with 25 μg of plasmid DNA (DCCMV-1,4GT) and selected in the presence of geneticine (500 mg/ml). The infectious medium containing amphotropic recombinant retroviruses was obtained by incubation for 16 h of subconfluent (70-80%) PA-1,4GT cells. Next the medium was complemented with polybrene and added for 4 hrs to B-78-H1 cells. Then cells were selected in the presence of geneticin (500 mg/ml) for 3 weeks.

Assay for Human 1,4GT mRNA Expression

Total RNA was isolated from 1,4GT transduced and control B-78-H1 cells using Chomczynski method. After denaturing gel electrophoresis RNA was transferred to the nylon (GeneScreen-Plus) and hybridized with specific probe (1.1 kb EcoRI fragment) radiolabeled with G-ATP using random priming (Promega Kit) according to producent instructions.

Analysis of B-78-H1 and B-78-H1-1,4GT cells proliferation

Transduced and control cells proliferation was analyzed using MTT assay (Abe et al. 1994). No differences in cell proliferation between two analyzed cell lines was seen.

Tumor Model

Eight to ten weeks old C57BL/6 x C3H mice were injected subcutaneously (s.c.) with 5 x 10^5 viable (as determined by trypan blue exclusion) control (mock transduced) or 1,4GT transduced B-78-H1 cells. Tumor growth and survival were monitored.

RESULTS

Expression of 1,4GT mRNA in Transduced B-78-H1 Cells

Expression of 1,4GT mRNA levels in PA-1,4GT and B-78-H1 cells is shown on Fig. 2.

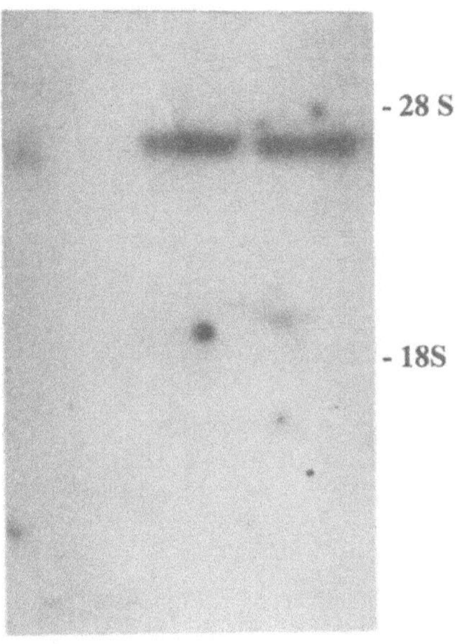

Figure 2. Northern blot of 1,4GT mRNA in: 1 - B-78-H1 cells; 2 - PA317 cell transfected with DCCMV-1,4GT vector; 3 - B-78-H1 cells transduced with 1,4GT cDNA.

Analysis of Tumor Growth Kinetics and Survival of Mice Injected with B-78-H1 and B-78-1,4GT Cells

Kinetics of tumor growth in C57BL/6 x C3H mice are demonstrated in Fig. 3A. All animals injected s.c. with B-78-H1 control cells developed tumors 3 weeks following injection. However, mice injected with B-78-1,4GT cells started to develop tumors two weeks later and all of them had tumors 5 weeks following injection.

Survival analysis demonstrated increased survival time of mice injected with B-78-1,4GT cells when compared with the mice injected with control cells (Fig. 3B). Characteristic three weeks lag period in survival curve between mice injected with control and transduced cells was observed. The lag period was seen until 9th week following injection. One week later all mice of control group and 80% of mice injected with B-78-1,4GT died.

a.

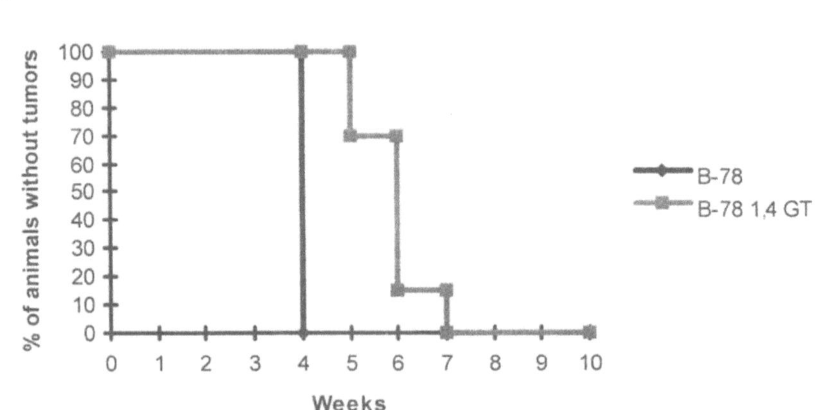

b.

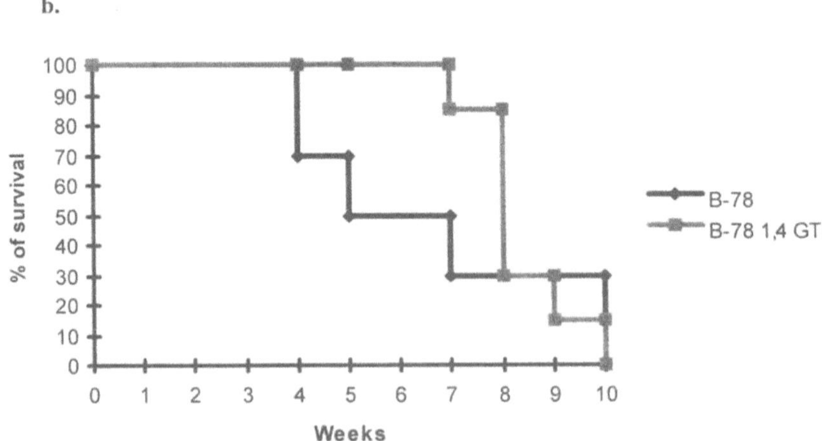

Figure 3. Analysis of tumor formation (a) and survival (b) of mice injected with 1,4GT transduced and control B-78-H1 melanoma cells.

DISCUSSION

Retroviral vectors are mostly based on Moloney leukemia virus and utilize the 5'LTR to promote production of the therapeutic protein mRNA. Such procedure yields from few up to 60% of target cells expressing protein of interest (Jaffe et al. 1993). Most of retroviral vectors incorporate a second internal promoter to produce two proteins from the same construct. In such vectors therapeutic protein is produced from mRNA derived from the viral (LTR) promoter and a protein conferring resistance from mRNA derived from internal promoter. The major problem with this strategy is that selection will yield resistant cells, which may not necessary express the therapeutic gene. To overcome this phenomenon chimeric protein was created by fusing sequence from the open reading frames of a resistance and therapeutic genes. However, products of such fusion-genes may present impaired function of either the therapeutic or the resistance protein due to improper tertiary conformation. Recently, another strategy to omit internal promoter has been developed. It relays on internal ribosome entry site (IRES) sequences which are used to link therapeutic and resistance genes and form so called dicistronic gene. IRES elements are 500-600 bp sequences that are characteristic 5' untranslated regions of picornaviruses including polio virus and encephalomyocarditis virus. IRES elements form tertiary structures which mediate cap independent translation of virus proteins. In consequence two proteins are produced from a single mRNA derived from dicistronic gene promoted by LTR. Finally, double copy vectors (DC) were design. Their unique feature is that therapeutic gene with internal promoter is inserted outside the transcriptional region of the vector namely within U3 region of the 3'LTR. It results in the duplication of the gene and its transposition to the 5'LTR what may finally lead to a 10 to 20-fold increase in transcripts and protein synthesis (Hatzantopoulos et al. 1989).

Recently, we have combined two retroviral vector construction strategies namely DC design and dicistronic gene concept and have developed DCCMV vector which have proven to be functional in variety of cell types providing high expression of therapeutic gene in all of the selected cells (Wiznerowicz et al. 1997). The limitation of the system was the relatively lower vector titer. In the present studies we have used DCCMV vector for *ex vivo* delivery of 1,4GT gene as a potential therapeutic agent.

Transfection of α1,3GT gene into murine melanoma resulted in alteration of cell surface carbohydrates and inhibition of melanoma metastatic property (Gorelic et al. 1995). Transfection of N-acetylglucosaminyl transferase III gene into murine melanoma cells affected ß1,6 branching of surface N-glycoproteins and suppressed melanoma lung metastasis (Yoshimura et al. 1995). In our studies transduction of 1,4GT gene into murine melanoma cells had no effect on their proliferation *in vitro*, however demonstrated inhibition of the tumor formation and growth *in vivo*. In addition animals injected with modified tumor cells displayed extended survival compared to those injected with control cells. It suggests that genetically modified melanoma cells were inducing non-specific anti-tumor immune response more efficiently than the control cells. However, nature of the mechanisms involved in the described phenomenon require further studies since 1,4GT was found within Golgi complex and on the cell surface (Evans et al. 1993).

Obtained results demonstrate that double copy dicistronic retroviral system presents a very powerful tool for delivery of glucosyltransferase genes into cancer cells.

ACKNOWLEDGEMENTS

This work was supported by The State Committee for Scientific Research (Warsaw) grants 203/S4/94/06p and 1067/PO5/96/10 and European Commission grant ERBC1PDCT940207.

REFERENCES

Abe, R., Ueo, H., Akiyoshi, T. 1994, Evaluation of MTT assay in agarose for chemosensitivity testing of human cancers: comparison with MTT Assay. *Oncology*, 51:416-425.

Evans, S.C., Lopez, L.C., Shur, B.D. 1993, Dominant negative mutation in cell surface ß1.4-Galactosyltransferase inhibits cell-cell and cell matrix interactions. *J. Cell Biol.* 120:1045-1057.

Friedmann, Th. 1992, A brief history of gene therapy. *Nat. Genet.*, 2:93-98.

Gorelik, E., Duty, L., Anaraki, F., Galili, U. 1995, Alterations of cell surface carbohydrates and inhibition of metastatic property of murine melanomas by a1,3 Galactosyltransferase gene transfection. *Cancer Res.* 55: 4168-4173.

Hatzantopoulos, P.A., Sullenger, B.A., Ungers, G., Gilboa, E. 1989 Improved gene expression upon transfer of the adenosine deaminase minigene outside the transcriptional unit of a retroviral vector. *Proc. Natl. Acad. Sci. USA* 86, 3519-3523.

Jacken, J., Schachter, H., Carchon, P., De Cock, B., Coddeville, B., Spik, G. 1994, Carbohydrate deficient glycoprotein syndrome type II: a deficiency in Golgi localized N-acetyl-glucosaminyltransferase II. *Arch. Dis. Child.* 71:123-127.

Jaffe, E.M., Dranoff, G., Cohen, L.K., Hauda, K.M., Clift, S., Marshall, F.F., Mulligan, R.C., Pardoll, D.M. 1993, High efficiency gene transfer into primary human tumor explants without cell selection. *Cancer Res.* 53:2221-2226.

Kornfeld, R., and Kronfeld, S. 1985, Assembly of asparagine-linked oligosaccharides. *Annu. Rev. Biochem.* 54:631-664.

Miller, A.D., Buttimore C. 1986, Redesign of retrovirus packaging lines to avoid recombination leading to helper virus production. *Mol. Cell. Biol.* 6:2895.

Mulligan, R.C. 1993, The basic science of gene therapy. *Science* 260:926-931.

Wiznerowicz, M., Fong, A., Mackiewicz, A., Hawley, R.C. 1997, Development of dicistronic double copy retroviral vectors for human gene therapy. *Gene Ther.,* submitted.

Yoshimura, M., Ihara, Y., Ohnishi, A., Ijuhin, N., Nishiura, T., Kanakura, Y., Matsuzawa, Y., Taniguchi, N. 1996, Bisecting N-acetlyglucosamine on K562 cells suppresses natural killer cytotoxicity and promotes spleen colonization. *Cancer Res.,* 56:412-418.

Yoshimura, M., Nishikawa, A., Ihara, Y., Tanigushi, S., Tanigushi, N. 1995, suppression of lung metastasis of B16 mouse melanoma by N-acetyloglucosaminyltransferase III gene transfection. *Proc. Natl. Acad. Sci. USA,* 92:8754-8758.

OLIGOSACCHARIDE EPITOPE DIVERSITY AND THERAPEUTIC POTENTIAL

Elizabeth F. Hounsell and David V. Renouf

Department of Biochemistry & Molecular Biology
University College London
Gower Street
London WC1E 6BT

INTRODUCTION

In the search for new therapeutics based on oligosaccharide-protein interactions (glycotherapeutics) several unique features about glycans must be borne in mind; i.e., their structural diversity; the recognition of epitopes on branched sequences with local conformation; and, multivalent presentation. These characteristics are variously important in, for example, the potential exploitation of a) high affinity interactions of glycosaminoglycans (proteoglycan oligosaccharides) with proteins, b) monoclonal antibody, mammalian lectin and microorganism recognition of mucin-type oligosaccharides, and c) the functions of both lipid-linked oligosaccharides (glycolipids) and glycoproteins (GPI anchored). In the first, diverse sequence determinants (reviewed in Hounsell, 1994; Hounsell, 1995; Hounsell and Bailey, 1997) tend to be displayed at multiple sites along a linear polymer; in the second a high degree of peptide substitution and oligosaccharide branching leads to crowding of potential ligands (Hounsell et al., 1996); and, in the last, cooperativity may involve lipid-lipid, oligosaccharide-oligosaccharide and oligosaccharide-protein mediated clustering. These different strategies of nature can be explored by NMR spectroscopy of functional motifs modelled by computer graphics in the context of multi-component macromolecular systems (Hounsell, 1994; 1995).

The field of this chapter thus covers areas so far not discussed in previous contributions in this book in the context of our work which has culminated in the characterisation of a series of neutral and anionic oligosaccharides (Hounsell, 1994; Hounsell and Bailey, 1997) which are all potential targets for therapeutic interactions in particular as tumour vaccines, anti-microbial reagents and inhibitors of multi-component networks involved in cell regulation in cancer (growth factor receptors) and immune regulation (cytokines and cell adhesion molecules). There is a legitimate question as to whether the large diversity of potential glycoconjugate epitopes (i.e., particular orientation of functional groups which interact in protein combining sites) all serve as recognition motifs, but the growing awareness of the diverse roles of glycosylation suggests that the more structures we characterise the more specific interactions can be ascribed to different oligosaccharides as high affinity ligands. The final therapeutic may be carbohydrate itself, or peptide mimic, or a ligand from an RNA combinatorial library (see previous chapters in this book by Robert Feldman and Ajit Varki).

GLYCOSAMINOGLYCANS (GAGs)

The paradigm for high affinity oligosaccharide-protein interactions is anti-thrombin III binding to heparin; a hexasaccharide, related to sequences in heparin which are the natural receptors presented in multivalent form, is now an accepted specific low molecular weight anti-coagulant with Kd around 10^{-8}M (Van Boeckel and Petitou, 1993). Heparin has also been known for several years to bind growth factors and several specific sequences are now being researched as bearing the epitopes for high affinity binding which could form the basis of inhibitors of the 6-20 oligomeric saccharide size. (Gallagher, 1995; Salmivirta et al., 1996). Their role can be either in controlling growth factor concentration at the extracellular level, interactions within the extracellular matrix (ECM), or specific recognition at the cell surface. It is now realised that GAG oligosaccharide sequences of proteoglycans (PGs) are found on many cell surface proteins with potential for exploitation in cell signalling events and anti-viral interactions (Price et al., 1995; Rider et al., 1994) as well as in anti-coagulation and growth factor regulation. In order to differentiate each action, specific structures within the polymer have to be characterised. Analysis includes the use of endoglycosidase digestion with enzymes of known specificity: e.g. for keratan sulphate (KS) the endo-β-galactosidase of *Bacteroides fragilis* (Scudder et al., 1986); for chondroitin sulphate (CS) and dermatan sulphate (DS) the chondroitinases; and for heparin (H or Hep) and heparan sulphate (HS) the heparin- and heparatin-ases. Fig 1 gives a general summary of Hep and HS structure and enzyme cleavage of the high molecular weight PGs from cow and pig intestinal heparin. Figs 2 and 3 show just one example of GAGs as ubiquitous cell surface components, on one of the splice variants of CD44. Chondroitin sulphate chains are added at particular serine (Ser) amino acids and the CD44 molecule will then interact with ECM molecules such as hyaluronan (Hyman et al., 1991). Other examples can be found in Esko and Zhang (1996) which charts the amino acid sequence motifs which support addition of Hep, HS, CS and DS saccharides to specific Ser hydroxyl groups.

O-LINKED GLYCOPROTEINS

GAGs are linked to the hydroxyl group (O) of Ser aminoacids via a Gal-Gal-Xyl-O- sequence. Classical terminology calls this O-glycosylation when the linkage sugar is GalNAcα1-O-Ser or Thr, originally defined as of "mucin-type" because of the multiple forms of this glycosylation that occur on the high molecular weight mucins which line the respiratory and gastrointestinal tracts. However, as with GAGs of PGs, the same sequences can occur in lower molecular weight oligosaccharide-protein glycoconjugates, the example in Fig 3 being CD8 where multiple O-glycosylation causes an extended

Heparin/Heparan sulphate.

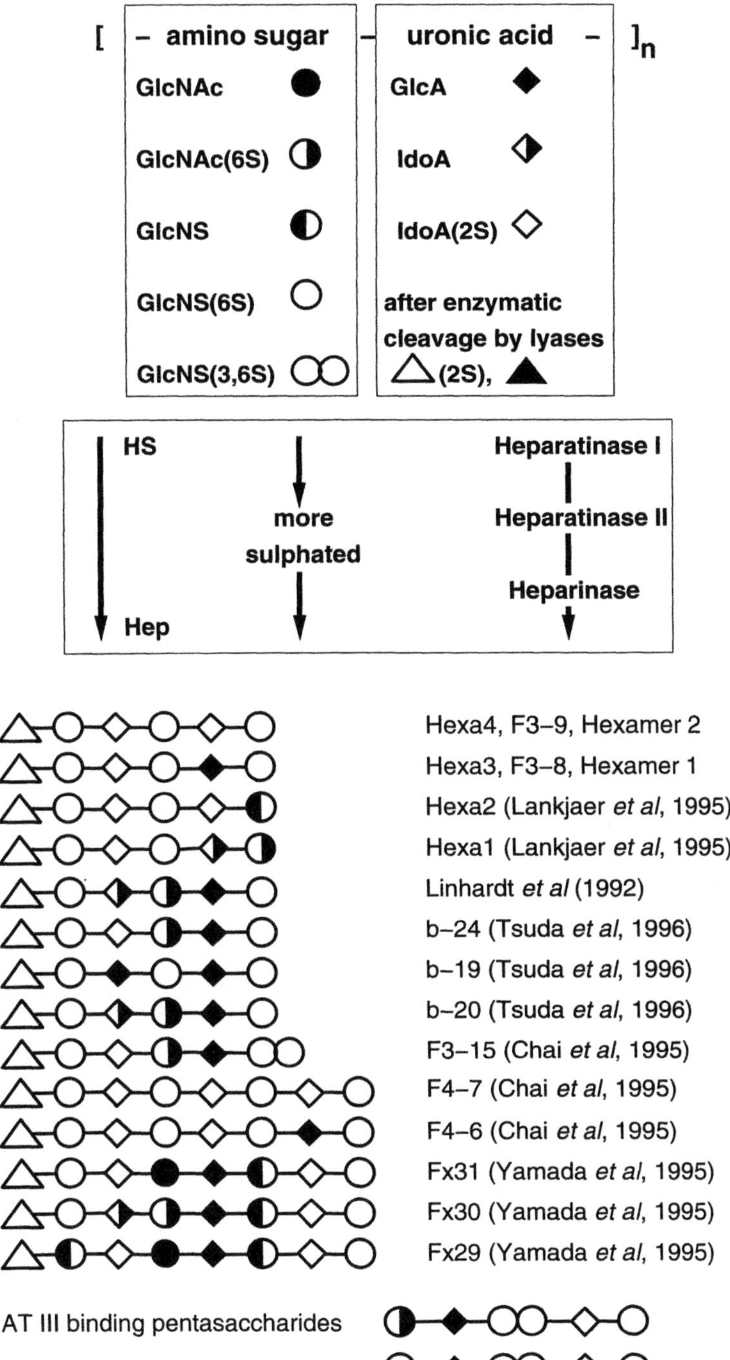

Figure 1. A generalised summary of the biosynthesis, structural definition, enzyme release and characterised hexa- and octa-saccharide sequences of heparin (H) and heparan sulphate (HS) GAGs. (See Hounsell and Bailey, 1997 for further details).

Proteoglycans: Keratan sulphate (KS), Heparin (Hep or H), Heparan sulphate (HS), Chondroitin sulphate (CS) and Dermatan sulphate (DS); ■

Glycoproteins: O-linked (Ser/Thr); ●
N-linked (Asn); ▼

Glycolipids, Phosphatidylinositol glycans (GPI), Peptidoglycolipids, Lipopolysaccharide.

Figure 2. The different types of glycoconjugates to be discussed in this article with symbols ■, ●,▼ relating to Fig 3. Protein glycosylation includes glycosaminoglycan (GAG) to the OH group of Ser amino acids of proteglycans (PG) and other glycoproteins, classical *O*-linked glycosylation having the GalNAcα1-Ser/Thr linkage (for GlcNAcβ1-Ser/Thr see the chapter by Gerald Hart in this volume and other *O*-glycosylation in Hounsell, et al., 1995) and *N*-glycosylation to the "N"-acetamido group of Asn in the consensus sequence Asn.Xaa.Ser/Thr. In addition are glycolipids which will not be discussed herein, but bear several of the recognition motifs discussed in this book and many more besides. Other important conjugates of oligosaccharide, lipid and protein are the GPI protein membrane anchors, diverse molecules of the surface of mycobacteria such as glycopeptidolipids (Hounsell, 1995 and see the work on trypanosomes etc of Steve Homans in this book, McConville and Ferguson, 1993) and the lipopolysaccharides of bacteria (eg. articles by Robert Feldman and Richard Moxon in this book).

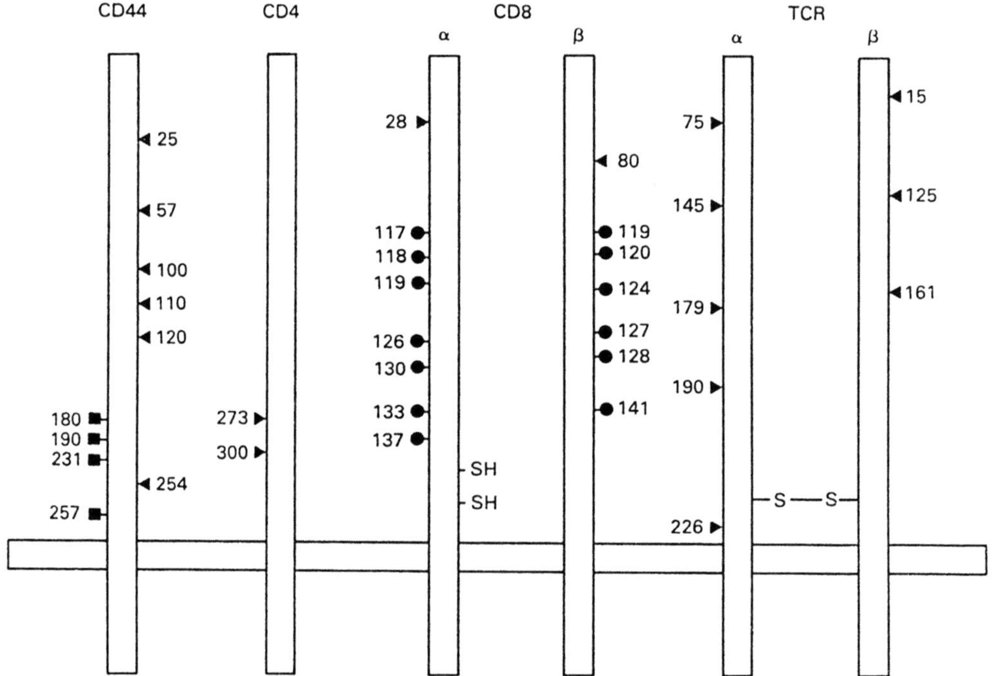

Figure 3. Examples of the usage of diverse glycosylation patterns in molecules of lymphocyte cell surfaces. Symbols relate to Fig 2. Only one of the several splice variants of CD44 are shown. Of interest too are the different splice variants of the TCRγδ chains having different N-glycosylation patterns (Hounsell and Davies, 1993 and with permission of the Annals of Rheumatic Disease).

protein sequence which probably functions to place the N-terminal sequences of CD8 out into the correct intracellular space for protein-protein interactions. One of the many general roles of oligosaccharides is now being appreciated as this spatial function, in addition to presenting multivalent ligands for oligosaccharide-to-protein binding (eg. see the discussions in this book on selectin binding; Barclay et al., 1993; Hounsell et al., 1996).

The commonest oligosaccharide sequence linked via GalNAc is that having Galβ1-3 (core 1) and with zero, one or two N-acetylneuraminic acid residues ±NeuAcα2-3Galβ1-3[±NeuAcα2-6]GalNAcα1-. This is present not only on cell surface glycoproteins such as CD8, but also on serum glycoproteins as well as in the mucins which line the respiratory and gastrointestinal tracts. Here they are known as tumour associated antigens e.g. Fig 4; (recently reviewed in Hounsell et al., 1996; Fukuda, 1996, Kim et al., 1996 and Young, et al., submitted): the company Biomira, (Edmonton, Canada) have in phase II clinical trials the sequence NeuAcα2-6GalNAc as an anti-cancer vaccine (reported by James Paulsen in this book); and Galβ1-3GalNAcα1- Ser/Thr antigen sequences are exposed in colon carcinoma (Campbell et al., 1995). Structural work on mucin oligosaccharides has revealed seven more core region sequences (Hounsell, 1994; Hounsell et al., 1996), each of which we believe will display an oligosaccharide or glycopeptide epitope for B cell (or possibly T cell; see below) responses (reviewed in Hounsell and Davies, 1993; Hounsell et al., 1996). The majority of mucin structures so far characterised have GalNAc, Gal, GlcNAc and Fuc, but there are also many anionic ones having sialic acid and sulphate. Purification and characterisation has been carried out by a combination of HPLC, MS and NMR (Hounsell, 1994; Hounsell, 1995; Davies and Hounsell 1996 a,b). Changes in oligosaccharides in cancer detection were first explored using lectins and polyclonal sera. With the advent of monoclonal antibodies (mAb) came the prospect of specific diagnosis and therapy. The antibody to the sequence NeuAcα2-3Galβ1-3[Fucα1-4]GlcNAcβ1- (SLe[a]) was perhaps the first mAb to be thought of as a therapeutic reagent (Fig 4). Related oligosaccharides are now known to function in endogenous recognition either on O- or N-linked chains of glycoproteins.

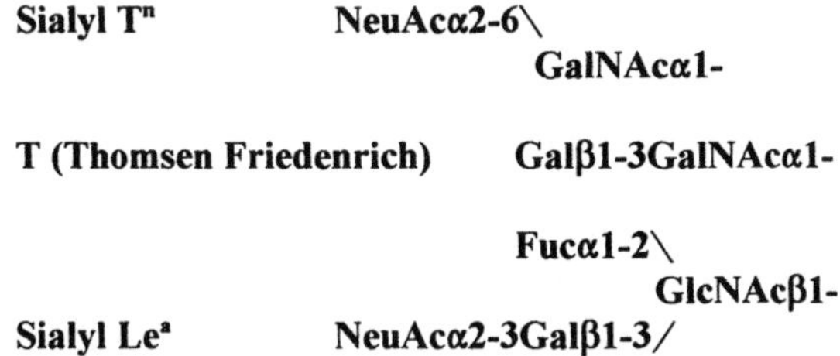

Figure 4. Some of the tumour-associated antigens of potential therapeutic value (reviewed in Fukuda, 1996; Kim et al., 1996; Young, et al., submitted).

RECOGNITION AND INTERACTIONS

Fig 3 shows a few examples of N-linked chain glycopeptide diversity, i.e. of CD8, CD4 and TCRαβ. CD8 and CD4 have very similar actions, mediating the recognition of the T cell receptor (TCR) to the structurally very closely related class I and class II MHC molecules, respectively, but they have very different peptide and glycosylation patterns. In addition TCRγδ have very different glycosylation profiles to TCRαβ and which also vary in different splice variants (reviewed in Hounsell and Davies, 1993). Thus when looking at glycoprotein conformation and recognition, it is important to consider the diversity of oligosaccharide sequence at the glycopeptide level, i.e., the distribution of each glycoform along the peptide sequence. We have recently carried out such an analysis of the receptor for epidermal growth factor (Smith et al.,1996 and see also the chapter by Pauline Rudd in this book on other CD molecules). In addition to B cell epitope and functional recognition at the glycopeptide level, evidence is accumulating that specific cytotoxic T cell responses can be raised to glycopeptide epitopes (Otvos et al., 1995; Haurum et al., 1995; Abdel-Motal et al., 1996; Dustin et al., 1996). Some of the complex discussion on T cell independent or T cell dependent MHC restriction of oligosaccharide determinants is to be found in the chapter by Feldman in this book. In general the role of N-linked glycosylation on protein conformation and interactions is beginning to be appreciated outside the field of the glycosylation experts, in for example, T cell recognition of MHC and superantigens (Fields et al., 1996; Garboczi et al., 1996) and in prion glycoproteins (Collinge et al., 1996).The latter are not only N-glycosylated but are also GPI anchored (discussed next).

Another exciting new aspect is the interactions of lipid, oligosaccharide, GAGs and lectins at the cell surface. A few examples are given in Fig 5 (compiled from Stelter et al., 1996, Chen et al., 1996; Holland et al., 1996; Massague, 1996; Niehrs, et al., 1996; Petty, et al., 1996) which shows how post-translational modifications work at the multicomponent level. The receptor for GDNF and the CR3 complement receptor have "C-type" lectin like domains which probably interact through the glycosylation of GPI anchored glycoproteins (GDNFRα, CD16, CD14, uPAR). In this way single passage transmembrane proteins can mediate cell signalling via oligosaccharide-protein and lipid-lipid interactions rather than trying to transfer a conformational change from the extracellular ligand-binding domain through the constrained hydrophobic transmembrane region to the cytoplasmic phosphorylation sites (Y or S/T amino acids). We have also shown that the GPI anchor can adopt a different conformation either when it is attached to lipid or after phospholipase C or D action. On cleavage both the glycoprotein and glycan minus lipid both appear to adopt a different conformation, resulting in the protein being longer reactive with antibodies (Barboni et al., 1995) or presumably its natural ligand at the cell surface. A proteoglycan has also been found linked to the membrane by a GPI anchor, glypican. Other receptors, such as those for TGFβ and FGF interact with the transmembrane GAG-containing protein, betaglycan (Fig 5) as well as ECM PGs.

The field of the glycosciences can lead the way in trying to understand such multicomponent systems because initially we looked at proteins as glycoproteins, furthermore, a lot of the early work on glycoconjugates was also on glycolipids, i.e., we have had to consider protein-protein, lipid-lipid, protein-water, oligosaccharide-water, oligosaccharide-protein and oligosaccharide-oligosaccharide interactions. We have also had to worry about the oligosaccharide chains as the dynamic component at the cell surface. Oligosaccharides generally do not fall against the proteins but are surrounded by water and this may dictate their role as a half-way house between the cell and its water and salt environment. The quantitative rather than qualitative expression of oligosaccharides is most likely to be a fine control mechanism in biochemistry, but if targets are chosen wisely there are many areas where we can exploit their interactions in new therapeutics.

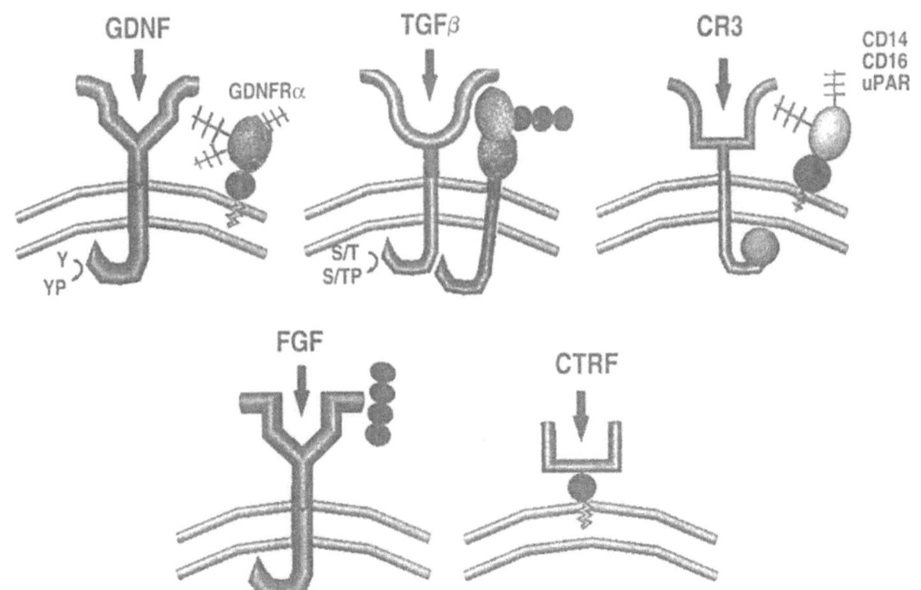

Figure 5. The interactions of the receptor for glial cell derived neurotrophic factor (GDNF) and its coreceptor (GDNFR α); transforming growth factor (TGF) receptor in association with transmembrane proteoglycan, betaglycan; and complement receptor 3 (CR3). CD16 is the immunoglobulin G receptor FCRIII γb. CD14 is the lipolysaccharide (LPS) receptor in the context of LPS binding protein (LBP). uPAR is the receptor for plasminogen activation (PA) found in urine (u). GDNFR/CR3 and TGFR signal intracellularly via, respectively, tyrosine (Y) or Ser/Thr (S/T) phosphorylation. The GDNFR does this in association with GPI anchored GDNFRα. The receptor for ciliary neurotrophic factor is also GPI anchored. Other transmembrance growth factors and their receptors interact through PGs of the ECM, e.g., fibroblast growth factor (FGF). Key: ⧻,N-linked glycan; ●●●●,glycosaminoglycan; ●∧∧ , GPI anchor.

ACKNOWLEGEMENT

The authors are grateful for the help of Gail Evans in preparing this chapter.

REFERENCES

Abel-Motal, U.M., Berg, L., Rosén, A., Bengtsson, M., Thorpe, C.J., Kihlberg, J., Dahmén, J., Magnusson, G., Karlsson, K-A., and Jondal, M., 1996, Immunization with glycosylated Kb-binding peptides generates carbohydrate-specific, unrestricted cytotoxic T cells. *Eur. J. Immunol.* 26:544-551.

Barboni, E., Rivero, B.P., George, A.J.T., Martin, S.R., Renouf, D.V., Hounsell, E.F., Barber, P.C., Morris, R.J., 1995, The glycophosphatidylinositol anchor affects the conformation of the Thy-1 protein. *J.Cell Sci.* 108:487-497.

Barclay, N.A., Birkeland, M.L., Brown, M.H., Beyers, A.D., Davis, S.J., Somoza, C., Williams, A.F. 1993, The Leucocyte Antigen Facts Book. Acad. Press Ltd, London.

Campbell, J.B., Finnie, I.A., Hounsell, E.F., Rhodes, J.A., 1995. Direct demonstration of increased expression of Thomsen-Friedenreich (TF) antigen in colonic adenocarcinoma and ulcerative colitis mucin and its concealment in normal mucin. *J. Clin Invest.* 95:571-576.

Chai W., Hounsell E.F., Bauer C.J., Lawson A.M. (1995) Characterisation by LSIMS and ^{1}H-NMR of tetra-, hexa- and octasaccharides of porcine intestinal heparin. *Carbohydrate Res.* 269:139-156.

Chen, X., Rubock, M.J. and Whitman, M., 1996, A transcriptional partner for MAD proteins in TGF-β Signalling. *Nature* 383:691-696.

Collinge, J., Sidle, K.C.L., Meads, J., Ironside, J. and Hill, A.F., 1996. Molecular analysis of prion strain variation and the aetiology of 'new variant' CJD. *Nature* 383:685-690.

Davies, M.J. and Hounsell, E.F. (1996a) Carbohydrate Chromatography: Towards Yoctomole Sensitivity. *Biomed. Chrom.* Vol.10 No.6 p285-289.

Davies, M. J. and Hounsell, E. F., (1996b). Comparison of separation modes for high-performance liquid chromatography of glycoprotein- and proteoglycan-derived oligosaccharides. *J Chromatogr.* 720:227-234.

Dustin, M.L., McCourt, D.W., and Kornfeld, S., 1996, A mannose 6-phosphate containing N-linked glycopeptide derived from lysosomal acid lipase is bound to MHC class II in B lymphoblastoid cell lines. *J. Immunol.* 156:1841-1847.

Esko, J.D. and Zhang, L., 1996, Influence of core protein sequence on glycosaminoglycan assembly. *Curr. Opin. Struct. Biol.* 6:663-670.

Fields, B.A., Malchiodi, E.L., Li, H., Ysern, X., Stauffacher, C.V., Schlievert, P.M., Karjalainen, K. and Mariuzza, R.A., 1996, Crystal structure of a T-cell receptor β-chain complexed with a superantigen. *Nature* 384:188-192.

Fukuda, M. 1996, Possible roles of tumour-associated carbohydrate antigens. *Cancer Res.* 56:2237-2244.

Gallagher, J.T. 1995, Heparan sulphate and protein recognition. Binding specificities and activation mechanisms. *Adv. Exp. Med. Biol.* 376:125-34.

Garboczi, D.N., Ghosh, P., Utz, U., Fan, Q.R., Biddison, W.E., and Wiley, D.C., 1996, Structure of the complex between human T-cell receptor, viral peptide and HLA-A2. *Nature,* 384:134-141.

Haurum, J.S., Tan, L., Arsequell, G., Frodsham, P., Lellouch, A.C., Moss, P.A.H., Dwek, R.A., McMichael, A.J., and Elliott, T. (1995) Peptide anchor residue glycosylation: effect on class I major histocompatibility complex binding and cytotoxic T lymphocyte recognition. *Eur. J. Immunol.* 25:3270-3276.

Holland, S.J., Gale, N.W., Mbamalu, G., Yancopoulos, G.D., Henkemeyer, M and Pawson, T. 1996, Bidirectional signalling through the EPH-family receptor nuk and its transmembrane Ligands. *Nature* 383:722.

Hounsell, E.F. and Davies, M.J., 1993, Role of protein glycosylation in immune regulation. *Annals Rheum. Dis.* 52, S22-S29.

Hounsell, E.F., Davies, M.J. and Renouf, D.V., 1996, *O*-linked protein glycosylation structure and function. *Glycoconjugate J.* 13:19-26.

Hounsell, E.F. and Bailey, D., 1997, Approaches to the structure determination of oligosaccharides and glycopeptides using NMR. In *Glycopeptides and related compounds: Synthesis, Analysis and Applications.* Eds. D.G. Large and C.D. Warren. Marcel Dekker Inc pp. 631-660.

Hounsell, E.F., 1995, ^{1}H-NMR in the Structural and Conformational Analysis of Oligosaccharides and Glycoconjugates. *In Prog. NMR Spec.,* Eds. J.W. Emsley, J. Feeney and L.H. Sutcliffe, Elsevier 27:445-474.

Hounsell, E.F., 1994, Physicochemical Analysis of Oligosaccharide Determinants of Glycoproteins. *Adv. Carb. Chem. Biochem.*, 50:311-350.

Hyman, R., Lesley, J., Schulte, R., 1991, Somatic cell mutants distinguish CD44 expression and hyaluronic acid binding. *Immunogenetics.* 33: (5-6): 392-5.

Kim, Y.S., Gum, J. jnr. and Brockhausen, I., 1996, Mucin glycoproteins in neoplasia. *Glycoconj. J.* 13: 693-707.

Larnkjær, A., Nykjær, A., Olivecrona, G., Thøgersen, H. and Østergaard, P.B., 1995, Structure of heparin fragments with high affinity for lipoprotein lipase and inhibition of lipoprotein lipase binding to α_2-macroglobulin-receptor/low-density-lipoprotein-receptor-related protein by heparin fragments. *Biochem. J.* 307:205-214.

Lehner, P.J. and Cresswell, P., 1996, Processing and delivery of peptides presented by MHC class I molecules. *Curr. Opin. Immun.*8, 59-67.

Linhardt, R.J., Wang, H-M., Loganathan, D. and Bae, L-H., 1992, Search for the heparin antithrombin III-binding site precursor. *J. Biol. Chem.* 267:2380-2387.

Massagué, J., 1996, Cross receptor boundaries. *Nature*, 382:29-30.

McConville, M. J. and Ferguson, M.A.J. 1993, The structure, biosynthesis and function of glycosylated phosphatidylinositols in the parasitic protozoa and higher eukaryotes. *Biochem. J.* 294:305-324.

Niehrs, C., 1996, Mad connection to the nucleus, *Nature*, 381:561-562.

Otvos, L. jr., Krivulka, G.R., Urge, L., Szendrei, G.I., Nagy, L., Xiang, Z.Q., Ertl, H.C.J., 1995, Comparison of the effects of amino acid substitutions and *β-N-* vs. *α-O-*glycosylation on the T-cell stimulatory activity and conformation of an epitope on the rabies virus glycoprotein. *Biochimica et Biophysica Acta*, 1267:55-64.

Petty, H.R. and Todd III R. F., 1996, Integrins as promiscuous signal transduction devices. *Trends Immunol. Today 17*, 5:209-211.

Price, P., Allcock, R.J.N., Coombe, D.R., Shellam, G.R., McCluskey, J.,1995, MHC proteins and heparan sulphate proteoglycans regulate murine cytomegalovirus infection. *Immunol. Cell Biol.* 73:308-315.

Rider, C.C., Coombe, D.R., Harrop, H.A., Hounsell, E.F., Bauer, C.J., Feeney, J., Mulloy, B., Mahmood, N., and Parish, C.R. 1994, Anti-HIV-1 activity of chemically modified heparin: correlation between binding to the V3 loop of gp120 and inhibition of cellular HIV-1 infection in vitro. *Biochem.* 33:6974-6980.

Salmivirta, M., Lidholt, K., and Lindahl, U., 1996, Heparan sulfate: a piece of information, *The FASEB J.* 10:1270-1279.

Scudder, P., Tang, P.W., Hounsell, E.F., Lawson, A.M., Mehmet, H. and Feizi, T. 1986, Isolation and characterisation of sulphated oligosaccharides released from bovine corneal keratan sulphate by endo-ß-galactosidase. *Eur. J. Biochem.*, 157:365-373.

Smith, K.D., Bailey D.H.,Davies M.J., Renouf, D.V., and Hounsell,E.F. 1996, Analysis of the glycosylation patterns of the extracellular domain of the epidermal growth factor expressed in Chinese hamster ovary fibroblasts. *Growth Factors* 13:1-12.

Stelter, F., Pfister, M., Berheiden, M., Jack, R., Bufler P. Engelmann, H. and Schutt, C., 1996, The myeloid differentiation antigen CD14 is N- and O-glycosylated. Contribution of N-linked glycosylation to different soluble CD14 isoforms. *Eur. J. Biochem*, 236:457-464.

Tsuda, H., Yamada, S., Yamane, Y., Yoshida, K., Hopwood, J.J. and Sugahara, K., 1996, Structures of five sulfated hexasaccharides prepared from porcine intestinal heparin using bacterial heparinase. Structural variant with apparent biosynthetic precursor-product relationships for the antithrombin III-binding site. *J. Biol. Chem.* 271:10495-10502.

Van Boeckel, CAA., Petitou, M. 1993, The unique antithrombin III binding domain of heparin: a lead to new synthetic antithrombotics. *Angew Chem Int. Ed* 32:1671-1690.

Yamada, S., Sakamoto, K., Tsuda, H., Yamane, Y., Yoshida, K and Sugahara, K., 1996, Structural studies on the octasaccharide fraction prepared from porcine intestinal heparin after flavobacterium heparinase digestion. *Proc. Milan 1996 Int. Carb Symp.*

Young, M., Davies, M.J. and Hounsell, E.F., 1996, Glycoprotein changes in tumours: a renaissance in clinical potential *Nature Med.* Submitted.

THE GROUP B STREPTOCOCCAL CAPSULAR CARBOHYDRATE: IMMUNE RESPONSE AND MOLECULAR MIMICRY

Robert G. Feldman,[1,2] Ger T. Rijkers,[2] Maaike E. Hamel,[2] S. David,[2] and Ben J.M. Zegers[2]

[1] Department of Infectious Diseases, Royal Postgraduate Medical School, London, UK.
[2] Department of Immunology, Het Wilhelmina Kinderziekenhuis, University Hospital for Children and Youth, Utrecht, The Netherlands.

INTRODUCTION

The group B streptococcus (GBS) is the commonest cause of lethal bacterial infection in the newborn infant.1 GBS possesses a sialylated carbohydrate capsule which is known to be a major virulence factor.[2] Mutant GBS which have a non-sialylated capsule are considerably less virulent, fix complement to a greater degree than the isogenic wild type and are more avidly opsonophagocytosed by polymorhonucleoctes.[2] Antibodies to GBS capsule have been shown to be protective but the majority of the population do not possess these antibodies. For example, 90% of the population lack antibodies to the type III capsule of GBS (GBS III).[3] Furthermore, 60% of seronegative individuals do not respond to experimental vaccination with purified GBS III.[4] There has been considerable interest over the last few years in the development of a vaccine which would prevent neonatal infection with GBS. An appropriate approach would seem to be to conjugate the capsular carbohydrates to proteins as has been successfully achieved for *Haemophilus influenzae* serotype b. It is envisaged that any such vaccine would be given early in pregnancy and the vaccine induced IgG would subsequently be placentally transported to the fetus. In order to facilitate the development of these new vaccines, it is important to understand the basis of the poor immune response to the native capsular carbohydrate in humans and to document the naturally occurring human antibody response.

The structure of GBS III was initially determined in 1980 and subsequently revised in 1987.[5,6] The accepted structure of the pentasaccharide repeating unit is [α-D-NeupNAc-(2→3) β-D-Galp-(1→4)- [→4] β-D-Glcp-(1→6)] β-D-GlcpNAc-(1→3) β-D-Galp-(1→). Immunisation with GBS III induces two populations of antibody: the major response is to a sialic acid dependent conformational epitope and the minor response is to a non-conformational, sialic acid independent backbone epitope.[7] Coincidentally, the structure of pneumococcal capsule type 14 (PPS14) is physically and immunologically identical to pneumococcal capsule type 14 (PPS14) is physically and immunologically identical to desialylated GBS III capsule.[5,8] The amount of antibody against the backbone epitope of

GBS III has been established by performing an ELISA with GBS III as the solid phase and inhibiting with PPS14. It is evident from these studies that antibodies to the backbone epitope are uncommon in a human population.[9] The importance of the sialic acid in maintaining the conformational epitope has been established and it is likely that the carboxylate group of the sialic acid hydrogen bonds to the 2-acetamido-2-deoxy-β-D-glucopyranose residue of the backbone.[7] This results in an alteration of the orientation of the sidechain β-D-galactopyranose. It appears that the major epitope is small and is located precisely at the junction of this same β-D-galactopyranose residue with the backbone of GBS III.

Using the lymphocyte *in vitro* culture method developed for GBS III it is possible to manipulate the culture conditions in order to investigate the T cell control of the response.[10] It is possible to detect anti-GBS III antibody producing B lymphocytes from both seropositive and seronegative donors.[10] The response has been shown to be T lymphocyte dependent, with no response in the absence of T lymphocytes nor in the presence of >40% T lymphocytes. Cells from 5 of 6 seropositive donors and 3 of 7 seronegative donors produced specific IgM antibody after culture with antigen. It has already been shown that the B lymphocyte response to GBS III is dependent upon T cells so we have attempted in this study to begin to understand the nature of the T cell control.

The method of B lymphocyte activation is not entirely understood although it has been shown that carbohydrate antigens can directly lead to early activation *in vitro* presumably by cross-linking surface immunoglobulins.[11] Any role which T cells may play in this process is even more obscure although there is some evidence from the murine model that suppressor T cells can specifically downregulate T cell independent (TI) antibody responses by idiotypic recognition.[11] There is no evidence though for any specific T cell helper role. It is also recognized that T cell help of the TI response is not restricted to autologous T cells.[12] This suggests that there is no direct role for MHC molecules unlike protein antigens which is consistent with the model outlined above. We have therefore investigated the ability of allogeneic T cells to augment the anti-GBS III response and examined the possible function of CD8[+] cells.

We also investigated the possibility of an alternative approach to generating antibody to GBS capsule, namely the mimicry of the capsular carbohydrate by a peptide. Two examples of peptides which mimic carbohydrate structures have been reported.[13,14] A dodecapeptide was isolated from a random display library which binds specifically to the lectin concanavalin A. The peptide contained a consensus sequence of Pro-Tyr-Pro which appears critical for the binding. The K_d of the peptide for the lectin was 46 μM compared to 89 μM for methyl α-D-mannopyranoside. Hoess *et al.* generated an octapeptide which binds to the monoclonal antibody, named B3, which is specific for Lewis[Y] antigen.[14] Furthermore, the peptide competes with the Lewis[Y] carbohydrate. Four amino acids within the octapeptide (Pro-Trp-Leu-Tyr) have been shown to be critical for the binding properties.

MATERIALS AND METHODS

Preparation of the GBS type III carbohydrate

This was prepared as described previously.[15] No detectable protein or nucleic acids are present in the preparation.

Isoelectric focusing of IgG and visualization of clonotypes

IgG preparations were dialyzed and concentrated to approximately 100 μg/ml specific antibody against 1% glycine in Centricon-10 concentrators (Amicon) before applying between 5 and 10 µl to an agarose immunoelectrophoresis gel, pH 3 to 10 (FMC). Focusing was performed as recommended by the manufacturers (25W constant power, 1000 V maximum) on a flat-bed apparatus (Pharmacia, Uppsala, Sweden). Immediately after focusing the gel was press-blotted onto a NY 13 nylon membrane (Schleicher and Schuell) for 30 minutes and subsequently blocked with 2% goat serum in PBS for 20 minutes. Specific clonotypes were detected by probing with 1 µg/ml biotinylated GBS and then a streptavidin-alkaline phosphatase conjugate diluted 1 in 1000 (Tago, Burlingame, CA). The duration of each step was 45 minutes at room temperature and all dilutions were in PBS plus 2% goat serum. Steps were separated by three washes in PBS plus 0.02% Tween-20 (Sigma). Nitro blue tetrazolium and 5-bromo-4-chloro-3-indolylphosphate were used as the alkaline phosphatase substrate (Promega).

Relative antibody affinity

The thiocyanate elution method was used as described previously.[9] In order to rank different mean antibody avidities, the molar concentration of thiocyanate necessary to inhibit the absorbance of each antibody in an enzyme immunoassay by 50% was estimated.

Preparation of peripheral blood lymphocytes

T cell and non-T cell fractions were prepared as described previously.[10] In some experiments T cells depleted of $CD8^+$ cells were utilized in place of the T cell fraction. 30×10^6 T cells were incubated with excess leu2 (anti-CD8) monoclonal antibody (Becton Dickinson, Mountain View, CA) for 30 minutes at 4 °C in Eagles minimal essential medium (MEM) supplemented with 10% fetal calf serum (Flow Laboratories, Irvine, UK). After two washes a 30-fold excess of ox erythrocytes coated with goat anti-mouse IgG (Tago) was added and centrifuged at 250 g for 7 minutes.[16] The cells were then resuspended by gentle vortexing and the non-rosetted cells were separated by density gradient centrifugation on Ficoll-Isopaque (Pharmacia). This fraction typically contained greater than 85% $CD4^+$ cells and less than 1% $CD8^+$ cells measured by flow cytometry (FACStar Plus, Becton Dickinson).

Culture and activation of B cells

Culture conditions have been described previously.[10] Briefly all cultures were performed in 2 ml of RPMI 1640 medium (Flow, Irvine, UK) supplemented with penicillin, streptomycin, 2 mM glutamine, 10% fetal calf serum and 25% T cell supernatant and incubated for 7 days at 37 °C, 100% relative humidity and 5% CO_2 in tissue culture tubes (Becton Dickinson). 1×10^6 monocyte depleted cells from the non-T fraction were added to each tube as well as irradiated T cells or CD8 depleted T cells and filter sterilized GBS type III polysaccharide. In some cases allogeneic T cells were employed which were isolated from a different donor as described.

Spot forming cell assay

This assay was performed as previously described.[10] Briefly cells producing specific antibody were detected by culturing them in microtiter wells precoated with GBS III carbohydrate. Local antibody concentration around cells is high and the presence of these antibodies is detected after washing away the cells with an antibody conjugate and insoluble substrate. The "spots" which are produced are counted and expressed finally as the number of spot forming cells (SFC) per million non-T cells in the starting culture.

Random phage display library

We have performed a single round of panning of a fUSE3 random hexapeptide display library in order to establish whether it will be possible to generate peptides which will bind to anti-GBS type III capsule antibody. To this end, amplified fUSE3 phage library was added to a microtiter well coated with affinity purified anti-GBS III IgG2. After extensive washing, phage was eluted with 1 mg/ml of purifed GBS III polysaccharide and the number of virions in the eluate were enumerated. In order to confirm this, a nylon membrane lift from the test plate was probed with a purifed IgG from a donor known to have high levels of anti-GBS III IgG (the specific IgG concentration used to probe the lift was 20 ng/ml). The donor for this IgG preparation was not related to the donor for the IgG used to coat the well used for panning. The colony giving heaviest staining and a non-staining control were picked from the plate and phage was produced from each of these. A microtiter tray was then coated with phage and probed with affinity purified anti-GBS III IgG2 antibody. A goat anti-human IgG alkaline phosphatase conjugated antibody was used to detect binding of human antibody to the phages bound to the solid phase of the microtiter plate.

Statistics

All results given are the mean and standard error of duplicate cultures.

RESULTS

Relative antibody avidity measurements

The ranking of avidity of IgG antibodies to GBS III obtained from randomly chosen donors is shown in table 1.

Table 1. Ranking of avidity of anti-GBS III IgG
from randomly chosen donors.

Donor	Thiocyanate concentration[1]
ACC	0.09
PD	0.1
AM	0.15
BL	0.2
MAB	0.25
BJMZ	0.3
CB	0.55
MH	0.75

[1]Molar concentration required to inhibit antibody binding
by 50% in an enzyme immunoassay.

Anti-GBS III IgG clonotypes

Examination of the clonotypes of four different donors designated MAB, MM, SJM and CB showed that the number of clonotypes visible on the blots varied between 8 and 15.

Response to CD8 depleted T cells

Figure 1 shows the effect of increasing numbers of T cells and $CD8^{+}$ depleted T cells in the same donor following B lymphocyte culture with 2.5×10^{-4} µg/ml GBS III antigen. As shown previously, increasing numbers of T cells above 10% resulted in a reduced response.[10] The number of $CD8^{+}$ depleted T cells added was such that comparable cultures of T cells and CD8[+] depleted T cells contained the same quantity of CD4[+] cells as measured by flow cytometry. The increasing and much higher response with CD8[+] depleted T cells was seen in 4 of 4 donors tested (table 2).

Table 2. Demonstration of the enhancing effect CD8 depleted T cells on three sero-positive (1,2 and 3) and one seronegative (4) donor. Data for donor 4 is derived from that shown in figure 1.

	%T	No Antigen	With Antigen	CD8 depleted T cells
Donor				
1	10	266±4	379±59	4584±536
2	10	560[1]	2677±1568	4433±2094
3	20	10±10	63±23	666±188
4	20	0±0	40±0	110±1

[1]No duplicate.

Comparison of autologous and allogeneic T cells

We found that specific SFC could also be induced by culturing with T cells obtained from separate non-related donors as shown in figure 2. For this donor the total number of IgM SFC per 10^6 non-T cells cultured was 1750 (±790) for 10% autologous added T cells and 3179 (±539) for 10% allogeneic T cells. We saw the lower but positive response using allogeneic T cells in two separate donors cultured with allogeneic T cells and on both occasions the total IgM producing SFC was lower with allogeneic than autologous T cells. One seronegative donor in which specific SFC were detected with autologous T cells, no specific SFC were seen using allogeneic T cells. Conversely, we have seen a response to

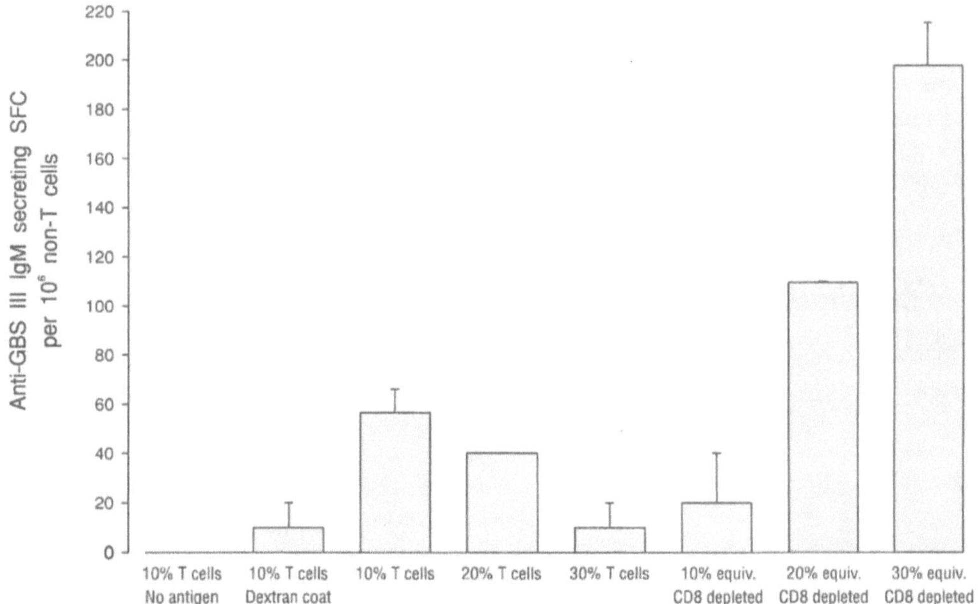

Figure 1. IgM anti-GBS type III spot forming cells from a seronegative donor showing the effect of increasing concentrations of added autologous T cells and CD8[+] depleted T cells. 2.5 × 10⁻⁴ µg/ml GBS III polysaccharide was added to each culture unless stated. Dextran coat denotes biotinylated dextran was used in place of biotinylated GBS III as the coating antigen in the SFC assay. Percentage equivalent CD8 depleted cells represents the equivalent number of CD4[+] cells shown to be present in each of the cultures where T cells were added.

GBS III antigen when allogeneic but not autologous T cells were incorporated into the culture (figure 3).

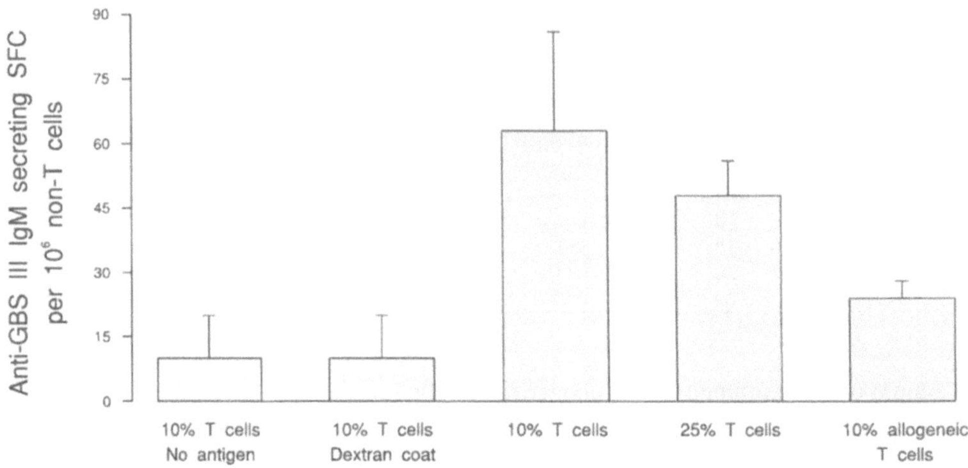

Figure 2. IgM anti-GBS type III spot forming cells from a seropositive donor (8 μg/ml anti-GBS III IgG) showing the effect of added autologous and allogeneic T cells. 2.5×10^{-4} μg/ml GBS III polysaccharide was added to each culture unless stated.

Random phage display library

Approximately 400 phage particles per 100 μl were eluted from the test well, coated with 10μg of affinity purified anti-GBS III IgG, compared to 300 particles per 100 μl from a control (non-coated) well. This indicated that approximately one quarter of the phages isolated had some specificity. The results of the enzyme immunoassay using the phages

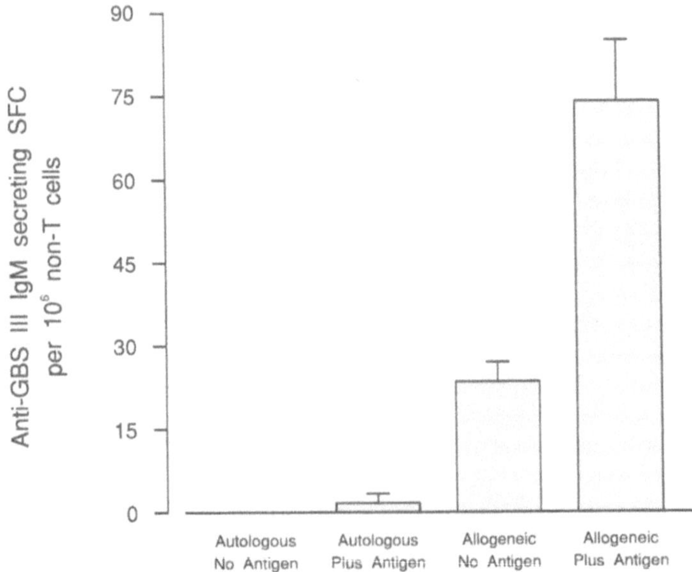

Figure 3. IgM anti-GBS type III spot forming cells from a seronegative donor showing a positive *in vitro* response to GBS type III antigen (2.5×10^{-4} μg/ml) when allogeneic (10%) but not autologous (10%) irradiated T cells were added to the culture.

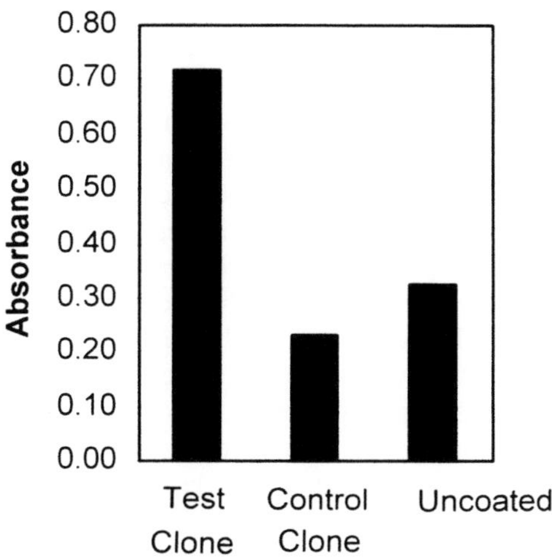

Figure 4. Enzyme immunoassay demonstrating binding of the test fUSE3 phage to anti-GBS III antibody.

which gave the highest and lowest signal in the immunoblots are shown in figure 2 and show that even with a single round of panning, the test clone has the ability to bind to anti-GBS III IgG antibodies from two separate sources.

DISCUSSION

As with the response to many other carbohydrates, naturally occurring antibodies to GBS III are highly restricted in nature. The IgG isotype response is >90% IgG2 in all sera examined[15] and the number of IgG clonotypes is also limited in number. This is more restricted than the response to other carbohydrates and significantly more restricted compared to anti-protein responses.[17,18] The ranking of avidties of specific IgG from various donors demonstrates, as one would expect, that there is a range of avidities. We have recently examined the IgG antibody kinetics one of these donors (MAB) using a resonant mirror biosensor. MAB was chosen as the relative avidity was representative in the ranking of the avidities. The average is 1.1 e-8 M and the mean half life is 2044 sec (submitted). In absolute terms this shows that human IgG antibody is at the low end of high affinity binding.[19]

With respect to the *in vitro* culture results, there appears to be a suppressive effect mediated by $CD8^+$ cells as demonstrated by an increased response with $CD8^+$ depleted T cells. The lack of response with high numbers of T cells (50%) may also be due to the predominant effect of suppressor T cells. It is notable that the augmented SFC response seen with increasing numbers of $CD8^+$ depleted T cells is opposite to that with increasing numbers of T cells. This difference must be due to the suppressive effect of $CD8^+$ T cells. $CD8^+$ cells have also been shown to induce suppression of antibody production to TI antigens in mice by a mechanism which is antigen specific signaled by idiotypic determinants on activated B cells rather than the antigen itself.[11] Similar mechanisms may operate for GBS III in the human, so that the suppressive T cells may determine whether an individual is responsive. Although it is likely that $CD4^+$ cells are responsible for the amplifying role, the role of other cells ($CD4^-$ and $CD8^-$) cannot be excluded. Indeed, there is evidence that lymphocytes which are CD4 and CD8 negative can augment the *in vitro* response to pneumococcal type 4 polysaccharide.[12] It is not clear why little or no response is seen with the equivalent of 10% $CD8^+$ depleted T cells (figure 1), however it is possible that some stimulatory cells are removed by CD8 depletion.

It is clear that not only are autologous T cells required for the *in vitro* response but that this role can be fulfilled by allogeneic T cells. These findings are compatible with T cell interactions which are not MHC restricted but we have not attempted to demonstrate the role of MHC molecules directly. As the total number of IgM secreting cells is also less when allogeneic T cells are added this excludes a non-specific mixed lymphocyte reaction.

The fact that allogeneic but not autologous cells can facilitate a response (and *vice versa*) suggests that observed responsiveness is dependent upon T cell control rather than any inability of the B cells to respond. However, the number of subjects studied to date are small and the data presented should be confirmed with larger numbers. Based on the visualization of clonotypes, the number of B cell clones in an individual capable of producing anti-GBS III antibodies is highly restricted so it could theoretically be possible for non-responders not to possess the appropriate germline genes. It nevertheless seems unlikely that this number of germline genes would be absent in such a large proportion of the population. This would be consistent with what has been observed in relation to the *Haemophilus influenzae* capsular carbohydrate (PRP): there is a predominant VH gene which is utilized in most antibodies to PRP but individuals lacking this gene are still able to produce a normal antibody response to PRP.[20] It has also been recognized that the side chain of the GBS capsular carbohydrates is similar to many glycosylation moieties on human glycoproteins and it is also possible therefore that GBS III may be recognized as self by some individuals. In this case, one would propose that GBS III specific B cells are actively suppressed by T cells and it would therefore be of interest to examine the *in vitro* culture of a non-responder using autologous CD8 depleted T cells.

We conclude that the culture system for B cell activation by GBS III enables the analysis of the role of regulatory T cells in the *in vitro* antibody response. Whether such studies will contribute to the understanding of the cellular basis of non-responsiveness following vaccination with GBS III remains to be established. In addition, the mechanism of both the amplifying and suppressive role of T cells still requires clarification. The restricted nature of the antibody response to GBS III is consistent with other TI responses where there is no MHC mediated T cell help. Preliminary experiments support the idea that it is possible to mimic GBS III using a short peptide. Further pannings or affinity maturations using random PCR will be required to enhance the affinity of binding of the isolated clone.

ACKNOWLEDGEMENTS

R.G. Feldman was funded by the Wellcome Trust, London, UK and Glaxo-Wellcome plc and S. David was funded by The Praeventiefonds, The Netherlands. We are indebted to Dr A. George, Department of Immunology, Royal Postgraduate Medical School for supplying the fUSE3 phage display library and for invaluable assistance with the antibody kinetics study.

REFERENCES

1. R.G. Feldman, Streptococcus agalactiae (group B streptococcus), *in:* "Congenital, Perinatal and Neonatal Infections," A. Greenough, J. Osborne, and S. Sutherland, eds., Churchill Livingstone, Edinburgh (1992).
2. M.R. Wessels, C.E. Rubens, V.J. Benedi, and D.L. Kasper, Definition of a bacterial virulence factor: sialylation of the group B streptococcal capsule, *Proc Natl Acad Sci.* 8983:86 (1989).
3. C.J. Baker and D.L. Kasper, Correlation of maternal antibody deficiency with susceptibility to neonatal group B streptococcal infection, *N Engl J Med.* 753:294 (1976).
4. C.J. Baker and D.L. Kasper, Vaccination as a measure for prevention of neonatal GBS infection, *Antibiot Chemother.* 281:35 (1985).

5. H.J. Jennings, K.G. Rosell, and D.L. Kasper, Structural determination and serology of the native polysaccharide antigen of type-III group B Streptococcus, *Can J Biochem.* 112:58 (1980).

6. M.R. Wessels, V. Pozsgay, D.L. Kasper, and H.J. Jennings, Structure and immunochemistry of an oligosaccharide repeating unit of the capsular polysaccharide of type III group B Streptococcus, *J Biol Chem.* 8262:262 (1987).

7. H.J. Jennings, C. Lugowski, and D.L. Kasper, Conformational aspects critical to the immunospecificity of the type III group B streptococcal polysaccharide, *Biochemistry.* 4511:20 (1981).

8. D.L. Kasper, D.K. Goroff, and C.J. Baker, Immunochemical characterization of native polysaccharides from group B streptococcus: the relationship of the type III and group B determinants, *J Immunol.* 1096:121 (1978).

9. R.G. Feldman, M.E. Hamel, M.A. Breukels, N.F. Concepcion, and B.F. Anthony, Solid-phase antigen density and avidity of antibodies detected in anti-group B streptococcal type III IgG enzyme immunoassays, *J Immunol Methods.* 37:170 (1994).

10. R.G. Feldman, G.T. Rijkers, and B.J. Zegers, Human B lymphocyte in vitro response to the group B streptococcal type III capsular carbohydrate, *Scand J Immunol.* 343:37 (1993).

11. P.J. Baker, Regulation of magnitude of antibody response to bacterial polysaccharide antigens by thymus-derived lymphocytes, *Infect Immun.* 3465:58 (1990).

12. A.W. Griffioen, E.A. Toebes, G.T. Rijkers, F.H. Claas, G. Datema, and B.J. Zegers, The amplifier role of T cells in the human in vitro B cell response to type 4 pneumococcal polysaccharide, *Immunol Lett.* 265:32 (1992).

13. K.R. Oldenburg, D. Loganathan, I.J. Goldstein, P.G. Schultz, and M.A. Gallop, Peptide ligands for a sugar-binding protein isolated from a random peptide library, *Proc Natl Acad Sci.* 5393:89 (1992).

14. R. Hoess, U. Brinkmann, T. Handel, and I. Pastan, Identification of a peptide which binds to the carbohydrate-specific monoclonal antibody B3, *Gene.* 43:128 (1993).

15. R.G. Feldman and A. Ferrante, Prevalence of anti-group B streptococcal type III capsular IgG antibodies in the United Kingdom and an analysis of their specific IgG subclasses, *J Infect Dis.* 883:162 (1990).

16. G.C. Mudde, C.J.M. Verberne, K. Groeneveld, and G.C. De Gast, Human tonsil B lymphocyte function I. The proliferative response to Staphylococcus aureus and pokeweed mitogen in relation to surface heavy chains mu and delta, *Clin Exp Immunol.* 709:56 (1984).

17. K.A. Knisley and L.S. Rodkey, Affinity immunoblotting. High resolution isoelectric focusing analysis of antibody clonotype distribution. *J Immunol Methods.* 79:95 (1986).

18. W.F. Riesen, F. Skvaril, and D.G. Braun, Natural infection of man with group A streptococci: levels, restriction in class, subclass and type, and clonal appearance of poly-saccharide group-specific antibodies. *Scand J Immunol.* 383:5 (1976).

19. R.J. Morris, Antigen-antibody interactions: how affinity and kinetics affect assay design and selection procedures, *in:* "Monoclonal antibodies. Production, engineering and clinical application," M.A. Ritter and H.M. Ladyman, eds., Cambridge University Press, Cambridge (1995).

20. M.G. Scott, D.L. Crimmins, D.W. McCourt, I. Zocher, R. Thiebe, H.G. Zachau, and M.H. Nahm, Clonal characterization of the human IgG antibody repetoire to Haemophilus influenzae type b polysaccharide, *J Immunol.* 4110:12 (1989).

INDEX

MIX
Papier aus verantwortungsvollen Quellen
Paper from responsible sources
FSC® C105338

If you have any concerns about our products,
you can contact us on
ProductSafety@springernature.com

In case Publisher is established outside the EU,
the EU authorized representative is:
Springer Nature Customer Service Center GmbH
Europaplatz 3, 69115 Heidelberg, Germany

Printed by Libri Plureos GmbH
in Hamburg, Germany